Dynamics for Engineers

Dynamics for Engineers

Soumitro Banerjee

Department of Electrical Engineering
Indian Institute of Technology, Kharagpur, India

John Wiley & Sons, Ltd

Telephone (+44) 1243 779777

Email (for orders and customer service enquiries): cs-books@wiley.co.uk
Visit our Home Page on www.wiley.com

Other Wiley Editorial Offices

John Wiley & Sons Inc., 111 River Street, Hoboken, NJ 07030, USA

Jossey-Bass, 989 Market Street, San Francisco, CA 94103-1741, USA

Wiley-VCH Verlag GmbH, Boschstr. 12, D-69469 Weinheim, Germany

John Wiley & Sons Australia Ltd, 42 McDougall Street, Milton, Queensland 4064, Australia

John Wiley & Sons (Asia) Pte Ltd, 2 Clementi Loop #02-01, Jin Xing Distripark, Singapore 129809

John Wiley & Sons Canada Ltd, 22 Worcester Road, Etobicoke, Ontario, Canada M9W 1L1

Wiley also publishes its books in a variety of electronic formats. Some content that appears in print may not be available in electronic books.

Library of Congress Cataloging-in-Publication Data

Banerjee, Soumitro.
Dynamics for engineers / Soumitro Banerjee.
p. cm.
Includes bibliographical references and index.
ISBN-13 978-0-470-86843-0 (cloth : alk. paper) – ISBN-10 0-470-86844-9 (pbk. : alk. paper)
1. Dynamics. 2. Mechanics, Applied. I. Title.

TA352.B35 2005
620.1′04 – dc22

2005043777

British Library Cataloguing in Publication Data

A catalogue record for this book is available from the British Library

ISBN-13 978-0-470-86843-0 (HB) 978-0-470-86844-7 (PB)
ISBN-10 0-470-86843-0 (HB) 0-470-86844-9 (PB)

Typeset in 11/13.6pt Times by Laserwords Private Limited, Chennai, India

This book is printed on acid-free paper responsibly manufactured from sustainable forestry in which at least two trees are planted for each one used for paper production.

Contents

Preface

The undergraduate curricula of almost all disciplines of engineering include some courses on modelling and analysis of dynamical systems. They include modelling approaches based on block diagram, signal flow graph, and so on, which obtain the system models in the form of transfer functions. Analysis of stability and other properties is then carried out with the help of root locus, Nyquist criterion and similar tools. At base, the Laplace transform remains the tool of analysis of the engineer. This restricts the exposure of the students to the behaviour of linear systems only.

Over the past few years, there has been an increasing realization that most of the physical systems are nonlinear, and linearity is a very special case. Most of the systems that an engineer has to deal with are nonlinear, and nonlinear dynamics pervades the engineer's workplace. But the training of the engineer often renders him/her hopelessly short of the challenges. This book is aimed at addressing this problem, first by introducing those methodologies of system modelling that make no reference to linearity, and then by developing an understanding of dynamics where linear system description is put in proper perspective – as local linear approximation in the neighbourhood of an equilibrium point.

The book is divided into two parts. In the first part, the methods and techniques for translating a physical problem into mathematical language by formulating differential equations are introduced. Some part of it draws from classical mechanics, but instead of working with particles and groups of particles as in classical mechanics textbooks, I have dealt mainly with electrical and mechanical systems so that the ideas developed can easily be applied in engineering.

The basic methodology of deriving differential equations follows from Newton's laws for mechanical systems and from Kirchoff's laws for electrical circuits. However, the Newtonian method in its pure form is not suitable for handling practical systems. In Chapter 2, the Newtonian formalism is introduced, and the practical problems of this method are illustrated. In Chapter 3, the application of Kirchoff's laws in derivation of dynamical equations for electrical circuits is considered. The mesh current method, the node voltage method and the more general graph theoretic methods are introduced. This chapter is particularly useful for students of the electrical sciences, and may be skipped by those of the other disciplines without breaking the continuity of exposition.

The Lagrangian method is introduced in Chapter 4, and its application in handling electrical, mechanical and electromechanical systems is illustrated. To show the basic unity of dynamical systems, an equivalence of the mechanical and the electrical systems is shown at relevant places. In order to make the model amenable to the solution techniques to be introduced in the latter part of the book, first-order differential equations must be obtained. Chapter 5 shows how the Lagrangian equations can be used to obtain the equations in first-order through the definition of conjugate momenta. The Hamiltonian formalism – which allows one to obtain the first-order equations directly – is generally applied to conservative systems. I have shown how this approach, with the inclusion of the dissipative term, can be made useful in handling engineering systems also.

Then I have introduced the Bond Graph methodology – a powerful technique for obtaining first-order differential equations for a wide variety of physical and engineering systems. This method algorithmizes the process of obtaining equations, so that once the bond graph of a system is formulated, computer programs can handle the job of obtaining equations and simulating the model. This method is widely applicable to engineering problems, but has not yet entered the mainstream engineering curriculum. Engineering students are often found to put off learning this powerful technique for some day "when there is time at hand". In this book, the basic elements of this method are introduced in a span of only 40 pages without getting into the nitty-gritty of modelling complicated systems, so that the reader can get a feel of bond graph–based system modelling without spending too much time on it.

The main advantage of the methods introduced in Part I is that they are equally applicable to both linear and nonlinear systems. I believe, these will constitute the basic modelling tools in the hands of the scientist and the engineer in future.

After the differential equations are obtained, one has to *solve them*, and from the solutions, one has to understand how a given system is going to behave in specific circumstances. The second part of the book is aimed at developing an intuitive understanding of the dynamics of physical systems. For this, the concepts of *state space* and *vector field* are introduced, and a geometric view of the dynamics in the state space is provided – since with the availability of computers and computer graphics that viewpoint has become visualizable and intuitively appealing.

The method of locally linearizing a nonlinear system through the Jacobian matrix is introduced, and the dynamics of linear systems are then analysed. There are many approaches to solving linear differential equations, and I have chosen the one that allows the relationship of the eigenvalues and eigenvectors with system dynamics to be highlighted. This gives a geometric view of the dynamics, and facilitates a smooth transition to the understanding of the dynamics of nonlinear systems. The special features of the dynamics of nonlinear systems, like limit cycles, high-period orbits and chaotic orbits, are then discussed.

Even though limit cycles find wide application in engineering wherever some oscillatory behaviour is desired (as in oscillators, power electronics, etc.), treatment of the stability of limit cycles is rarely found in dynamics or control textbooks. An approach to this problem, developed in nonlinear dynamics, has now reached sufficient maturity to deserve being taught at the undergraduate level. In this book, I have illustrated the powerful method of obtaining discrete-time models or "maps", developed by the French mathematician Henri Poincaré, and have discussed the various ways a limit cycle can lose stability. In that process, I have included an exposition on the dynamics of discrete-time systems, which is finding increasing application in engineering.

The content of this book is suitable for teaching undergraduate students of all branches of engineering at the second- or third-year level. Though the scope of the topic is much wider, the material presented in this book is limited to the extent that can be taught in a one-semester course – which, I feel is a proper supplement of a control systems course. Some parts of it can also be integrated in an existing course on control theory.

The book addresses dynamics problems coming from a wide range of engineering disciplines, and can be used by mechanical engineers, electrical engineers, aeronautical engineers, civil engineers, and so on. For example, an electrical engineer who is not interested in mechanical systems may read Chapters 1 and 3 to pick up the methods of obtaining differential equations specific to electrical systems, and then may proceed on to Chapter 7 onward to develop ideas of dynamics. For those who have to deal with electromechanical systems like relays, motors and other electrically actuated mechanical systems, Chapters 4, 5 and 6 will be particularly useful. The students of mechanical, aeronautical and allied engineering disciplines, on the other hand, may skip Chapter 3.

This book can also be used by those who are primarily interested in linear systems, because the methods of obtaining differential equations are equally applicable to linear systems. Moreover, Chapters 9 and 10 specifically deal with solving linear differential equations and developing a visual impression of linear dynamics. Courses with such leaning may leave Chapters 11 and 12 as options or for a later study.

It should be kept in mind that this book is not meant to be an exhaustive treatise on system modelling. The purpose of Part I is to acquaint the engineering students with the general modelling approaches, which they can later pursue to any level of detail following the lead provided at the end of each chapter. This is also not a typical nonlinear dynamics book, and includes only those aspects of nonlinear dynamics which a twenty-first century engineer should be exposed to. I have deliberately chosen not to burden the students with too much material.

The subject matter of this book was developed through the teaching of the subject "Dynamics of Physical Systems" at the Indian Institute of Technology, Kharagpur, India. I am indebted to the vision of Prof. Y. P. Singh who was instrumental in introducing this subject as an important component in the curriculum of the Bachelor's degrees in Electrical Engineering and Energy Engineering at the IIT, Kharagpur. In the course of writing the book, valuable feedback was received from the students to whom this subject was taught. They contributed in many ways in giving it the shape of a concrete course material. Among them, Mr. Manish Agarwal deserves special mention, for he helped in formulating many of the exercise problems.

I am particularly indebted to Prof. Amalendu Mukherjee of the Mechanical Engineering Department of our institute, who went through the whole manuscript and made many valuable suggestions. Mr. Ashoke Mukherjee and Prof. G. P. Rao helped me in improving the language. I am indebted to the Continuing Education Cell of IIT, Kharagpur, for providing the financial support for the preparation of the manuscript. I also thank Ms. Wendy Hunter of John Wiley & Sons for her continuous support at times of difficulty. Last, but not the least, I thank my daughter Anita, my son Kiran and my wife Manua for being patient when I could not spare time for my family while preparing the manuscript.

I invite the students and teachers who will use the book to send me comments and suggestions for its further improvement.

Soumitro Banerjee
Department of Electrical Engineering
Indian Institute of Technology, Kharagpur, India

Part I

Obtaining differential equations for physical systems

Everything in nature is continuously changing. However static and unchanging some of the things may look, they are all changing – some fast and some slowly. And any system whose status changes with time is called a *dynamical system*.

The study of dynamical systems has an intrinsic value for scientists who attempt to understand how nature functions. For engineers, it is the bread and butter. Everything he has to deal with is a dynamical system. He has to design them, operate them and he has to predict how a given system is going to behave in a particular circumstance.

Dynamical systems are described by *differential equations* – whose solutions show how the variables of the system depend on the independent variable time. Hence the thrust of the following chapters will be to formulate the differential equations for different types of systems.

Dynamics for Engineers S. Banerjee

1

Introduction to System Elements

1.1 Introduction

Any electrical, mechanical or electromechanical system is composed of some *elements* that interact with one another to produce the dynamics of the total system. To model the dynamics of the system, therefore, it is necessary to understand the dynamical properties of the individual elements.

Though, in general, the elements of a practical engineering system may be quite complex, with complicated (often nonlinear) dynamical properties, we can take the first step by considering a few discrete elements often found in practical systems, and idealizing them into linear components. Often, the component parts of a system may be spatially extended, and if one looks into what goes on inside that component, the problem may become unmanageable. What one does, instead, is to take "lumped representation" of the component where one focuses attention on the way it interacts with other elements through the "end points", and ignores the behaviour at the points inside it. In the following section, we shall discuss some such system elements.

The behaviour of certain electrical quantities is mathematically identical to that of certain mechanical quantities. This enables us to establish an equivalence between electrical and mechanical systems. These equivalences are also discussed in the following sections.

1.1.1 The inertial element

The property of inertia is to resist change in velocity. A moving point mass or a rotating rigid body are examples of the inertial element in mechanical systems. If a force $\mathbf{f}$ is applied on a translational mass that moves with momentum $\mathbf{p}$, then the

Dynamics for Engineers S. Banerjee

fundamental property of the inertial element is given by the relation

$$\mathbf{f} = \frac{\mathrm{d}\mathbf{p}}{\mathrm{d}t}.$$

With **q**, **v** and **a** representing the position, velocity and acceleration respectively, this relation can be written as

$$\mathbf{f} = m\frac{\mathrm{d}^2\mathbf{q}}{\mathrm{d}t^2} = m\frac{\mathrm{d}\mathbf{v}}{\mathrm{d}t} = m\mathbf{a},$$

which defines the mass m as the slope of the (linear) graph of the magnitude $f = |\mathbf{f}|$ versus $a = |\mathbf{a}|$ (Fig. 1.1(a)). In SI units, the unit of mass is kg, and that of force is Newton (or kg m/s^2).

For the rotation of a rigid body about a fixed axis, the same relationship obtains between the applied torque **f** and angular momentum, given by

$$\mathbf{f} = I\frac{\mathrm{d}\boldsymbol{\omega}}{\mathrm{d}t} = I\boldsymbol{\alpha},$$

where $\boldsymbol{\omega}$ and $\boldsymbol{\alpha}$ are the angular velocity and acceleration respectively. The unit of moment of inertia is kg m^2 and that of torque is N m (Newton meters).

For translational motion in rectangular coordinate system, the directions of the vectors are obtained in a straightforward manner. But what represents the direction of the vectors in rotational motion? The vectors in the relation above are obtained by the right-hand rule, with the vector pointing in the direction of the thumb (Fig. 1.1(c)).

Inertial element in the electrical domain is the inductance, whose dynamical property is given by

$$E = L\frac{\mathrm{d}i}{\mathrm{d}t} = L\frac{\mathrm{d}^2q}{\mathrm{d}t^2}.$$

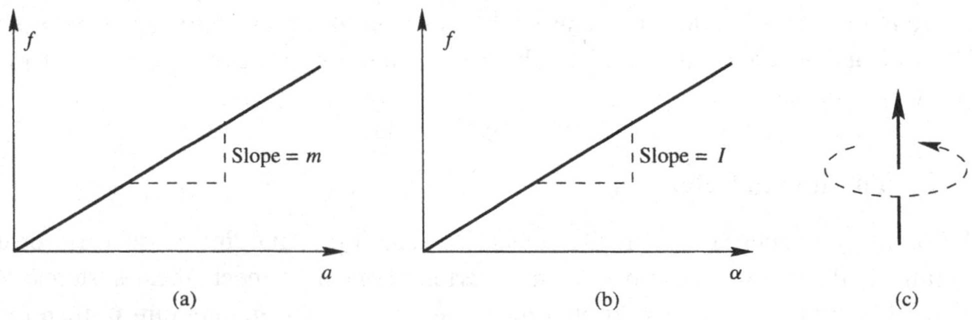

Figure 1.1 The law of inertia in (a) translational and (b) rotational motion, where the vectors point in the direction shown in (c) by the right-hand corkscrew rule.

Thus the electrical equivalent of mechanical force is voltage or electromotive force, and the electrical equivalent of position is charge. The analogy also indicates that the energy stored in an inductor, $\frac{1}{2}L\dot{q}^2$, is the electrical equivalent of kinetic energy. The unit of inductance is Henry (or V s/A). The property of inertia is to resist change in speed and the property of inductance is to resist change in current.

1.1.2 The compliant element

The property of a compliant element is to resist change in the separation between its end points, that is, to resist compression or stretching. If a translational spring is given a relative displacement of q between the two ends, it produces a force f given by

$$f = kq = k\int v\,\mathrm{d}t,$$

where k is the spring constant (or stiffness). Thus k represents the slope of the graph, assumed to be linear, between the force and the relative displacement, and its unit is N/m. Fig. 1.2(a) shows the characteristics of the hard and soft springs, as well as that of the linear spring represented by the equation above. For torsional springs, the same relationship applies, with f representing the torque, q representing the relative angular displacement and k representing the torsional stiffness.

The corresponding electrical element is the capacitance, whose dynamical relation between the voltage across the capacitor e and the charge in the capacitor q (Fig. 1.2(b)) is given by

$$e = \frac{1}{C}q = \frac{1}{C}\int i\,\mathrm{d}t.$$

The unit of charge is Coulomb (or A s) and that of capacitance is Farad (or A s/V).

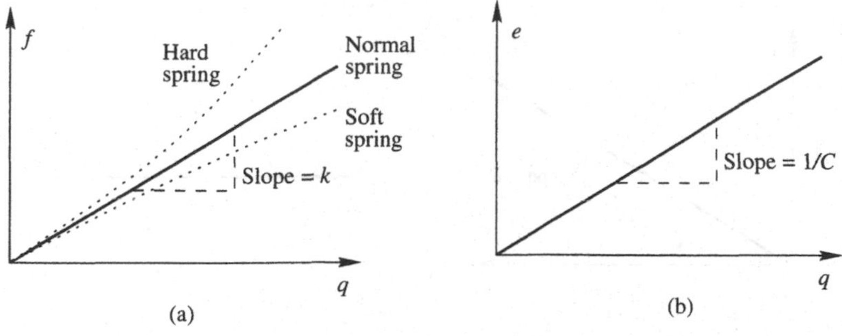

Figure 1.2 (a) The spring element in mechanical translational motion and (b) the capacitance in electrical circuit.

The potential energy stored in a spring is given by $\frac{1}{2}kq^2$, and the electrical equivalent of potential energy is the energy stored in a capacitor, given by $\frac{1}{2C}q^2$.

1.1.3 The resistive element

The electrical resistance is represented by the Ohm's law, which is a linear approximation of the relationship between the voltage across it and the current flowing through it:

$$e = Ri = R\frac{\mathrm{d}q}{\mathrm{d}t}.$$

There are, in the main, two types of mechanical friction or damping:

1. Viscous friction
2. Coulomb friction.

Viscous friction is generated when two surfaces separated by a liquid slide against each other. The damping force due to friction opposes the motion, and depends on the nature of fluid flow between the surfaces. The relationship between the relative velocity of sliding and the generated damping force is quite complex, but can be approximated to a linear relationship (Fig. 1.3) given by

$$f = R.v = R\frac{\mathrm{d}q}{\mathrm{d}t}.$$

Therefore viscous friction is similar in nature to electrical resistance. In rotational motion, the same relation applies between the relative angular velocity of the two surfaces and the torque created by friction.

Coulomb friction is generated when two dry surfaces slide against each other. Suppose a solid body is resting on a dry surface and a force is applied to move it.

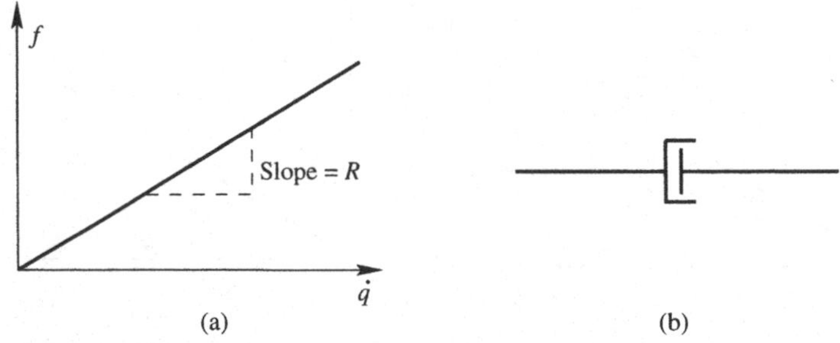

Figure 1.3 (a) The characteristics of viscous friction, and (b) the representation of a viscous damper.

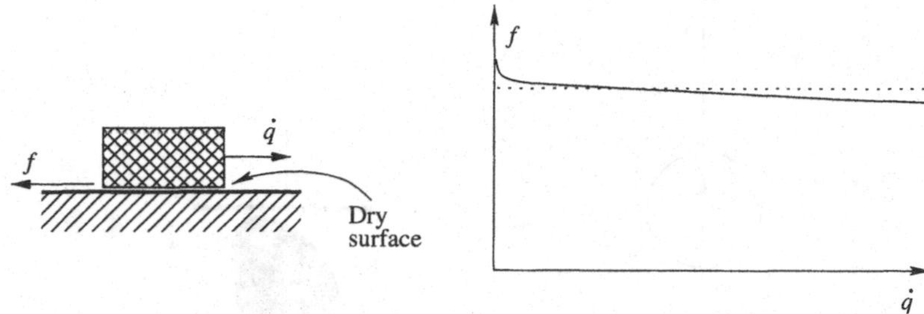

Figure 1.4 The characteristics of Coulomb friction.

As the force is increased gradually, the body does not move till a critical value of force is reached. Upto that point, the applied force equals the *static friction force*. At the critical juncture, just before the beginning of the sliding motion, the static friction force attains a maximum value. As soon as sliding occurs, the nature of the friction force changes, and now it is the *kinetic friction force*. At the beginning of motion, it has a value slightly less than the maximum value of static friction, and decreases with the increase of the relative velocity (Fig. 1.4).

For the purpose of this book, we will mainly assume friction elements to be of the viscous type. The viscous damper will be denoted by the symbol shown in Fig. 1.3(b). Friction elements with nonlinear characteristics can be modelled by considering R to be a variable parameter, and by giving it a suitable functional form.

1.1.4 The voltage source and externally impressed force

The voltage source and externally impressed force represent the *source of effort* in electrical and mechanical system respectively. What they apply on a system – mechanical force in case of mechanical system and electromotive force in case of electrical system – are independent variables, not affected by the rest of the system. The notation of the voltage source is shown in Fig. 1.5(a).

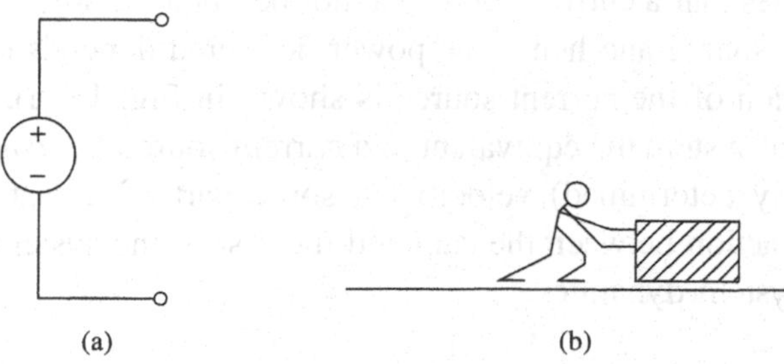

Figure 1.5 The voltage source (a) and the mechanically impressed force (b).

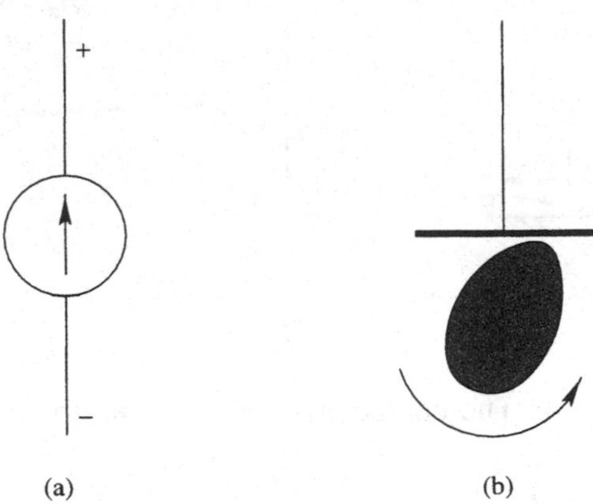

Figure 1.6 (a) The current source in an electrical circuit, and (b) the cam in a mechanical system.

However, the current through a voltage source may vary depending upon the rest of the circuit. Since the power delivered by the voltage source is the product of the voltage and the current, it follows that the power flowing from a source to the rest of the system depends not only on the characteristics of the source but also on the system configuration.

Likewise, the agent applying an external force in a mechanical system has to move along with the mass on which the force is applied (Fig. 1.5(b)), and hence the power delivered depends both on the character of the applied force and the rest of the system.

1.1.5 The current source and externally impressed sources of flow

The correct source in an electrical circuit has the property that the current through it is an independent variable and does not depend on the rest of the circuit (this immediately implies that a current source cannot be open-circuited). But the voltage across the current source and hence the power delivered depends on the rest of the circuit. The notation of the current source is shown in Fig. 1.6(a).

In a mechanical system the equivalent of a current source is a *cam* which imposes a given (externally determined) velocity on some part of a system (Fig. 1.6(b)). The force of interaction between the cam and the rest of the system is variable, and depends on the system dynamics.

1.2 Chapter Summary

Translational or rotational motion in a mechanical system is characterized by three types of quantities:

- inertia (or the tendency to resist acceleration),
- compliance (or the tendency to resist compression and stretching),
- friction (or the tendency to resist motion).

In an electrical system, the dynamics or the change in the variables like voltage, current, and so on, are governed by three types of quantities:

- inductance (or the tendency to resist change in current),
- capacitance (or the tendency to resist change in voltage),
- resistance (or the tendency to resist the flow of charge).

These are quantities that are internal to a system. There are also quantities that are given by things external to the system – the applied force and cam in a mechanical system, and the voltage source and the current source in an electrical system.

There is an equivalence between some mechanical quantities and some electrical quantities, in the sense that their dynamical character is the same. For the sake of convenience, the equivalence between mechanical quantities and electrical quantities is summarized in Table 1.1.

Table 1.1 Equivalence of variables and parameters in mechanical and electrical systems

Mechanical quantity	Electrical quantity
Displacement	Charge
Velocity	Current
Force	Voltage
Mass	Inductance
Spring constant	1/capacitance
Damping coefficient	Resistance
Potential energy	Energy in capacitor
Kinetic energy	Energy in inductor

Further Reading

C. Nelson Dorny, *Understanding Dynamic Systems: Approaches to Modeling, Analysis and Design*, Prentice Hall, Englewood Cliffs, New Jersey, 1993.

R. L. Woods and K. L. Lawrence, *Modeling and Simulation of Dynamic Systems*, Prentice Hall, Upper Saddle River, New Jersey, 1997.

2

Obtaining Differential Equations for Mechanical Systems by the Newtonian Method

The basic methodology for formulation of differential equations was provided by Isaac Newton (1642–1727). He discovered, following Galileo's thoughts, that the application of a force changes the velocity of a body. The rate of change of velocity, or acceleration, is proportional to the applied force. And the mass of the body appears in the constant of proportionality as $F = ma = m\frac{dv}{dt}$.

Following Newton, D'Alembert (1717–1783) proposed an equivalent but slightly different viewpoint. He saw the entity called *mass* as having the property of actively trying to maintain status quo. If left to itself, a mass maintains its own velocity with respect to any inertial frame of reference. But if it is not allowed to maintain its inertial status by the application of an external force, an opposing tendency or "inertial force" develops in it, whose magnitude is given by the rate of change of momentum (or mass × acceleration), which acts in opposition to the applied force. These two are equal in magnitude, and the body exists in the unity of the two opposing tendencies.

That gives us a way of writing an equation. And since acceleration is the double derivative of position, the resulting equation would be a differential equation. Thus, following Newton, a simple methodology of understanding the dynamics of any system emerged; just look for the opposing tendencies in a system since motion or change in any system is the result of these opposing tendencies.

► **Example 2.1** Consider the mass-spring system in Fig. 2.1. The externally applied force is F. Let the position of the mass m be measured such that it has value zero at the unstretched position of the spring, and is positive in the direction of the applied force.

Dynamics for Engineers S. Banerjee

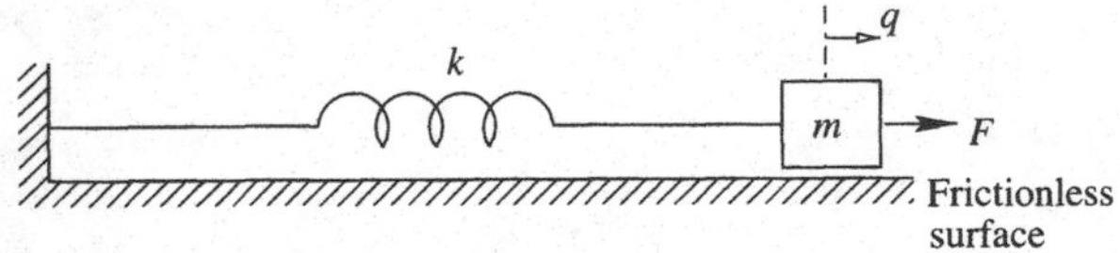

Figure 2.1 The mass-spring system of Example 2.1.

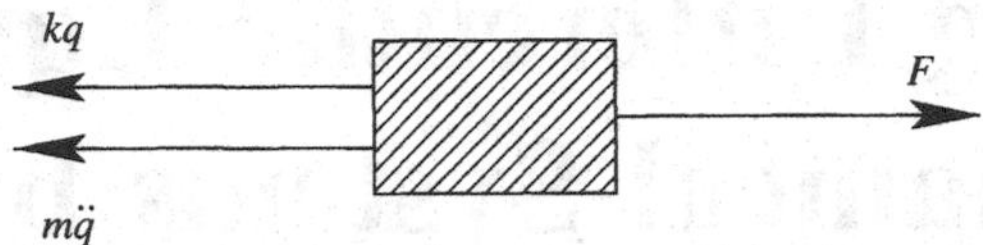

Figure 2.2 The free body diagram of the system of Example 2.1 following D'Alembert's principle.

For understanding the balance of forces, one draws what is known as the free body diagram (FBD) – where the forces acting on each mass point are shown separately. In this system, there is only one mass point, and its FBD is shown in Fig. 2.2. When the elongation of the spring is q, the force on the mass exerted by the spring is kq, where k is the spring constant. Therefore the total force on the mass is $F - kq$. Owing to this force the mass moves, and the rate of change of momentum acts in opposition to the applied force. Equating these two opposing tendencies, we get

$$\frac{\mathrm{d}}{\mathrm{d}t}(m\dot{q}) = F - kq$$

or

$$m\ddot{q} + kq - F = 0.$$

This is the differential equation that describes the dynamics of the mass. ◀

2.1 The Configuration Space

The positional status of any system can be uniquely specified with the help of a few real numbers. If we have one particle (or a solid body), its position can be specified with a vector $\mathbf{r}$ consisting of three real numbers representing the three coordinates. If we have two solid bodies making up a system, we would need two vectors $\mathbf{r}_1$ and $\mathbf{r}_2$ consisting of six position coordinates. Likewise, for a system with N mass points, one requires N vectors $\mathbf{r}_j$ (j varying from 1 to N) consisting of $3N$ coordinates.

An interesting feature is that there is no fundamental distinction between the x-coordinate of one body and the z-coordinate of another. All coordinates have equal value in defining the configuration of the whole system. It is therefore convenient to express the positional status or configuration of a system as a collection of $3N$ real numbers $(x_1, x_2, x_3, \ldots, x_{3N})$, each number representing a position coordinate of one of the participating bodies.

Geometrically, one can visualize a $3N$-dimensional space with the configuration coordinates x_i as the axes. The configuration of a system at any instant would be represented by a *point* in this space. This is the *configuration space*. Dynamics of a system would then be represented by movement of the point in the configuration space.

2.2 Constraints

Most physical systems have *constraints* that restrict the movement of the point in the configuration space. The bob of a pendulum is constrained to move on the surface of a sphere. A ball sliding down a plane is constrained to remain on the plane. People riding a roller coaster are constrained to move along given trajectories.

In these cases, the constraints can be expressed by algebraic equations of the form

$$f(x_1, x_2, x_3, \ldots, x_{3N}, t) = 0$$

or, equivalently,

$$f(\mathbf{r}_1, \mathbf{r}_2, \mathbf{r}_3, \ldots, \mathbf{r}_N, t) = 0. \tag{2.1}$$

A constraint of the form (2.1) or reducible to that form, is called a *holonomic* constraint.

► **Example 2.2** For a spherical pendulum, the equation of constraint is

$$x_1^2 + x_2^2 + x_3^2 - l^2 = 0.$$

If it is a planar pendulum, there would be another constraint given by $x_3 = 0$. These equations are of the form (2.1) without time dependence. If the point of suspension is moved as a function of time in x_1 direction, the constraint equation becomes

$$[x_1 + f(t)]^2 + x_2^2 + x_3^2 - l^2 = 0.$$

Here the constraint equation is a function of both x_j and time. ◄

Holonomic constraints can be further subdivided into two types. A constraint that does not depend explicitly on time is called *scleronomic*. The equation of a scleronomic constraint would be of the form

$$f(\mathbf{r}_1, \mathbf{r}_2, \mathbf{r}_3, \ldots, \mathbf{r}_N) = 0.$$

The constraints that depend on time and hence are not of the above form are called *rheonomic*. They have to be expressed in the form (2.1).

In a three-dimensional configuration space an algebraic equation of the form (2.1) represents a surface. The configuration point would be constrained to remain on this surface throughout the dynamical motion. If two such constraints exist, the configuration point would be constrained to remain on the line obtained by the intersection of the two surfaces. This argument applies to higher dimensional configuration space also, though the surfaces would then be difficult to visualize. It turns out that every holonomic constraint reduces the dimension of the system by one. The existence of holonomic constraints, therefore, simplifies system modelling problems a great deal.

There are constraints that cannot be described by algebraic equations of the form (2.1). These are called *non-holonomic* systems. Billiard balls are constrained to move within the boundaries of the table. Here the coordinates representing the location of the balls are bounded within some range of real numbers. The constraint is therefore expressible as an *inequality*, which is a signature of non-holonomic systems. Non-holonomic systems also include cases where the *velocities* of the elements are related by an algebraic equation, as[1]

$$f(\dot{\mathbf{r}}_1, \dot{\mathbf{r}}_2, \dot{\mathbf{r}}_3, \ldots, \dot{\mathbf{r}}_N, t) = 0. \tag{2.2}$$

► **Example 2.3** (Rosenberg 1977, p. 29) Two objects cannot occupy the same position at any time. To illustrate that this obvious fact imposes a non-holonomic constraint, consider two particles moving along a straight line, and let their positions at any instant of time be given by their distance from some fixed point on that line (Fig. 2.3). Since their positions

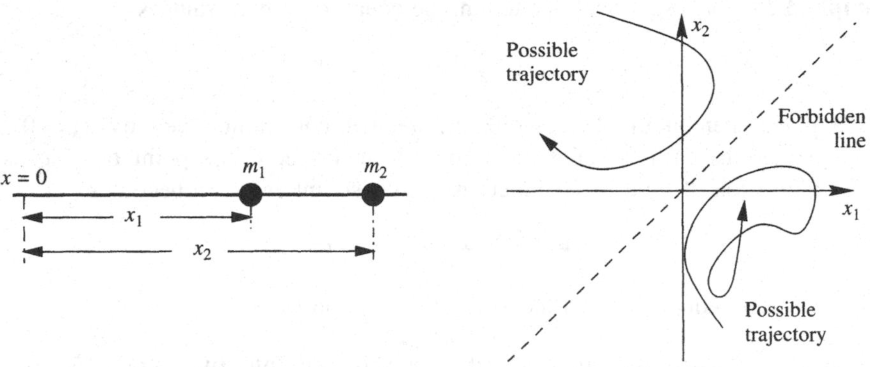

Figure 2.3 Two particles moving on a straight line. Two possible trajectories are shown in the configuration space.

[1]It is to be noted that a holonomic constraint equation of the form (2.1) can be differentiated term-wise to obtain a relationship between the velocities. Such an equation relating the velocities would not imply that the system is non-holonomic, because it can be integrated to the form (2.1). A differential constraint of the form (2.2) is non-holonomic if it is not integrable.

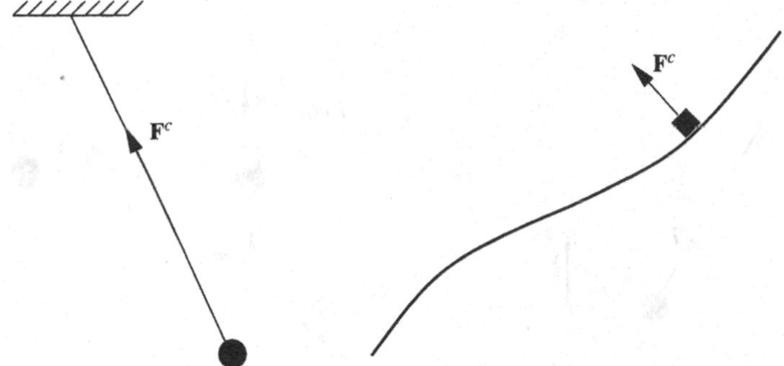

Figure 2.4 Forces of constraint.

can never coincide, the constraint is given by $x_1 \neq x_2$, that is, the line $x_1 = x_2$ in the configuration space is a forbidden line. Therefore the trajectory is constrained to be in one of the two halves of the configuration space divided by the forbidden line. The system will remain in the half where the initial condition lies, for it can never cross the forbidden line. ◀

It is therefore clear that a non-holonomic constraint imposes a restriction on the *region* of configuration space that can be accessed by the system. Such a constraint does not reduce the dimension of the system.

Fortunately, most engineering systems are holonomic in nature, and that allows us to work with low-dimensional models. From now onward, we shall confine ourselves to holonomic systems. Whenever we come across non-holonomic systems we shall have to obtain the dynamical equations with the full set of configuration variables.

Newton's second law says that whenever there is any alteration of the state of motion, there must be a force associated with it. Since holonomic constraints alter the natural trajectories of systems, it follows that they give rise to forces. These are called forces of constraint. The tension in the string of a pendulum is an example of such a constraint force. If a body moves on a surface, the constraint force is the force of reaction of the surface on the body (Fig. 2.4).

2.3 Differential Equations from Newton's Laws

The above shows that the motion of any system would be a consequence of two kinds of forces: the applied force (denoted by $\mathbf{F}$) and the constraint force (denoted by $\mathbf{F}^c$). We can write the Newton's equation for each mass point as

$$m_j \ddot{\mathbf{r}}_j - \mathbf{F}_j = \mathbf{F}_j^c, \tag{2.3}$$

where $\mathbf{F}_j$ and $\mathbf{F}_j^c$ are the given force and the constraint force on the jth mass point. This is the basic equation of the Newtonian formalism.

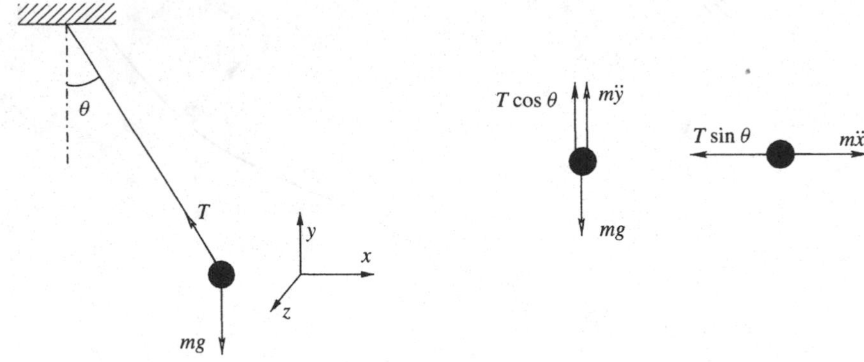

Figure 2.5 The simple pendulum considered in Example 2.4, and the free body diagrams for the x and y coordinates.

▶ **Example 2.4** Consider the simple planar pendulum shown in Fig. 2.5. Here the tension of the string T represents the constraint force that makes the bob move in a circular path. The force due to gravity is mg. The total force in the y direction is $mg - T\cos\theta$, and that in the x direction is $T\sin\theta$. There is no component of the force in the z direction. Therefore the differential equations in the three directions become

$$m\frac{d^2x}{dt^2} = T\sin\theta,$$

$$m\frac{d^2y}{dt^2} = mg - T\cos\theta,$$

$$m\frac{d^2z}{dt^2} = 0.$$

It is clear that this purely Newtonian formulation of the equation is inconvenient for practical application because the constraint force T enters the equations. It varies from point to point and its magnitude is not known a priori. ◀

2.4 Practical Difficulties with the Newtonian Formalism

We thus see that obtaining differential equations by the Newtonian method in its pure form has a few practical problems. First, all the constraint forces in a given system have to be included in the model, and in a complicated system there may be a large number of constraint forces that are not easily quantifiable. For example, in a system of pulleys, all the masses are attached to the rest of the system through strings and the forces in the strings constitute the set of constraint forces. Accounting for all of them would be a cumbersome exercise.

Second, in an interconnected system, the mass points interact with each other through springs or frictional elements. The forces exerted by these elements on each mass are to be categorized as "given forces". We need to take into account all these forces, and the method becomes unmanageable with the increase in the number of interacting elements.

Third, for an N-body problem one would have to obtain $3N$ differential equations. We have seen that the existence of holonomic constraints may reduce the number of configuration coordinates. For the simple system of Example 2.4, the number of configuration coordinates can be reduced to one by adopting a polar coordinate system. But for more complicated systems the Newtonian method does not offer any direct way of utilizing the advantage offered by the holonomic constraints.

In Chapter 4, we will see how these practical difficulties of the Newtonian method were overcome, resulting in a powerful method of obtaining mathematical models of dynamical systems.

2.5 Chapter Summary

For a system consisting of N mass points, the positional status is specified by $3N$ real numbers. A space with these $3N$ real numbers as coordinates is called the configuration space. Constraints in a given system may restrict the motion of the configuration point in the configuration space, and in doing so, give rise to constraint forces. The Newton's law, applied to each mass point, gives

$$m_j\ddot{\mathbf{r}}_j - \mathbf{F}_j = \mathbf{F}^c_j, \tag{2.4}$$

where $\mathbf{F}_j$ and $\mathbf{F}^c_j$ are the given force and the constraint force on the jth mass point. For N mass points there will be N such equations. Summing up the N equations, we get

$$\sum_{j=1}^{N}(m_j\ddot{\mathbf{r}}_j - \mathbf{F}_j) = \sum_{j=1}^{N}\mathbf{F}^c_j. \tag{2.5}$$

This method of formulation of the differential equations encounters three practical problems. First, the constraint forces get into the formulation; second, all the forces of interaction between elements have to be accounted; and third, it ignores the advantage offered by holonomic constraints, that is, the possibility of reduction of system dimension.

Further Reading

H. Goldstein, *Classical Mechanics*, Addison Wesley, 1980.
R. M. Rosenberg, *Analytical Dynamics of Discrete Systems*, Plenum Press, New York, 1977.

Problems

1. Are the following systems holonomic or non-holonomic? Write down the equations (or in equations) that define the constraints.

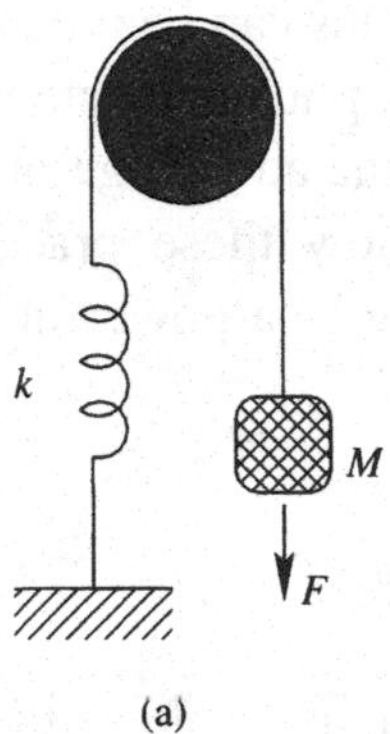

(a)

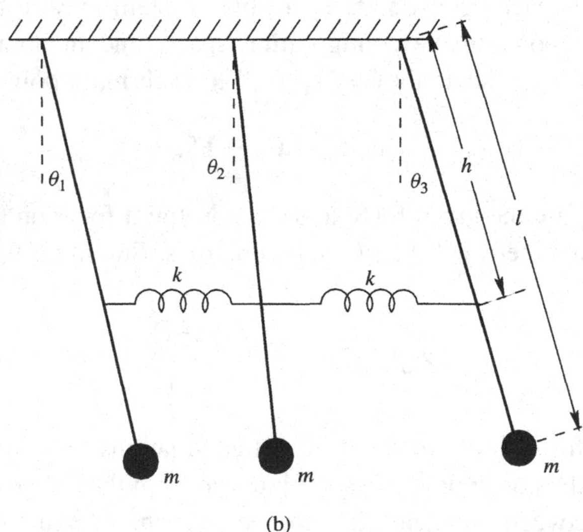

(b)

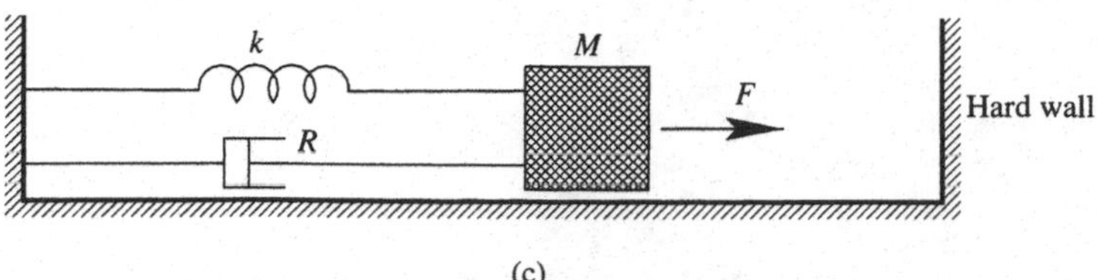

(c)

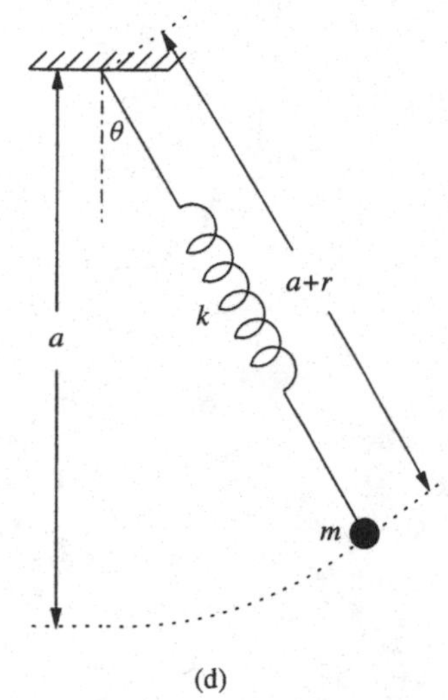

(d)

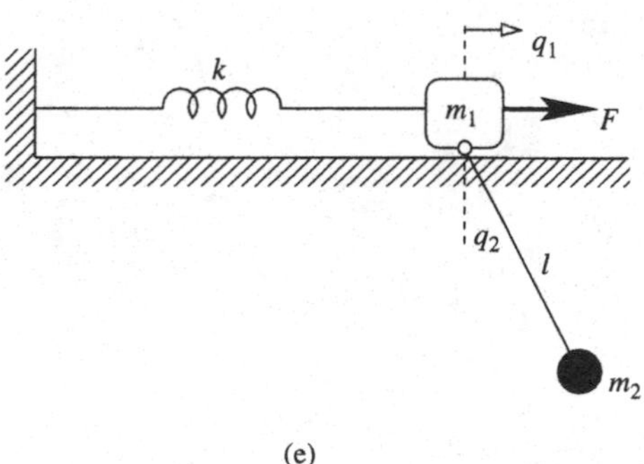

(e)

2. For the above problems, try to obtain the dynamical equations using the Newton's laws. Comment on the observations in Section 2.4.

3

Differential Equations of Electrical Circuits from Kirchoff's Laws

In electrical circuits, the elements like inductors, capacitors, resistors and sources are connected in specific ways that establish some relationships between the variables. Kirchoff's laws express the fundamental relations between the voltages and currents, which can be used to formulate differential equations.

3.1 Kirchoff's Laws about Current and Voltage

The points in a circuit that can be at different voltages are called *nodes*. The Kirchoff's current law (KCL) establishes the relationship between the currents entering or going out of nodes. It states that *the algebraic sum of the currents in all the branches that converge in a common node is equal to zero.*

The Kirchoff's voltage law (KVL) relates the voltages across the branches that form a loop. It states that *the algebraic sum of the voltages between successive nodes that form a closed path or loop is equal to zero.*

► **Example 3.1** In the circuit shown in Fig. 3.1, there are four nodes 0, 1, 2 and 3. If the currents in the branches 1-2, 1-3 and 1-0 are denoted by i_{12}, i_{13}, i_{10} respectively, since the three branches are connected to node 1, the KCL gives

$$i_{12} + i_{10} + i_{13} = 0.$$

Notice the directions of the currents. A current should have one sign if it enters the node, and the opposite sign if it is leaving the node.

Dynamics for Engineers S. Banerjee

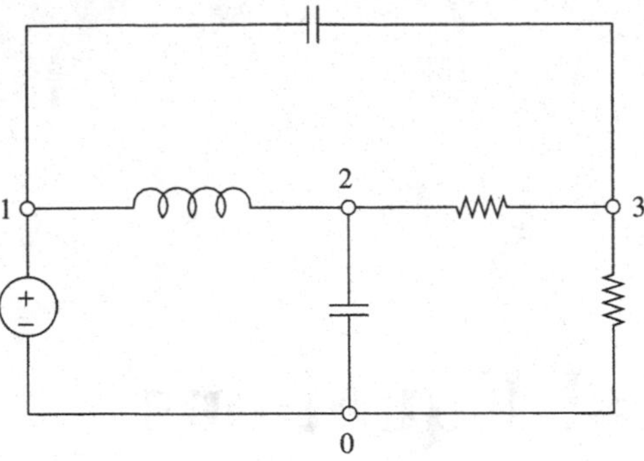

Figure 3.1 The circuit pertaining to Example 3.1.

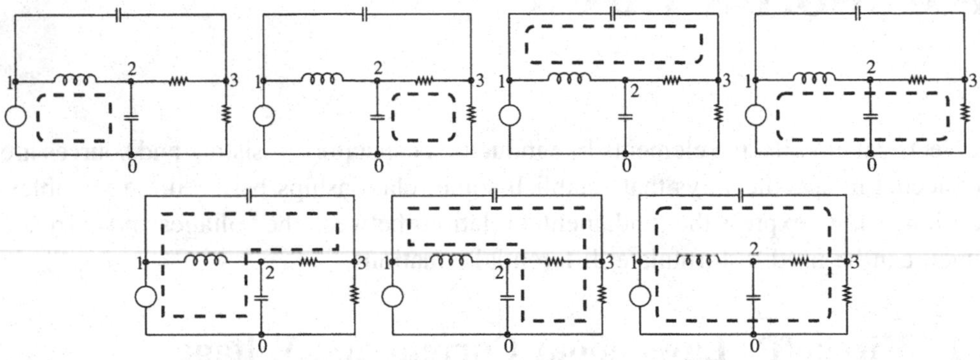

Figure 3.2 The possible loops or closed paths in the circuit of Example 3.1.

Similarly, the KCL equations for the other nodes will be

$$i_{21} + i_{20} + i_{23} = 0,$$
$$i_{31} + i_{30} + i_{32} = 0,$$
$$i_{01} + i_{02} + i_{03} = 0.$$

One can easily identify many closed paths in this circuit: 1-2-0-1, 0-2-3-0, 1-2-3-1, 1-2-3-0-1, 1-3-2-0-1, 0-2-1-3-0, 0-1-3-0 (see Fig. 3.2). KVL applied to these loops will yield the equations

$$v_{12} + v_{20} + v_{01} = 0,$$
$$v_{02} + v_{23} + v_{30} = 0,$$
$$v_{12} + v_{23} + v_{31} = 0,$$
$$v_{12} + v_{23} + v_{30} + v_{01} = 0,$$

$$v_{13} + v_{32} + v_{20} + v_{01} = 0,$$

$$v_{02} + v_{21} + v_{13} + v_{30} = 0,$$

$$v_{01} + v_{13} + v_{30} = 0.$$

◄

The basic method of using these equations in obtaining differential equations is as follows. If there is a capacitor in a branch, the relationship of its current and voltage is given by

$$\frac{\mathrm{d}v}{\mathrm{d}t} = \frac{i}{C}. \tag{3.1}$$

Likewise, if there is an inductor in any branch, its constitutive relation is

$$\frac{\mathrm{d}i}{\mathrm{d}t} = \frac{v}{L}. \tag{3.2}$$

The right-hand sides of these equations can then be expressed in terms of the variables of our choice using the KCL and KVL equations to obtain first-order differential equations.

Two questions immediately crop up:

1. What will be the right choice of the variables?
2. The total number of KCL and KVL equations are far larger than what are actually required. Indeed, many will not be independent equations. How can we save unnecessary labour by selecting the minimum number of independent equations?

Equations (3.1) and (3.2) suggest that the voltages across capacitors and the currents through inductors should form the natural choice of variables in whose terms the differential equations are to be obtained. The KCL and KVL equations are then to be used to express the other currents and voltages in terms of these variables.

To answer the second question, two approaches are generally adopted: the mesh current method and the node voltage method. These methods help in obtaining the minimum number of independent KCL and KVL equations in most cases.

3.2 The Mesh Current and Node Voltage Methods

A mesh is a closed path that does not contain any other loop within it. The meshes in a circuit are therefore the basic windows. The mesh current method is applicable only to planar circuits where there are no branches that pass over or under any other branches. In this method, one assumes that the currents flowing in the meshes are independent. Therefore, out of all the possible KVL equations for a given circuit, only the ones related to the meshes contain the information necessary for forming the differential equations. If a mesh contains an inductor, the resulting $L\frac{di}{dt}$ term gives a differential equation.

In the node voltage method, one identifies the nodes to which at least three branches are connected. One of the nodes can be assumed to be at zero potential (ground). KCL Equations can then be written relating the currents entering the remaining $(n-1)$ nodes in terms of the voltages of the nodes. If a branch connected to a node contains a capacitor, one gets an equation involving $C\frac{dv}{dt}$.

► **Example 3.2** In the circuit of Fig. 3.3, there are three closed paths, but only two meshes. Therefore, the independent mesh currents are i_1 and i_2 as shown in the figure. The independent variables are the current through the inductor i_1 and the voltage across the capacitor v_c. The independent KVL equations at the two meshes in terms of these variables are

$$E - L\frac{di_1}{dt} - v_c = 0,$$

$$v_c - i_2 R = 0.$$

The first one gives the differential equation in terms of i_1:

$$\frac{di_1}{dt} = E/L - v_c/L.$$

The other equation is not a differential equation. The differential equation involving v_c has to be obtained from a KCL equation.

In order to apply the node voltage method, we proceed as follows. First identify the nodes, which in this case are marked 0, 1 and 2. Out of these one node (say, node 0) can be assumed to be at ground potential, which leaves us with two nodes 1 and 2. Out of these, node 1 has only two branches connected, and so the corresponding KCL equation is trivial. That leaves us with only node 2 at which the KCL equation will be

$$i_1 - C\frac{dv_c}{dt} - \frac{v_c}{R} = 0.$$

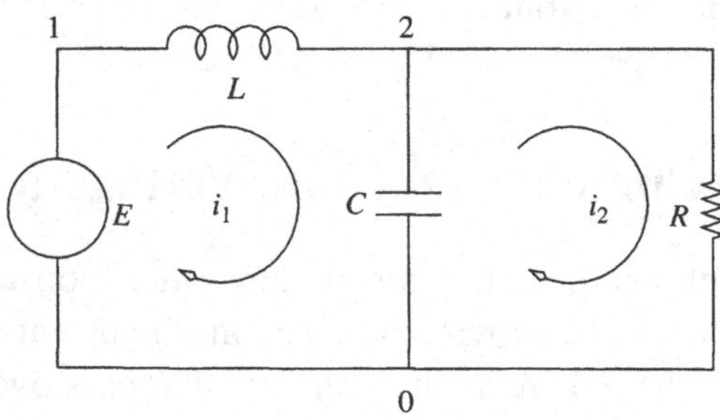

Figure 3.3 The circuit pertaining to Example 3.2.

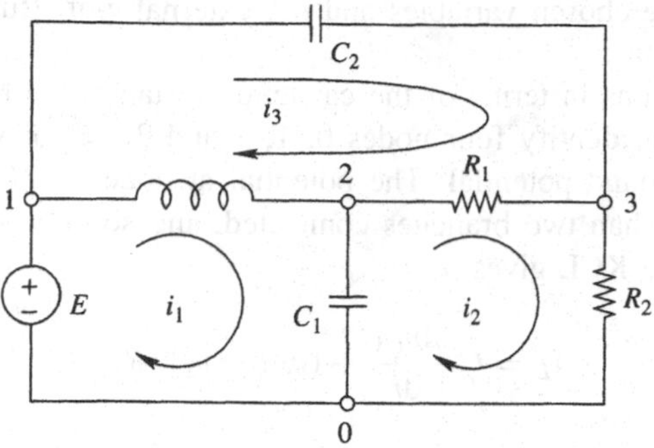

Figure 3.4 The meshes in the circuit of Example 3.1.

This straightaway gives the differential equation in terms of the capacitor voltage as

$$\frac{dv_c}{dt} = \frac{1}{C}i_1 - \frac{1}{RC}v_c.$$ ◀

► **Example 3.3** In the circuit considered in Example 3.1, there are three open windows or meshes as shown in Fig. 3.4. Therefore, the independent KVL equations are

$$v_{12} + v_{20} + v_{01} = 0,$$
$$v_{02} + v_{23} + v_{30} = 0,$$
$$v_{13} + v_{32} + v_{21} = 0.$$

The variables are the voltages across the capacitors v_{c1} and v_{c2}, and the current through the inductor i_L. We need also to specify which directions of these variables are assumed to be positive. Let the direction of v_{c1} be positive when the upper plate is at a higher potential with respect to the lower one; let the direction of v_{c2} be positive when the left plate is at a higher potential than that at the right; and let the direction of the current through the inductor be positive when it flows from left to right, that is,

$$i_L = i_1 - i_3. \tag{3.3}$$

Writing the KVL equations in terms of the circuit parameters, we get

$$E - L\frac{d(i_1 - i_3)}{dt} - v_{c1} = 0,$$
$$v_{c1} - R_1(i_2 - i_3) - R_2 i_2 = 0,$$
$$v_{c2} - R_1(i_3 - i_2) - L\frac{d(i_3 - i_1)}{dt} = 0.$$

By substituting (3.3), we get the first equation as

$$\frac{di_L}{dt} = \frac{E}{L} - \frac{v_{c1}}{L}, \tag{3.4}$$

which is in terms of the chosen variables and the external emf. But the other two do not yield differential equations.

To obtain the equations in terms of the capacitor voltages, we have to apply the node voltage method. We can identify four nodes 0, 1, 2 and 3, out of which the node 0 may be assumed to be at ground potential. The potential at node 1 is known; it is E. All the other nodes have more than two branches connected, and so they yield independent KCL equations. At node 2 the KCL gives

$$i_L = C_1 \frac{\mathrm{d}v_{C1}}{\mathrm{d}t} + (v_{C1} - v_3)/R_1$$

or

$$\frac{\mathrm{d}v_{C1}}{\mathrm{d}t} = \frac{1}{C_1} i_L - \frac{1}{R_1 C_1} v_{C1} - \frac{1}{R_1 C_1} v_{C2} + \frac{1}{R_1 C_1} E. \tag{3.5}$$

At node 3, the KCL gives

$$(v_2 - v_3)/R_1 - v_3/R_2 + C_2 \frac{\mathrm{d}v_{C2}}{\mathrm{d}t} = 0$$

or

$$\frac{\mathrm{d}v_{C2}}{\mathrm{d}t} = -\frac{1}{R_1 C_2} v_{C1} + \frac{1}{C_2}\left(\frac{1}{R_1} + \frac{1}{R_2}\right)(E - v_{C2}). \tag{3.6}$$

(3.4), (3.5) and (3.6) are the differential equations for the system. ◄

The above method of obtaining the differential equation is suitable for simple circuits. However, for complicated circuits it becomes difficult to keep track of the necessary and redundant equations. Keeping track of the signs of the voltages and currents also becomes difficult. And, as it has been mentioned earlier, the mesh current method fails if the circuit is non-planar. We therefore need a systematic method of analysing circuits.

There is another reason for the need of a systematic analysis. We have so far assumed that each inductor or capacitor will lead to a first-order differential equation. This is true for the situations where each capacitor voltage and inductor current are independent. However, such situations may also arise where the voltage across one capacitor is dependent (expressed as an algebraic equation) on other variables.

► **Example 3.4** Such a situation is illustrated in Fig. 3.5, where the resistance R_1 of Example 3.1 is replaced by a capacitance.

Here, the loop shown has only capacitors and the voltage source, leading to the KVL equation:

$$v_{01} + v_{13} + v_{32} + v_{20} = 0$$

or

$$E - v_{C2} + v_{C3} - v_{C1} = 0,$$

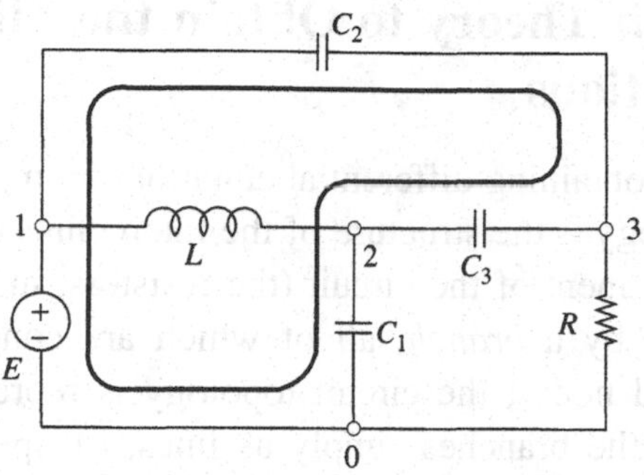

Figure 3.5 The circuit pertaining to Example 3.4.

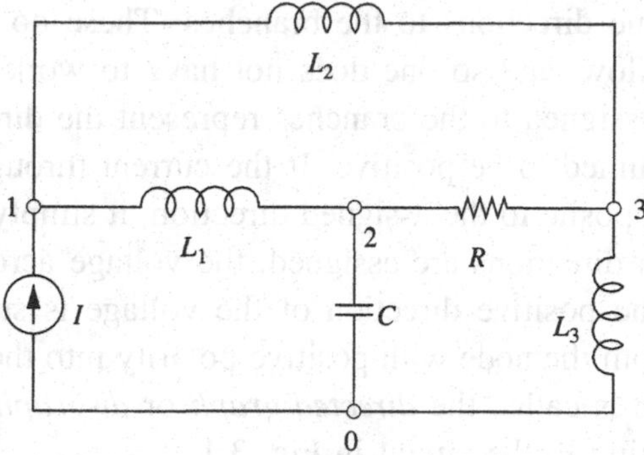

Figure 3.6 The circuit with only inductors and current sources connected to a node.

which makes one of the capacitor voltages linearly dependent on the others and the applied voltage. As a result, one of the capacitor voltages no longer remains an independent variable. ◀

A similar situation arises if, at any node in the circuit, the connected branches contain only inductors and current sources. An example of such a situation is shown in Fig. 3.6. The KCL applied to node 1 gives

$$I = i_{12} + i_{13},$$

which makes one of the inductor currents linearly dependent on the other inductor current and the current source.

In that situation, it becomes important to follow a systematic method to choose the minimum number of independent variables.

3.3 Using Graph Theory to Obtain the Minimal Set of Equations

The general method of obtaining differential equations from Kirchoff's laws makes use of the circuit topology – the structure of the interconnection between elements. In this method, each element of the circuit (the resistors, inductors, capacitors and sources) is represented by a *branch*, all of which are connected through *nodes*. Using the branches and nodes, the circuit topology is represented by means of a *graph* – which depicts the branches simply as lines, irrespective of what the element associated with a branch is. Thus, the circuit in Fig. 3.1 is represented by the graph in Fig. 3.7.

The next stage is to assign identifiers to the nodes and branches. In Fig. 3.8, the nodes have been identified with numbers and the branches with letters. Then, one has to assign some directions to the branches. These do not represent actual directions of current flow, and so one does not have to work out these directions a priori. The arrows assigned to the branches represent the directions in which the flow of current is assumed to be positive. If the current through a branch actually flows in a direction opposite to the assigned direction, it simply assumes a negative sign. Once the branch directions are assigned, the voltage across the branches can be assigned a sign; the positive direction of the voltage is such that current in a given branch flows from the node with positive polarity into the node with negative polarity. Such a graph is called the *directed graph* or *di-graph*. Fig. 3.8 shows the directed graph pertaining to the circuit in Fig. 3.1.

A *loop* within a graph is a set of branches that form a closed path. The loop in the context of a graph is therefore the same as that in a circuit discussed earlier. The loops in Fig. 3.8 are formed by the branches a-b-c, c-d-e, b-d-f, a-f-e, a-f-d-c, b-c-e-f and a-b-d-e.

A *cutset* is a set of branches in a graph, which, when cut off, will divide the graph into two disconnected pieces. Imagine you have a pair of scissors with which

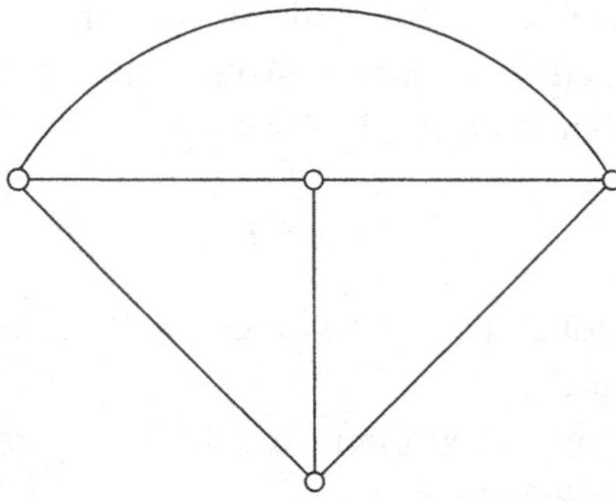

Figure 3.7 The graph of the circuit in Fig. 3.1.

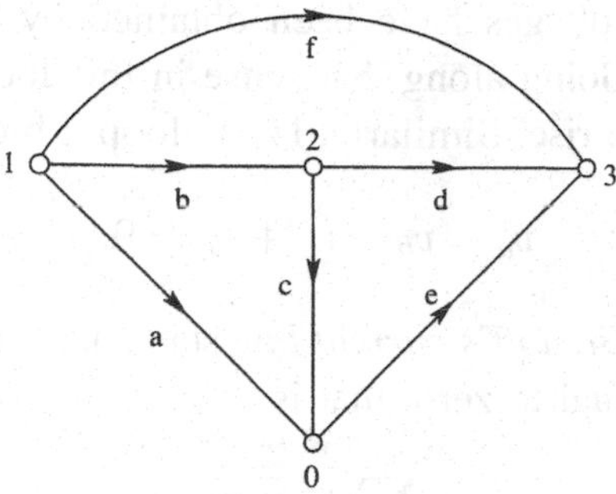

Figure 3.8 The directed graph of the circuit in Fig. 3.1.

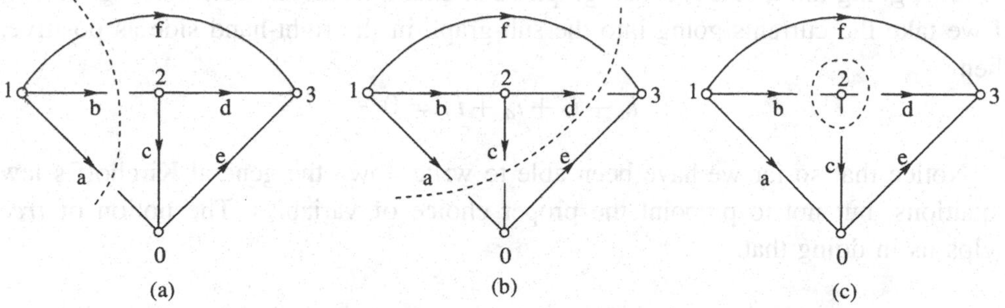

Figure 3.9 Three examples of cutsets of the graph of Fig. 3.8.

you can cut the branches of the graph in Fig. 3.8. In how many different ways can you cut the branches to form two disconnected sub-graphs? Figure 3.9 shows some examples of such cutsets.

3.3.1 Kirchoff's laws relating to loops and cutsets

The general form of Kirchoff's laws is related to the loops and cutsets in a graph.

Kirchoff's voltage law states that the sum of the voltages in the branches forming a loop is equal to zero, that is,

$$\sum_{\text{loop}} v_n = 0.$$

In order to obtain the correct sign of the voltages, one has to travel along the loop in a specific sense, clockwise or counterclockwise. For example, referring to Fig. 3.8, for the loop b-d-f, KVL will say

$$v_b + v_d - v_f = 0,$$

where the signs of the voltages have been obtained by traversing the loop in a counterclockwise sense. Going along that sense in this loop, v_b and v_d are voltage drops, and v_f is a voltage rise. Similarly, in the loop a-b-d-e, the KVL equation is

$$v_a - v_b - v_d + v_e = 0.$$

The general form of *Kirchoff's current law* states that the sum of currents in the branches of a cutset is equal to zero, that is,

$$\sum_{\text{cutset}} i_n = 0.$$

Here again, in order to get the signs correct, one has to assign positive signs to the currents going into one of the sub-graphs. For example, for the cutset in Fig. 3.9(b), if we take the currents going into the sub-graph in the right-hand side as positive, then

$$i_a + i_c + i_d + i_f = 0.$$

Notice that so far we have been able to write down the general Kirchoff's law equations, but not to pinpoint the proper choice of variables. The notion of *tree* helps us in doing that.

3.3.2 Tree and co-tree

A *tree* is a maximal set of branches in a graph that does not contain any loop. The set is maximal in the sense that if any branch is added to a tree, a loop will form.

One can identify a tree within a graph by starting from any one of the branches and adding neighbouring branches until one reaches a point where the addition of any more branch will cause a loop to close. Obviously the tree is not unique, and one can identify many trees within any graph. Figure 3.10 shows some of the trees corresponding to Fig. 3.8. It is an interesting fact that although the tree is

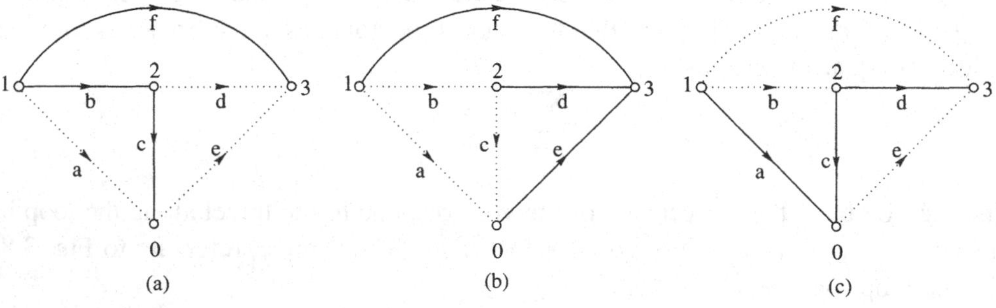

Figure 3.10 Three possible trees of the graph of Fig. 3.8.

non-unique, the number of branches in a tree is always the same for a given graph. For planar circuits, this number is one less than the number of nodes.

For every tree, the set of the remaining branches in a graph is called a *co-tree*. Thus, the union of the tree and co-tree is the graph itself.

3.3.3 The independent KCL and KVL equations

We had earlier seen that the KCL applied to all the cutsets and the KVL applied to all the loops yield many more equations than necessary, as some are linear combinations of other equations. Having identified a tree (any one) and its co-tree in a graph, we are now in a position to pinpoint the minimum number of independent KCL and KVL equations. But for that, we will need to choose those cutsets and loops that yield independent equations.

A *basic cutset* is that cutset that contains only one tree branch. If there are t number of tree branches, there can be the same number of basic cutsets. It can be shown that the KCL equations corresponding to the basic cutsets are independent[1]. If the tree in Fig. 3.10(a) is chosen, the basic cutsets are as shown in Fig. 3.11.

A *basic loop* is a loop that contains only one co-tree branch. It can be shown that the basic loops give the independent KVL equations. For the same tree as in Fig. 3.10(a), the basic loops are shown in Fig. 3.12.

3.3.4 The choice of the state variables

We have said in Section 3.1 that the voltages across capacitors and the currents through the inductors in a circuit form a natural choice of variables in whose terms the differential equations can be obtained. We have also seen that sometimes the storage elements may not be independent, and such cases occur if there is a loop

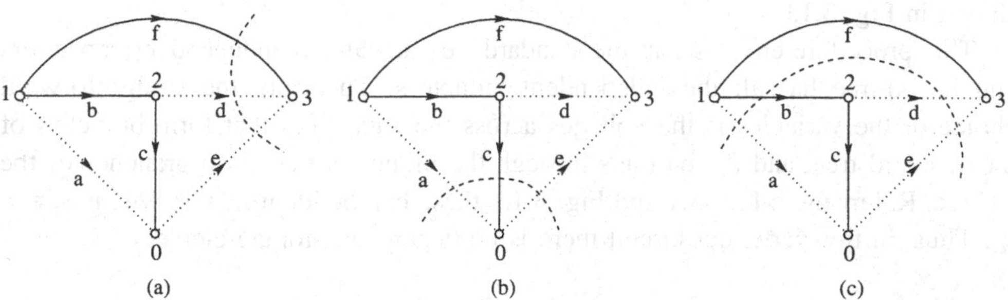

Figure 3.11 Three possible basic cutsets if the tree in Fig. 3.10(a) is chosen.

[1]For the proof, see any graph theory textbook.

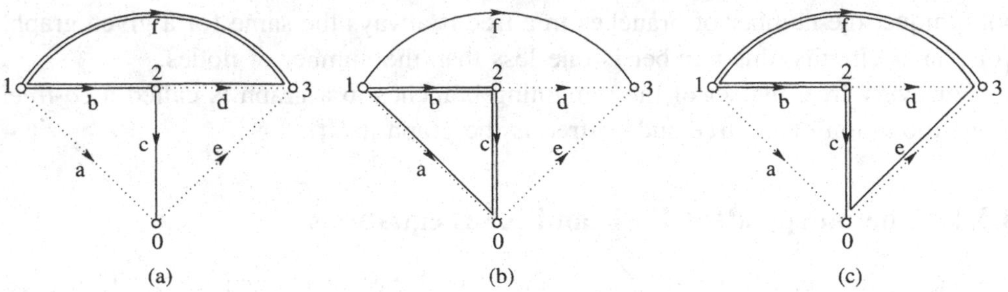

Figure 3.12 Three possible basic loops if the tree in Fig. 3.10(a) is chosen.

with only capacitors and voltage sources, or if there is a cutset with only inductors and current sources. In that case, we have to tackle the problem of identifying the independent storage elements.

We have also seen that for any given circuit, there can be plurality in the choice of the tree. The question is, does any particular choice of the tree offer advantage in obtaining the minimum number of independent equations? The answer is yes. In order to identify that particular tree, one has to go by the following order of priorities:

1. It should include *all* the voltage sources.
2. It should include a maximum number of capacitors.
3. It should include a maximum possible number of resistors.
4. It may then include the necessary number of inductors to complete the tree.

A tree chosen according to the above order of priorities is called a *standard tree*. For the circuit in Fig. 3.1, one can easily identify the standard tree as the one shown in Fig. 3.13.

This procedure ensures that the standard tree has all the independent capacitors and its co-tree has all the independent inductors. Therefore the straightforward choice of the variables is the voltages across the capacitors that form branches of the standard tree, and the currents through the inductors that form branches of the co-tree. Referring to Fig. 3.1 and Fig. 3.13, these can be identified as v_{C1}, v_{C2}, and i_L. Thus, in this particular circuit there is no dependent storage element.

3.3.5 Derivation of differential equations

There will be one first-order differential equation for each variable. We take the variables one by one.

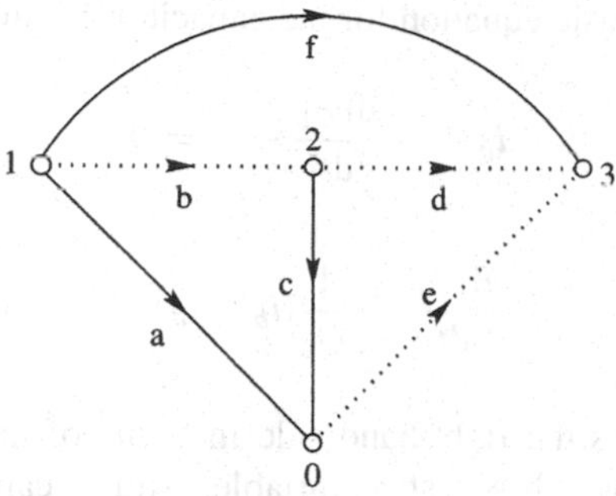

Figure 3.13 The standard tree of the circuit in Fig. 3.1.

- If the variable is a capacitor voltage, identify the basic cutset containing that capacitor. The differential equation is given by the KCL equation for that basic cutset.
- If the variable is an inductor current, identify the basic loop containing that inductor. The KVL equation for that basic loop will yield the desired differential equation.

In both cases, the final stage involves some substitution and algebraic manipulation to express the right-hand side in terms of the other variables and the sources. Let us illustrate the procedure with the help of a few examples.

► **Example 3.5** For the circuit in Fig. 3.1, the differential equation for v_{C1} is obtained from the basic cutset involving capacitor C_1, as shown in Fig. 3.14. The KCL equation for that cutset is

$$i_b - i_c - i_d = 0.$$

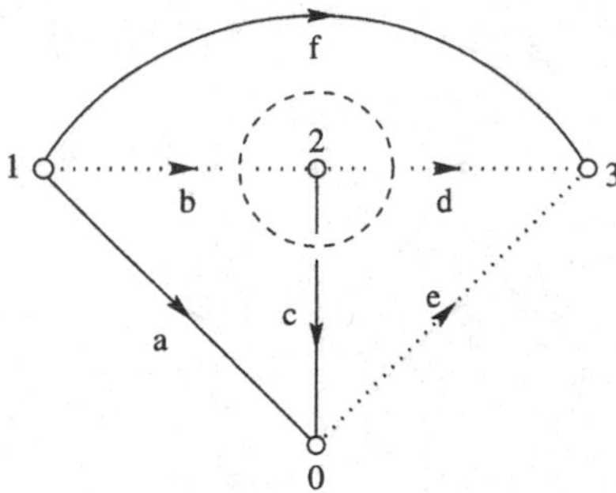

Figure 3.14 The basic cutset that includes capacitor C_1 in the standard tree of Fig. 3.13.

First, we substitute the basic equation for the capacitor C_1, and bring the derivative term to the left-hand side:

$$i_b - C_1 \frac{dv_{C1}}{dt} - i_d = 0$$

or

$$\frac{dv_{C1}}{dt} = \frac{1}{C_1}(i_b - i_d).$$

Next, we have to express the right-hand side in terms of the chosen variables and the sources. Note that $i_b = i_L$, which is a state variable. And i_d can be expressed as follows:

$$i_d = v_d / R_1.$$

Now notice the basic loop that involves R_1, whose KVL gives

$$\begin{aligned} v_d &= v_c - v_a + v_f \\ &= v_{C1} - E + v_{C2}. \end{aligned}$$

Substituting, the differential equation becomes

$$\begin{aligned} \frac{dv_{C1}}{dt} &= \frac{1}{C_1}(i_L - v_d/R_1) \\ &= \frac{1}{C_1} i_L - \frac{1}{R_1 C_1} v_{C1} - \frac{1}{R_1 C_1} v_{C2} + \frac{1}{R_1 C_1} E. \end{aligned} \quad (3.7)$$

The equation for v_{C2} comes from the cutset shown in Fig. 3.15. Following the same procedure, its KCL equation gives

$$i_f + i_d + i_e = 0,$$

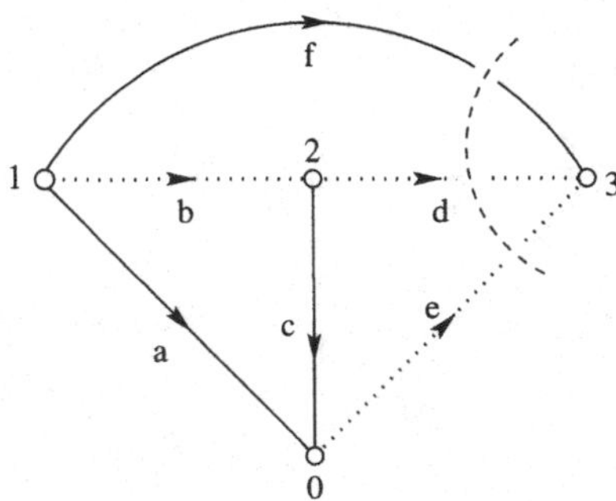

Figure 3.15 The basic cutset that includes capacitor C_2 in the standard tree of Fig. 3.13.

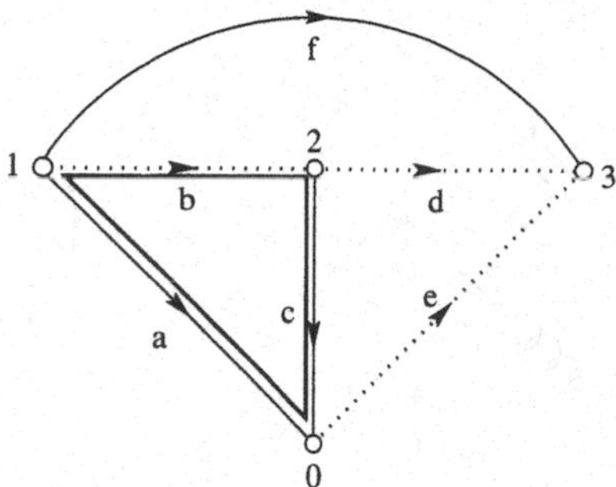

Figure 3.16 The basic loop that includes the co-tree inductor L in the standard tree of Fig. 3.13.

which in turn gives

$$\frac{dv_{C2}}{dt} = \frac{1}{C_2}(-i_d - i_e),$$
$$= \frac{1}{C_2}(-v_d/R_1 - v_e/R_2).$$

Here v_d is obtained from the basic loop containing R_1, and v_e is obtained from the basic loop containing R_2. This gives

$$\frac{dv_{C2}}{dt} = \frac{1}{C_2}\left[-\frac{v_{C1} - E + v_{C2}}{R_1} - \frac{v_{C2} - E}{R_2}\right],$$
$$= -\frac{1}{R_1 C_2}v_{C1} - \frac{1}{C_2}\left(\frac{1}{R_1} + \frac{1}{R_2}\right)v_{C2} + \frac{1}{C_2}\left(\frac{1}{R_1} + \frac{1}{R_2}\right)E. \quad (3.8)$$

The equation for i_L is given by the basic loop involving the co-tree inductor (Fig. 3.16). The corresponding KVL equation gives

$$v_b + v_c - v_a = 0.$$

Using the basic relation for the inductor, this is expressed as

$$\frac{di_L}{dt} = \frac{1}{L}(-v_c + v_a),$$
$$= -\frac{1}{L}v_{C1} + \frac{1}{L}E. \quad (3.9)$$

Equations (3.7), (3.8), and (3.9) are the desired differential equations of the system. ◀

► **Example 3.6** Let us consider the circuit in Fig. 3.17(a). Its directed graph and the standard tree, obtained following the steps outlined earlier, are given in Fig. 3.17(b). The

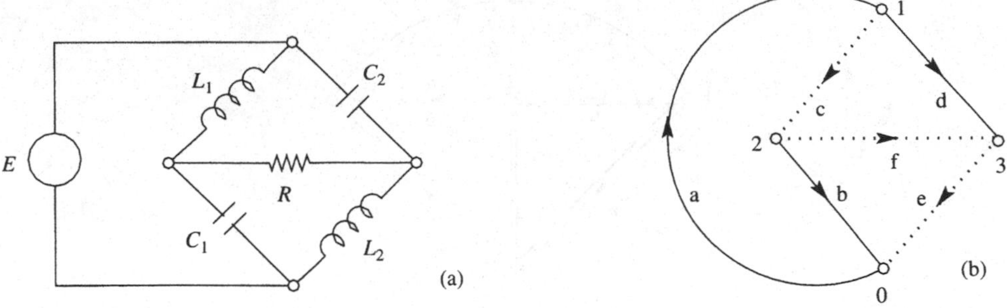

Figure 3.17 The circuit and its directed graph corresponding to Example 3.6, showing the standard tree.

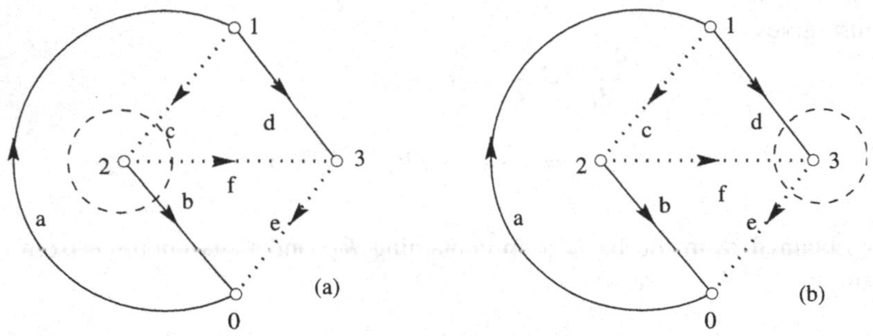

Figure 3.18 The basic cutsets containing the capacitors, corresponding to Example 3.6.

variables are the voltages across the tree capacitors v_{C1} and v_{C2}, and the currents through the co-tree inductors, i_{L1} and i_{L2}.

The differential equations for the capacitor voltages are to be obtained from the basic cutsets involving the capacitors. These basic cutsets are shown in Fig. 3.18. Let us first take the cutset in Fig. 3.18(a). Its KCL equation is

$$i_b + i_f - i_c = 0$$

or

$$C_1 \frac{dv_{C1}}{dt} + v_f/R - i_{L1} = 0.$$

The value of v_f is to be obtained from the basic loop involving the resistance R, which gives

$$\begin{aligned} v_f &= v_d + v_a + v_b \\ &= v_{C2} - E + v_{C1}. \end{aligned}$$

Note that, as given by the arrow, v_a has been assumed positive when node 0 is at higher potential than node 1. But E acts in the opposite direction, hence the negative sign. Substituting, we get

$$\frac{dv_{C1}}{dt} = \frac{1}{C_1} i_{L1} - \frac{1}{RC_1} v_{C1} - \frac{1}{RC_1} v_{C2} + \frac{1}{RC_1} E. \tag{3.10}$$

For the second cutset in Fig. 3.18(b), the KCL equation is

$$i_d + i_f - i_e = 0$$

or

$$C_2 \frac{dv_{C2}}{dt} + v_f/R - i_{L2} = 0$$

or

$$\frac{dv_{C2}}{dt} = \frac{1}{C_2}, i_{L2} - \frac{1}{RC_2} v_{C1} - \frac{1}{RC_2} v_{C2} + \frac{1}{RC_2} E. \tag{3.11}$$

The differential equations involving i_{L1} and i_{L2} are to be obtained from the KVL equations for the basic loops involving these inductors, which are shown in Fig. 3.19. For the loop in Fig. 3.19(a), the KVL equation is

$$v_a + v_c + v_b = 0$$

or

$$-E + L_1 \frac{di_{L1}}{dt} + v_{C1} = 0$$

or

$$\frac{di_{L1}}{dt} = \frac{1}{L_1} E - \frac{1}{L_1} v_{C1}. \tag{3.12}$$

For the other loop in Fig. 3.19(b), the KVL equation is

$$v_a + v_d + v_e = 0$$

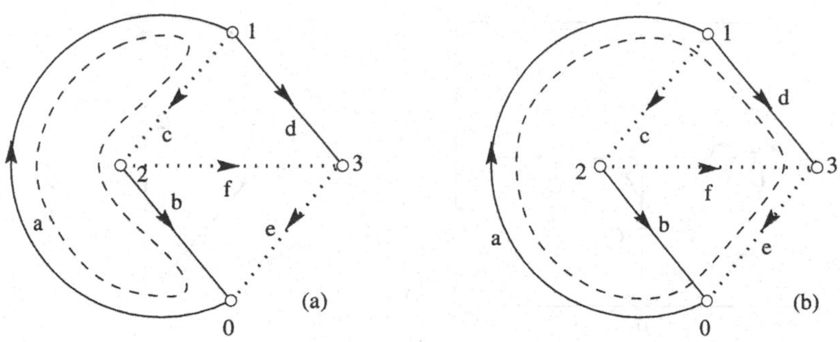

Figure 3.19 The basic loops containing the inductors, corresponding to Example 3.6.

or

$$-E + v_{C2} + L_2 \frac{di_{L2}}{dt} = 0$$

or

$$\frac{di_{L2}}{dt} = \frac{1}{L_2} E - \frac{1}{L_2} v_{C2}. \tag{3.13}$$

Thus, (3.10), (3.11), (3.12) and (3.13) are the required differential equations. ◀

▶ **Example 3.7** In this example, we consider the circuit in Fig. 3.20(a). We have shown in Example 3.4 that all the storage elements are not independent, as there exists a closed path with only the voltage source and capacitances. In identifying the standard tree, we start from the branch containing the voltage source, then add branch c, then branch f. At this stage, we find that addition of any other branch will close a loop, and so it is not possible to include all the capacitors in the standard tree. The resulting standard tree is shown in Fig. 3.20(b). Accordingly, the independent variables for the system are chosen as the voltages across the tree capacitors v_{C1} and v_{C2}, and the current through the co-tree inductor, i_L.

For the tree capacitor C_1, the basic cutset contains branches b, c and e. Thus, the KCL equation is

$$i_b - i_c - i_e = 0$$

or

$$C_1 \frac{dv_{C1}}{dt} - i_L - C_3 \frac{dv_e}{dt}.$$

Now, v_e is obtained in terms of the other state variables and sources from the basic loop involving branch e. Thus,

$$\begin{aligned} v_e &= v_c - v_a + v_f \\ &= v_{C1} - E + v_{C2}. \end{aligned}$$

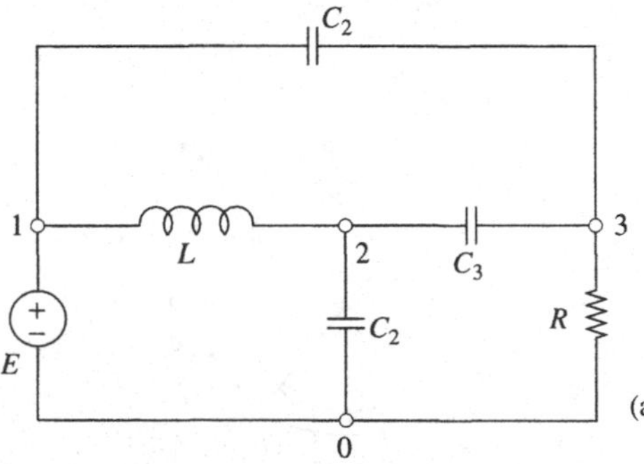

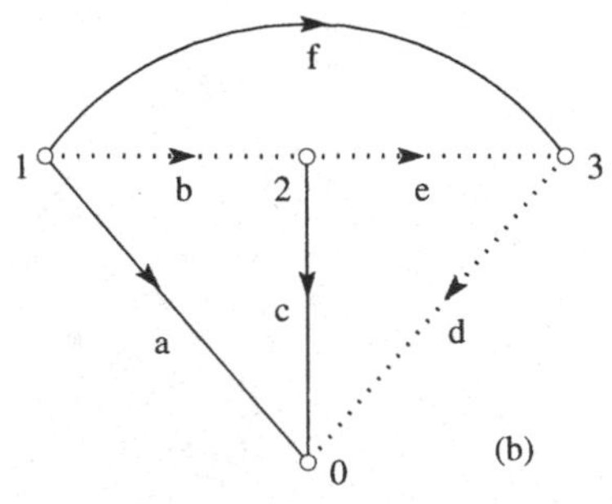

Figure 3.20 The circuit and its directed graph corresponding to Example 3.7, showing the standard tree.

Substituting, we get

$$C_1\frac{dv_{C1}}{dt} = i_L - C_3\frac{d}{dt}(v_{C1} - E + v_{C1}),$$
$$= i_L - C_3\frac{dv_{C1}}{dt} - C_3\frac{dv_{C2}}{dt} + C_3\frac{dE}{dt}$$

or

$$(C_1 + C_3)\frac{dv_{C1}}{dt} = i_L - C_3\frac{dv_{C2}}{dt} + C_3\frac{dE}{dt}. \tag{3.14}$$

For the tree capacitor C_2, the basic cutset contains the branches f, e and d, and the KCL gives

$$i_f + i_e - i_d = 0$$

or

$$C_2\frac{dv_{C2}}{dt} = -C_3\frac{dv_e}{dt} + \frac{v_d}{R},$$
$$= -C_3\frac{d}{dt}(v_{C1} - E + v_{C1}) + \frac{-v_f + v_a}{R},$$
$$= -C_3\frac{dv_{C1}}{dt} - C_3\frac{dv_{C2}}{dt} + C_3\frac{dE}{dt} - \frac{1}{R}v_{C2} + \frac{1}{R}E$$

or

$$(C_2 + C_3)\frac{dv_{C2}}{dt} = -C_3\frac{dv_{C1}}{dt} + C_3\frac{dE}{dt} - \frac{1}{R}v_{C2} + \frac{1}{R}E. \tag{3.15}$$

Notice that in (3.14) and (3.15), both $\frac{dv_{C1}}{dt}$ and $\frac{dv_{C2}}{dt}$ appear, and the expressions for each can be obtained by solving these two equations.

For the inductor, the basic loop contains branches a, b and c, and the KVL equation is

$$v_b + v_c - v_a = 0$$

or

$$\frac{di_L}{dt} = -\frac{1}{L}v_{C1} + \frac{1}{L}E. \tag{3.16}$$

The right-hand side is just in terms of the chosen variables and the sources. ◀

▶ **Example 3.8** In this example, we consider the circuit in Fig. 3.21(a). As shown in Fig. 3.21(b), there is a cutset that contains only inductors and the current source. This means that one of the inductors is not independent.

In identifying the standard tree, we note that in this circuit there is one capacitor and one resistor. So the standard tree should include them. The tree can be completed by including any one of the inductors. Let us choose the standard tree shown in Fig. 3.22. This gives the state variables as i_{L1}, i_{L3} and v_C.

For the inductor L_1, the differential equation is obtained from the KVL equation of the basic loop b-d-f:

$$v_b + v_d - v_f = 0$$

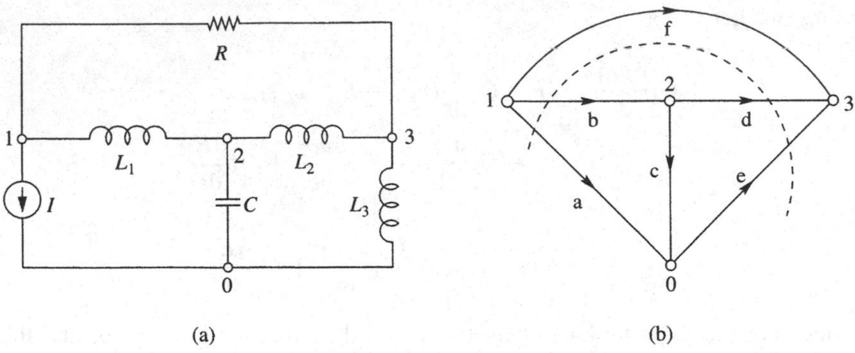

Figure 3.21 (a) The circuit corresponding to Example 3.8. The cutset in (b) shows that all the inductor currents are not independent.

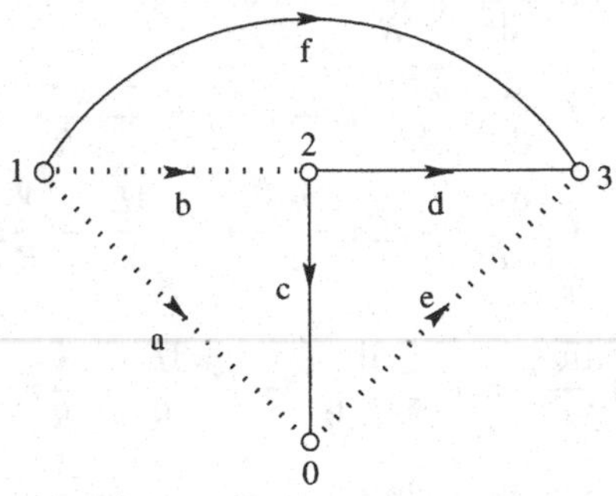

Figure 3.22 The standard tree pertaining to Example 3.8.

or

$$L_1 \frac{di_{L1}}{dt} = -L_2 \frac{di_{L2}}{dt} + Ri_R.$$

In our chosen standard tree, i_{L2} is not a state variable. It can be expressed in terms of the other variables by the KCL equation at the basic cutset shown in Fig. 3.21(b):

$$i_a + i_b - i_d - i_e = 0$$

or

$$i_{L2} = I + i_{L1} - i_{L3}.$$

The current i_R is to be obtained from the basic cutset involving that branch:

$$i_a + i_b + i_f = 0$$

or

$$i_R = -I + i_{L1}.$$

Substituting, we get

$$L_1\frac{\mathrm{d}i_{L1}}{\mathrm{d}t} = -L_2\frac{\mathrm{d}}{\mathrm{d}t}(I + i_{L1} - i_{L3}) + R(-I + i_{L1}),$$

$$(L_1 + L_2)\frac{\mathrm{d}i_{L1}}{\mathrm{d}t} = Ri_{L1} - RI + L_2\frac{\mathrm{d}i_{L3}}{\mathrm{d}t} - L_2\frac{\mathrm{d}I}{\mathrm{d}t}. \tag{3.17}$$

Similarly, for the inductor L_3, the basic loop c-d-e gives

$$v_e = -v_c + v_d,$$

$$L_3\frac{\mathrm{d}i_{L3}}{\mathrm{d}t} = -v_C + L_2\frac{\mathrm{d}i_{L2}}{\mathrm{d}t},$$

$$= -v_C + L_2\frac{\mathrm{d}}{\mathrm{d}t}(I + i_{L1} - i_{L3})$$

or

$$(L_2 + L_3)\frac{\mathrm{d}i_{L3}}{\mathrm{d}t} = -v_C + L_2\frac{\mathrm{d}i_{L1}}{\mathrm{d}t} + L_2\frac{\mathrm{d}I}{\mathrm{d}t}. \tag{3.18}$$

The expressions for $\frac{\mathrm{d}i_{L1}}{\mathrm{d}t}$ and $\frac{\mathrm{d}i_{L3}}{\mathrm{d}t}$ are obtained by solving these two equations.

Finally, the equation for v_C is obtained from the basic cutset a-c-e:

$$i_c = i_e - i_a,$$

$$C\frac{\mathrm{d}v_C}{\mathrm{d}t} = i_{L3} - I,$$

$$\frac{\mathrm{d}v_C}{\mathrm{d}t} = \frac{1}{C}i_{L3} - \frac{1}{C}I.$$

◄

It may be noted that systems that have dependent storage elements yield differential equations containing the derivatives of the external voltage or current sources. Such configurations are undesirable in practical systems, because if the sources are independent, they can also undergo sudden changes – as in a square wave or when a dc source is switched on or off. At the transition points the derivatives assume infinite value, which means that the elements of the circuit may be subjected to very high stress, possibly causing failure.

3.4 Chapter Summary

The Kirchoff's laws can be used to obtain differential equations for electrical circuits. Because of the constitutive relations $v = L\,\frac{\mathrm{d}i}{\mathrm{d}t}$ and $i = C\,\frac{\mathrm{d}v}{\mathrm{d}t}$, they normally yield one first-order differential equation per storage element. For simple systems, one can obtain the differential equations following the mesh current method and the node voltage method. For more complicated systems, and for systems with dependent storage elements (where a loop has only capacitances and voltage sources, or a cutset has only inductances and current sources), one has to use the graph theoretic technique to obtain the minimum number of independent differential equations.

In this method, one has to draw the directed graph and locate the standard tree. The voltages across the tree-branch capacitances and the currents through the co-tree-branch inductances form a minimum set of variables for obtaining the differential equations. The differential equation for a tree capacitance is obtained from the basic cutset containing that branch, and that for a co-tree inductance is obtained from the basic loop containing that branch.

Further Reading

A. Ioinovici, *Computer-Aided Analysis of Active Circuits*, Marcel Dekker, New York, 1990.

C. K. Tse, *Linear Circuit Analysis*, Addison Wesley, Longman Ltd., Harlow, England, 1998.

Problems

1. Obtain the differential equations for the following circuits:

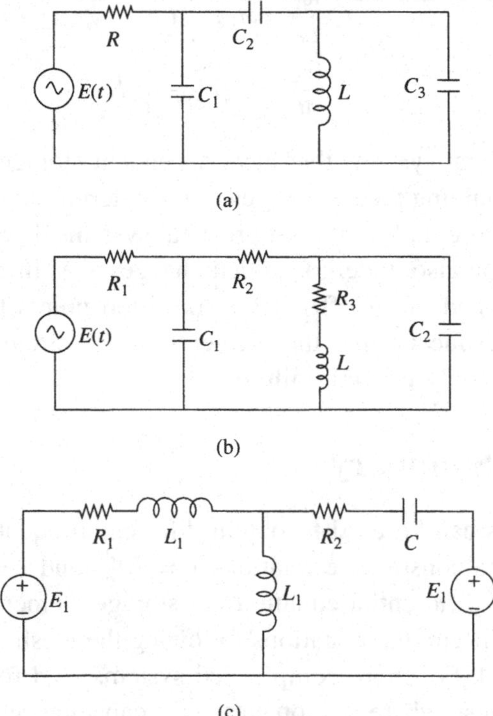

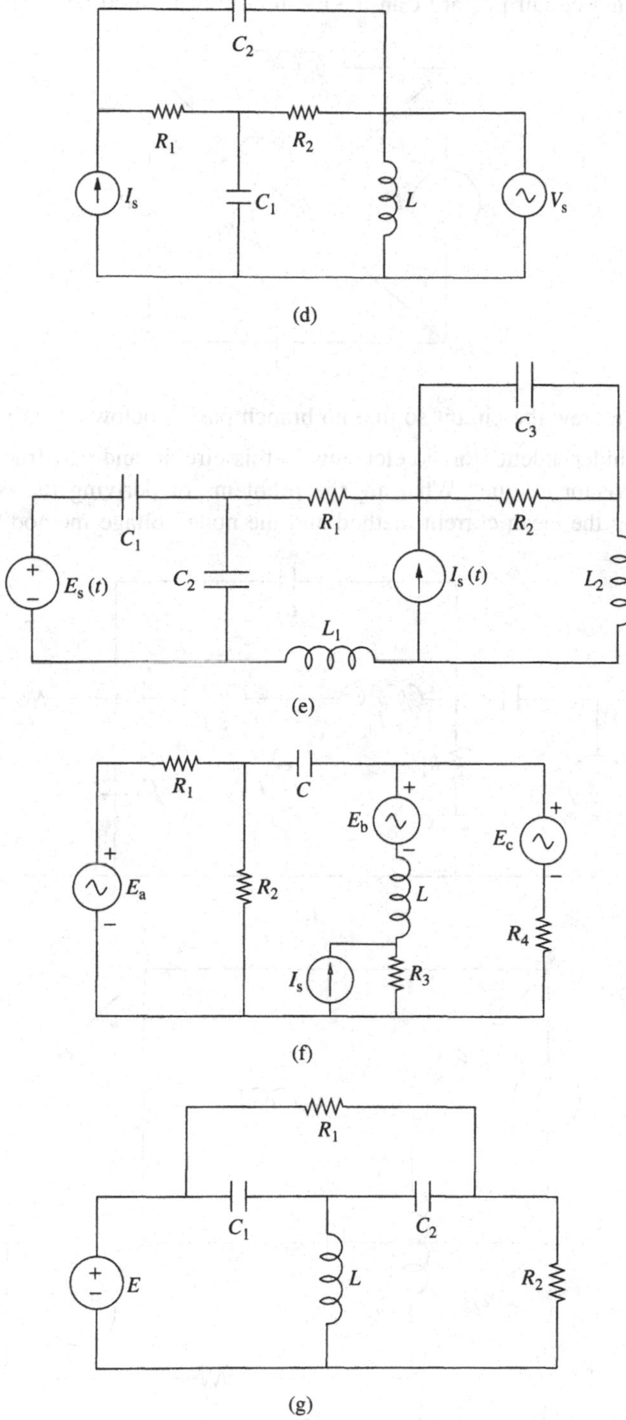

C_2
R_1
R_2
I_s
C_1
L
V_s
(d)
C_3
C_1
R_1
R_2
$E_s(t)$
C_2
$I_s(t)$
L_2
L_1
(e)
R_1
C
E_b
E_c
E_a
R_2
L
R_4
I_s
R_3
(f)
R_1
C_1
C_2
E
L
R_2
(g)

2. Is the following circuit planar? Can the mesh current method be applied?

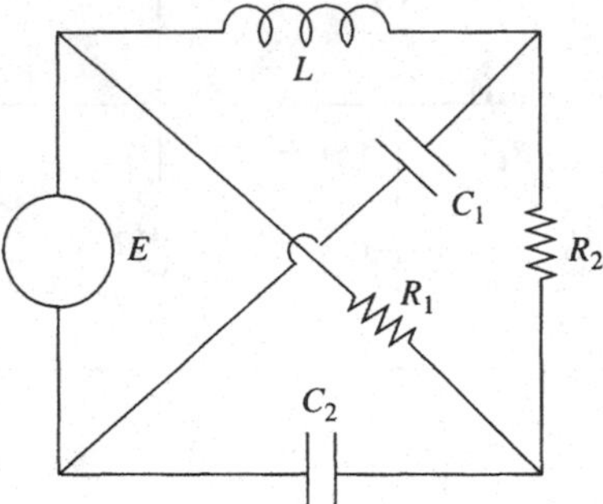

Hint: Try to redraw the circuit so that no branch passes below or above another branch.

3. Identify the independent storage elements in this circuit, and construct the state vector. Is the state vector unique? What are the problems of deriving the equations of these systems using the mesh current method and the node voltage method?

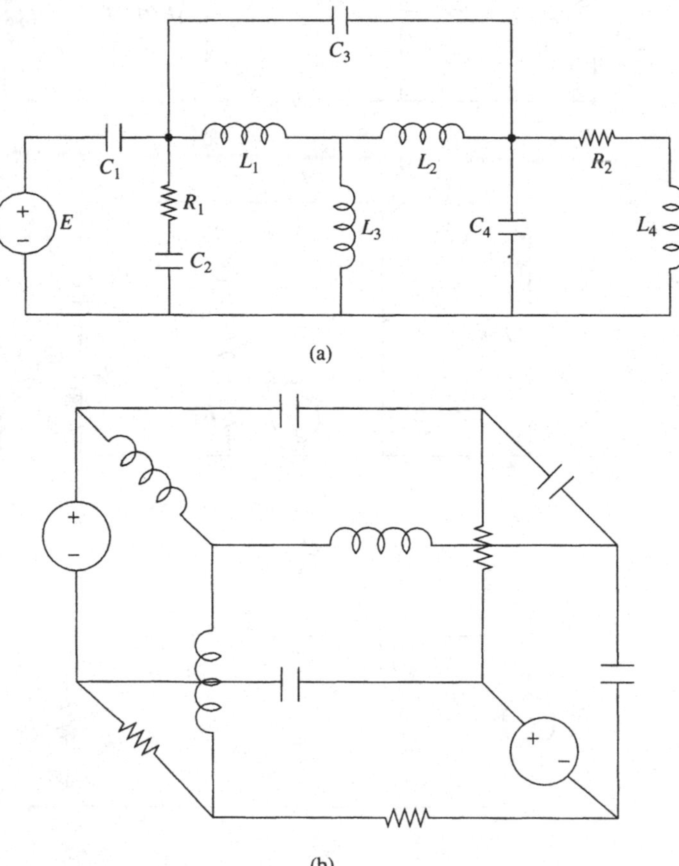

(a)

(b)

4

The Lagrangian Formalism

We have seen that in the Newtonian approach one has to take account of two kinds of forces: the given force (denoted by **F**) and the constraint force (denoted by $\mathbf{F}^c$). The given force includes the externally impressed forces, and the forces of interaction between mass points through springs and frictional elements. The Newtonian equation for each mass point is

$$m_j \ddot{\mathbf{r}}_j - \mathbf{F}_j = \mathbf{F}_j^c, \tag{4.1}$$

where $\mathbf{F}_j$ and $\mathbf{F}_j^c$ are the given force and the constraint force respectively, on the jth mass point. If there are N number of mass points there will be N such equations. Summing up the N equations, we get

$$\sum_{j=1}^{N} (m_j \ddot{\mathbf{r}}_j - \mathbf{F}_j) = \sum_{j=1}^{N} \mathbf{F}_j^c. \tag{4.2}$$

This is the basic Newtonian equation for a system comprising of N mass points.

About a century after Newton, Lagrange (1736–1813) showed that it is much more convenient to formulate differential equations in terms of the two basic forms of energies contained in a system: the kinetic energy and the potential energy. While there may be a large number of interacting forces in a system, there are only two forms of energy. Thus, if the central concept of Newtonian mechanics can be expressed in terms of the energies, the formulation would be much simpler. In the following sections, we shall explore this conceptual breakthrough to develop a powerful method of formulating differential equations of physical systems.

Dynamics for Engineers S. Banerjee

4.1 Elements of the Lagrangian Approach

4.1.1 Motivation

It is clear from Chapter 2 that if one tries to write the dynamical equations of any system in the Newtonian way, one faces a few practical difficulties:

1. All the constraint forces have to be included in the formulation. This would be difficult, for in most cases the constraint forces are not easily quantifiable. Moreover, in a complicated system there may be a large number of constraint forces. Accounting for all of them is a cumbersome exercise.

2. In a system with interconnected elements, the components interact with each other through forces that are to be categorized under "given forces" $\mathbf{F}_j$. These include the force of interaction between masses through springs or frictional elements. We need to take into account all these forces, and the method becomes unmanageable with the increase in the number of interacting elements.

   It would naturally be more convenient to express the forces in terms of some quantity of lesser complexity. The energies in a system, that is, the kinetic energy and the potential energy, are *scalar* quantities that are easily expressible in terms of the system coordinates. It would therefore be convenient to express the dynamical equations in terms of the energies of the system.

3. For an N-body problem, one would have to obtain $3N$ differential equations though we have seen that the existence of holonomic constraints reduce the number of configuration coordinates. It is desirable to utilize the advantage offered by the holonomic systems, that is, the number of system equations should be reduced as far as possible. This can be achieved by expressing the system equations in terms of a new coordinate system defined on the constraint surface.

The methodology for overcoming these problems was developed through the work of the post-Newtonian scientists like Bernoulli, D'Alembert, Lagrange, Hamilton and others. This will be taken up in the next section.

4.1.2 The concept of admissible motions

We take up the first problem, that is, elimination of the constraint forces from the system model. We notice that in most cases the constraint force does no work because it is perpendicular to the directions of motion allowed by the constraint. Let us illustrate it with three examples.

► **Example 4.1** (a) A pendulum's constraint allows motions that are always orthogonal to the tension in the string. Hence, the work done by the constraint force is zero.

(b) In a pulley, the constraint forces act on two bodies along the string. Since the direction of the constraint force and the possible displacement are along the same line, the work done by the constraint force exists for both the masses individually. However, since they are equal and opposite, the sum is always zero.

(c) For a body sliding along a frictionless surface, the force of reaction is always perpendicular to the direction of motion. Hence, the work due to the constraint force vanishes. If there is friction at the surface, the force of reaction would not be perpendicular to the motion. In that case we shall consider the perpendicular component of the reaction force as the force due to constraint and include the force due to friction (acting tangentially to the surface) in the category of given force $\mathbf{F}_j$. ◄

Notice that in the Example 4.1(a), if the point of suspension of the pendulum moves with time, the constraint force will no longer be orthogonal to possible directions of motion. The same thing happens if the surface in Example 4.1(c) moves with time. Still we would like to obtain a general framework in which the constraint force can be eliminated. We do this by considering the *geometrically admissible displacements* compatible with the constraints, for which the work due to constraint forces would be identically zero.

For example, a geometrically admissible displacement for a spherical pendulum can be any displacement tangential to the spherical constraint surface. But if it is a pendulum with an oscillating support, then the admissible displacement will be in relation to a given position of the pendulum. If we take a still photograph of any position of the pendulum, then any admissible displacement from that position will have to be perpendicular to the chord along which the constraint force acts. Clearly, the work done by the constraint force along an admissible displacement would always be zero.

A geometrically admissible quantity is denoted by the symbol δ. Thus, an admissible displacement is written as $\delta\mathbf{r}$. The work done by the constraint force at an admissible displacement is given by $\mathbf{F}^c\delta\mathbf{r}$.

Then we convert (4.2) into an equation involving work by multiplying with the admissible displacement

$$\sum_{j=1}^{N}(m_j\ddot{\mathbf{r}}_j - \mathbf{F}_j)\cdot\delta\mathbf{r}_j = \sum_{j=1}^{N}\mathbf{F}_j^c\cdot\delta\mathbf{r}_j. \tag{4.3}$$

The right-hand side of this equation would be zero, eliminating the constraint forces from our formulation. We thus obtain the equation

$$\sum_{j=1}^{N}(m_j\ddot{\mathbf{r}}_j - \mathbf{F}_j)\cdot\delta\mathbf{r}_j = 0. \tag{4.4}$$

4.1.3 The generalized coordinates

We now turn to the problem of minimizing the number of coordinates. In (4.2), each $\mathbf{r}_j$ had three components making the dimension of the configuration space $3N$. We have earlier seen that each holonomic constraint defines a surface in this $3N$-dimensional space. By taking the intersection of all such constraint surfaces, we get a space of dimension $3N - h$ (h being the number of holonomic constraints) on which the configuration point must lie. We define a new coordinate system on this $(3N - h)$ dimensional space where q_i are the independent coordinates. These are called *generalized coordinates*. This procedure has the effect of flattening out the constraint surface into a new $3N - h = n$ dimensional space whose axes are the generalized coordinates (Fig. 4.1).

Thus, the minimum number of independent coordinates in any system, consistent with the constraints, is also the number of generalized coordinates. To be exact, *any set of coordinates* $\{q_1, q_2, q_3, \ldots, q_n\}$ *is called a set of generalized coordinates of a system if and only if the number n of its members is necessary and sufficient to define the configuration or positional status of the system uniquely.*

There is no unique choice of the generalized coordinates, since any convenient set of variables that can specify the positional configuration of a system can serve as the generalized coordinates. These need not be the Cartesian coordinates of the mass points. In some cases of mechanical systems it may be convenient to express the system configuration in terms of polar coordinates r and θ. In systems consisting of two or more inertial elements, the position of some of the masses may be expressed

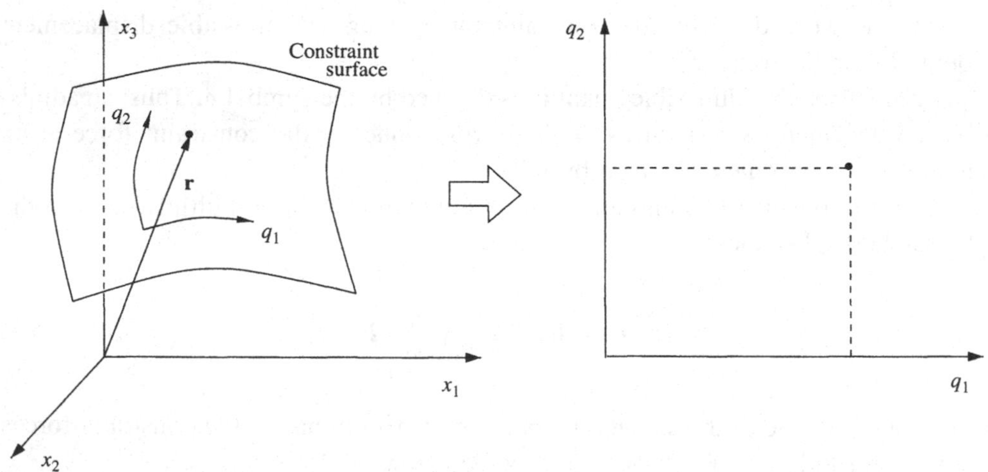

Figure 4.1 Generalized coordinates defined on the surface of constraint.

in rectangular coordinates and some others in polar coordinates. We shall see later that in electrical circuits, the charges flowing in loops are the natural choice for the generalized position coordinates.

In such a representation, the generalized velocities are merely the time derivatives of the chosen position coordinates. Thus, if a mass is represented by the polar coordinates r and θ, then the generalized velocities are just $\dot{r}$ and $\dot{\theta}$, and not the actual velocities in the radial and circumferential directions (the velocity in the circumferential direction is $r\dot{\theta}$).

The transformation of the system representation from $3N$ coordinates to a set of $3N - h = n$ generalized coordinates can be achieved by expressing the old vectors x_k in terms of the new coordinates q_i in the form

$$x_k = x_k(q_1, q_2, q_3, \ldots, q_n, t),$$

where k takes values between 1 and $3N$.

It will however be convenient to put together the coordinates of each mass point to form vectors $\mathbf{r}_j$, which can then be expressed as functions of the new coordinates:

$$\mathbf{r}_1 = \mathbf{r}_1(q_1, q_2, q_3, \ldots, q_n, t),$$
$$\vdots$$
$$\mathbf{r}_N = \mathbf{r}_N(q_1, q_2, q_3, \ldots, q_n, t)$$

or, in short,

$$\mathbf{r}_j = \mathbf{r}_j(q_i, t),$$

where j varies from 1 to N and i varies from 1 to n.

By differentiation with respect to t and by applying the chain rule, we get the velocities of the mass points in terms of the generalized coordinates as

$$\dot{\mathbf{r}}_j = \frac{\mathrm{d}\mathbf{r}_j}{\mathrm{d}t} = \sum_{i=1}^{n} \frac{\partial \mathbf{r}_j}{\partial q_i}\dot{q}_i + \frac{\partial \mathbf{r}_j}{\partial t}.$$

Note that the last term exists only in rheonomic systems where the constraint equations have explicit dependence on time. This term is independent of $\dot{q}_i$, and $\dot{q}_k$ is also independent of $\dot{q}_i$. Partially differentiating this equation with respect to the generalized velocities $\dot{q}_i$ we therefore get

$$\frac{\partial \dot{\mathbf{r}}_j}{\partial \dot{q}_i} = \frac{\partial \mathbf{r}_j}{\partial q_i}. \tag{4.5}$$

We shall need these equations for future reference.

Likewise, an admissible displacement $\delta\mathbf{r}_j$ in the old set of coordinates would be related to admissible displacements δq_i in the new generalized coordinates by

$$\delta\mathbf{r}_j = \sum_{i=1}^{n} \frac{\partial \mathbf{r}_j}{\partial q_i}\delta q_i. \tag{4.6}$$

This equation does not contain any time variation term δt because by definition an admissible displacement involves changes in space coordinates only.

This completes our conversion of the larger $3N$-dimensional coordinate system to the smaller n-dimensional coordinate system. We shall use these relationships later to formulate the dynamical equations in the generalized coordinate system.

4.1.4 Dynamical equations in terms of energies

Now we take up the third objective and start from the equation (4.4) derived earlier. Let us reproduce it here for the sake of convenience:

$$\sum_{j=1}^{N}(m_j\ddot{\mathbf{r}}_j - \mathbf{F}_j)\,\delta\mathbf{r}_j = 0.$$

Substituting the expression for admissible displacement from (4.6), we get (4.4) in terms of generalized coordinates as

$$\sum_{j=1}^{N}(m_j\ddot{\mathbf{r}}_j - \mathbf{F}_j)\sum_{i=1}^{n}\frac{\partial \mathbf{r}_j}{\partial q_i}\delta q_i = 0. \tag{4.7}$$

The summation over i can be brought forward, giving

$$\sum_{i=1}^{n}\left[\sum_{j=1}^{N}(m_j\ddot{\mathbf{r}}_j - \mathbf{F}_j)\frac{\partial \mathbf{r}_j}{\partial q_i}\right]\delta q_i = 0 \tag{4.8}$$

or

$$\sum_{i=1}^{n}\left[\sum_{j=1}^{N}\left(m_j\ddot{\mathbf{r}}_j\frac{\partial \mathbf{r}_j}{\partial q_i} - \mathbf{F}_j\frac{\partial \mathbf{r}_j}{\partial q_i}\right)\right]\delta q_i = 0. \tag{4.9}$$

Now consider the time derivative

$$\frac{\mathrm{d}}{\mathrm{d}t}\left(\dot{\mathbf{r}}_j\frac{\partial \mathbf{r}_j}{\partial q_i}\right) = \ddot{\mathbf{r}}_j\frac{\partial \mathbf{r}_j}{\partial q_i} + \dot{\mathbf{r}}_j\frac{\partial \dot{\mathbf{r}}_j}{\partial q_i}$$

or

$$\ddot{\mathbf{r}}_j\frac{\partial \mathbf{r}_j}{\partial q_i} = \frac{\mathrm{d}}{\mathrm{d}t}\left(\dot{\mathbf{r}}_j\frac{\partial \mathbf{r}_j}{\partial q_i}\right) - \dot{\mathbf{r}}_j\frac{\partial \dot{\mathbf{r}}_j}{\partial q_i}.$$

Substituting into (4.9), we get

$$\sum_{i=1}^{n}\left[\sum_{j=1}^{N} m_j \frac{\mathrm{d}}{\mathrm{d}t}\left(\dot{\mathbf{r}}_j \frac{\partial \mathbf{r}_j}{\partial q_i}\right) - \sum_{j=1}^{N} m_j \dot{\mathbf{r}}_j \frac{\partial \dot{\mathbf{r}}_j}{\partial q_i} - \sum_{j=1}^{N} \mathbf{F}_j \frac{\partial \mathbf{r}_j}{\partial q_i}\right] \delta q_i = 0. \tag{4.10}$$

Now let us analyse each term in (4.10). We observe that the kinetic energy T is given by

$$T = \sum_{j=1}^{N} \frac{1}{2} m_j \dot{\mathbf{r}}_j^2.$$

Differentiating with respect to the generalized velocities, we get

$$\frac{\partial T}{\partial \dot{q}_i} = \sum_{j=1}^{N} m_j \dot{\mathbf{r}}_j \frac{\partial \dot{\mathbf{r}}_j}{\partial \dot{q}_i}.$$

By (4.5) we get

$$\frac{\partial T}{\partial \dot{q}_i} = \sum_{j=1}^{N} m_j \dot{\mathbf{r}}_j \frac{\partial \mathbf{r}_j}{\partial q_i}.$$

Again, differentiating the expression for kinetic energy with respect to the generalized coordinates, we get

$$\frac{\partial T}{\partial q_i} = \sum_{j=1}^{N} m_j \dot{\mathbf{r}}_j \frac{\partial \dot{\mathbf{r}}_j}{\partial q_i}.$$

In the coordinate transformation, we can resolve the given forces on the mass points ($\mathbf{F}_j$) along the generalized coordinates as

$$Q_i = \sum_{j=1}^{N} \mathbf{F}_j \frac{\partial \mathbf{r}_j}{\partial q_i}. \tag{4.11}$$

These forces are called the *generalized forces*. Therefore, by (4.11) the last term in (4.10) turns out to be nothing but the force along the jth generalized coordinate.

Substituting these for the three terms in (4.10), we have

$$\sum_{i=1}^{n}\left\{\frac{\mathrm{d}}{\mathrm{d}t}\left(\frac{\partial T}{\partial \dot{q}_i}\right) - \frac{\partial T}{\partial q_i} - Q_i\right\} \delta q_i = 0.$$

Since the generalized coordinates are independent, any admissible displacement along the ith coordinate (δq_i) would be independent of admissible displacements along other generalized coordinates. Therefore, the summation in the above equation can vanish if the terms in the curly bracket are zero. Hence,

$$\frac{\mathrm{d}}{\mathrm{d}t}\left(\frac{\partial T}{\partial \dot{q}_i}\right) - \frac{\partial T}{\partial q_i} - Q_i = 0. \tag{4.12}$$

This is the Lagrangian equation in its most general form. Here the generalized force Q_i contains all the given forces in the system acting along the ith coordinate.

The given forces may be of three categories:

1. Forces of interaction between different mass points of the system. These happen generally by means of springs or elements of spring-like characteristics.

2. External forces, including the force due to gravity.

3. Forces due to friction.

Forces of the first two categories can be derived from a scalar "potential function" denoted as V. The force due to gravity can be obtained from the gravitational potential; the force exerted by a spring can be obtained from the potential energy stored in the spring. We shall show in Section 4.5 that externally impressed forces (including electromotive forces) can also be obtained from a suitably defined potential function. The generalized forces are derived from the potential as

$$Q_i = -\frac{\partial V}{\partial q_i}.$$

Systems where the generalized forces are obtainable from a scalar potential are called *conservative systems*. In such systems, the partial derivative of the potential function with respect to each coordinate gives the generalized force along that coordinate (see Fig. 4.2).

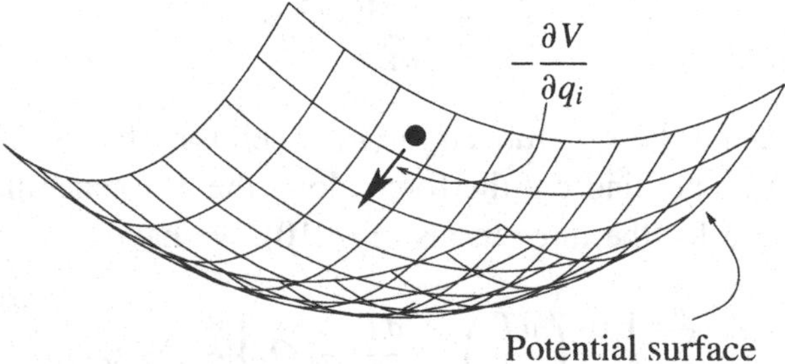

Figure 4.2 The potential function can be viewed as a surface whose gradient gives the generalized force.

For conservative systems (i.e. where forces are only of the first and second categories), (4.12) can be written as

$$\frac{\mathrm{d}}{\mathrm{d}t}\left(\frac{\partial T}{\partial \dot{q}_i}\right) - \frac{\partial T}{\partial q_i} + \frac{\partial V}{\partial q_i} = 0. \tag{4.13}$$

To make this equation further compact, we define a new function called the *Lagrangian* function[1] as

$$\mathcal{L} = T - V.$$

Since for most systems V is independent of the generalized velocities $\dot{q}_i$,

$$\frac{\partial \mathcal{L}}{\partial \dot{q}_i} = \frac{\partial T}{\partial \dot{q}_i}. \tag{4.14}$$

Hence, (4.13) becomes

$$\frac{\mathrm{d}}{\mathrm{d}t}\left(\frac{\partial \mathcal{L}}{\partial \dot{q}_i}\right) - \frac{\partial \mathcal{L}}{\partial q_i} = 0. \tag{4.15}$$

This is the Lagrangian equation for conservative systems.[2]

4.2 Obtaining Dynamical Equations by Lagrangian Method

As shown earlier, in the Lagrangian method, one does not have to worry about the constraint forces or the forces of interaction between elements of the system. Nor does one have to worry about the change of coordinates. The equations can be derived following a few simple steps given below.

1. Identify a minimum set of generalized coordinates consistent with the constraints.

2. Express the kinetic and potential energies in terms of the generalized coordinates and the generalized velocities, and obtain the Lagrangian function from these.

---

[1] We denote the Lagrangian by script $\mathcal{L}$ to distinguish it from the symbol of inductance, L.

[2] As we shall see later, there can be systems where the potential is dependent on the generalized velocities. In such systems (4.14) will not be valid, and so we will have to express the Lagrangian equation as

$$\frac{\mathrm{d}}{\mathrm{d}t}\left(\frac{\partial T}{\partial \dot{q}_i}\right) - \frac{\partial \mathcal{L}}{\partial q_i} = 0.$$

3. Partially differentiate $\mathcal{L}$ with respect to q_i and $\dot{q}_i$.

4. Write the Lagrangian equation (4.15) for each generalized coordinate.

Let us illustrate these steps with a few examples.

▶ **Example 4.2** We shall start with the simple example of a mass-spring system as shown in Fig. 4.3. The mass is constrained to move in only one direction and hence the position of the whole system can be defined by one generalized coordinate q, consistent with the constraint. We take $q = 0$ at the unstretched position of the spring.

The motion of the mass is governed by Newton's law: mass × acceleration = force or

$$m\frac{\mathrm{d}^2 q}{\mathrm{d}t^2} = -F. \tag{4.16}$$

Here F, the restoring force exerted by the spring, is negative because it acts in a direction opposite to q. This is the system equation, derived in the Newtonian way.

Now we obtain the equation in the Lagrangian way. When the spring is stretched, a distance q, the potential energy stored in the spring is

$$V = \frac{1}{2}kq^2, \tag{4.17}$$

where k is the spring constant.

The kinetic energy is given by

$$T = \frac{1}{2}m\dot{q}^2. \tag{4.18}$$

Hence, the Lagrangian function is

$$\mathcal{L} = \frac{1}{2}m\dot{q}^2 - \frac{1}{2}kq^2.$$

Now we differentiate $\mathcal{L}$ by $\dot{q}$ and q to get respectively

$$\frac{\mathrm{d}\mathcal{L}}{\mathrm{d}\dot{q}} = m\dot{q},$$

$$\frac{\mathrm{d}\mathcal{L}}{\mathrm{d}q} = -kq = -F.$$

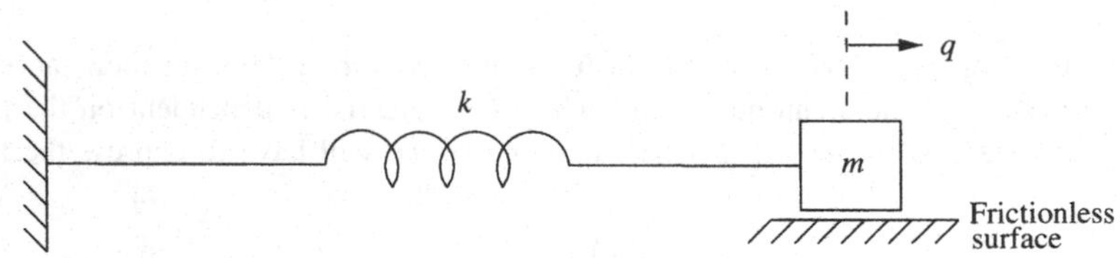

Figure 4.3 The one-dimensional mass-spring system.

Hence, the Lagrangian equation (4.15) gives

$$m\ddot{q} + F = 0,$$

which is the same as the Newtonian equation (4.16). ◄

In this example, no particular advantage of the Lagrangian methodology is visible. To illustrate the advantage, let us take the following examples.

► **Example 4.3** Consider the simple pendulum shown in Fig. 4.4. Let the mass of the bob be m, the length of the chord be l and at any point of time the angle of the chord with the vertical be θ. We notice that the position of the mass is given by three coordinates x, y and z. However, due to the constraint on its motion, the configuration of the system is uniquely given by the angle θ. This angle, therefore, becomes the only configuration coordinate. Moreover, in the Lagrangian formulation, we do not bother about the tension in the string, which is the force of constraint. The speed of the bob in the tangential direction is $l\dot{\theta}$, and so the kinetic energy becomes

$$T = \frac{1}{2}m\left(\dot{\theta}l\right)^2.$$

The potential energy measured with respect to the point of suspension is

$$V = -mgl\cos\theta.$$

Therefore, the Lagrangian function becomes

$$\mathcal{L} = \frac{1}{2}m\dot{\theta}^2l^2 + mgl\cos\theta.$$

Taking the partial derivatives

$$\frac{\partial\mathcal{L}}{\partial\dot{\theta}} = ml^2\dot{\theta},$$

$$\frac{\partial\mathcal{L}}{\partial\theta} = -mgl\sin\theta,$$

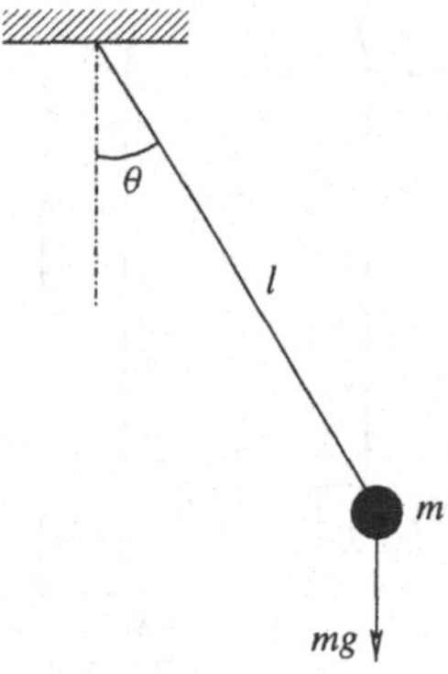

Figure 4.4 The simple pendulum.

and substituting into the Lagrangian equation we get

$$ml^2\ddot{\theta} + mgl\sin\theta = 0$$

or

$$\ddot{\theta} + \frac{g}{l}\sin\theta = 0.$$

Compare this method of obtaining the differential equation with that in Example 2.4. ◀

▶ **Example 4.4** We consider the motion of a pulley as shown in Fig. 4.5. Here, if you try to formulate the dynamical equation by the conventional method, you will have to worry about the tension in the string and all that. In contrast, by the Lagrangian approach we would simply note that though there are two masses, the configuration can be exactly defined with only one generalized coordinate q. The position of the other mass is $(l - q)$ where l is the length of the string. We would then write V, T and $\mathcal{L}$ in terms of q and $\dot{q}$ as

$$\begin{aligned} V &= -M_1 g q - M_2 g(l-q), \\ T &= \frac{1}{2}(M_1 + M_2)\dot{q}^2, \\ \mathcal{L} &= \frac{1}{2}(M_1 + M_2)\dot{q}^2 + M_1 g q + M_2 g(l-q). \end{aligned}$$

The derivatives with respect to q and $\dot{q}$ are

$$\frac{\mathrm{d}\mathcal{L}}{\mathrm{d}q} = (M_1 - M_2)g,$$

$$\frac{\mathrm{d}\mathcal{L}}{\mathrm{d}\dot{q}} = (M_1 + M_2)\dot{q}.$$

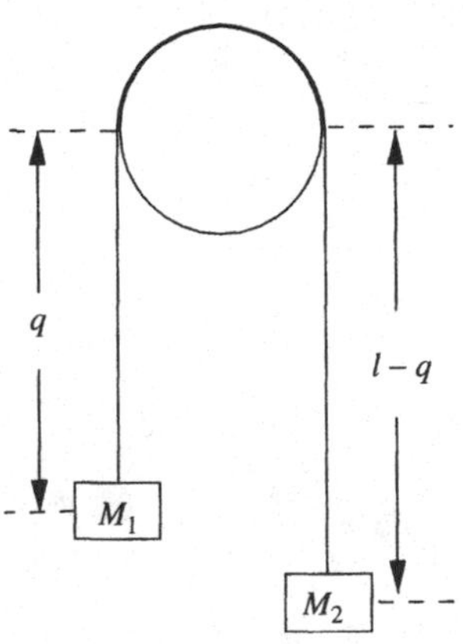

Figure 4.5 Dynamics of the pulley.

Thus, the Lagrangian equation gives

$$(M_1 + M_2)\ddot{q} - (M_1 - M_2)g = 0.$$

This is the dynamical equation for the system in Fig. 4.5. ◀

▶ **Example 4.5** To illustrate the advantages of the Lagrangian method, we now consider the pendulum with an oscillating support. Let the coordinates be as shown in Fig. 4.6. If the support were static, the coordinates of the bob would be given by

$$x = l\cos\theta,$$

$$y = l\sin\theta,$$

which give

$$\dot{x} = -l\sin\theta\dot{\theta},$$

$$\dot{y} = l\cos\theta\dot{\theta}.$$

If the position of the support is time-variable, given by $y_s(t)$, then the position of the bob is $(x, y + y_s)$. Therefore, the kinetic energy is given by

$$\begin{aligned} T &= \frac{1}{2}m\left[(\dot{y} + \dot{y}_s)^2 + \dot{x}^2\right], \\ &= \frac{1}{2}m\left(\dot{x}^2 + \dot{y}^2 + 2\dot{y}\dot{y}_s + \dot{y}_s^2\right), \\ &= \frac{1}{2}m\left(l^2\sin^2\theta\dot{\theta}^2 + l^2\cos^2\theta\dot{\theta}^2 + 2l\cos\theta\dot{\theta}\dot{y}_s + \dot{y}_s^2\right), \\ &= \frac{1}{2}m\left(l^2\dot{\theta}^2 + 2l\cos\theta\dot{\theta}\dot{y}_s + \dot{y}_s^2\right). \end{aligned}$$

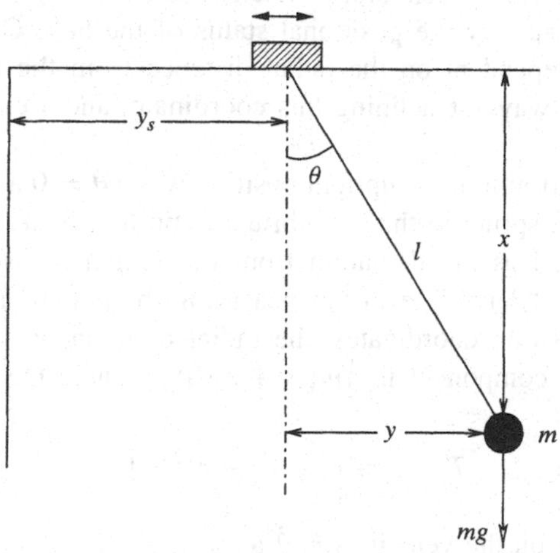

Figure 4.6 The pendulum with an oscillating support.

The potential energy is

$$V = -mgl\cos\theta.$$

Therefore, the Lagrangian function is

$$\mathcal{L} = \frac{1}{2}m\left(l^2\,\dot{\theta}^2 + 2l\cos\theta\,\dot{\theta}\dot{y}_s + \dot{y}_s^2\right) + mgl\cos\theta.$$

This gives

$$\frac{\partial\mathcal{L}}{\partial\dot{\theta}} = ml^2\dot{\theta} + ml\dot{y}_s\cos\theta,$$

$$\frac{\mathrm{d}}{\mathrm{d}t}\left(\frac{\partial\mathcal{L}}{\partial\dot{\theta}}\right) = ml^2\ddot{\theta} + ml\ddot{y}_s\cos\theta - ml\dot{\theta}\dot{y}_s\sin\theta,$$

$$\frac{\partial\mathcal{L}}{\partial\theta} = -ml\dot{\theta}\dot{y}_s\sin\theta - mgl\sin\theta.$$

The Lagrangian equation becomes

$$ml^2\ddot{\theta} + ml\ddot{y}_s\cos\theta - ml\dot{\theta}\dot{y}_s\sin\theta + ml\dot{\theta}\dot{y}_s\sin\theta + mgl\sin\theta = 0$$

or

$$\ddot{\theta} + \frac{g}{l}\sin\theta + \frac{1}{l}\cos\theta\,\ddot{y}_s = 0.$$

If the motion of the support is sinusoidal, given by

$$y_s = A\cos\omega t,$$

then the differential equation becomes

$$\ddot{\theta} + \frac{g}{l}\sin\theta - \frac{A}{l}\omega^2\cos\theta\ \cos\omega t = 0.$$ ◄

► **Example 4.6** For the spring-pendulum system shown in Fig. 4.7, we need two configuration coordinates to specify the positional status of the bob. One coordinate is θ and the other coordinate is dependent on the radial distance from the suspension to the bob. There are many possible ways of defining this coordinate, and we choose one of them as follows:

The spring is at equilibrium in the upright position when $\theta = 0$ and *elongation* $= mg/k$. Let the total length of the spring in the equilibrium condition be a. Let one of the position coordinates (r) be defined as the deviation from the length a. We take the equilibrium position of the bob as the "datum" level for measuring the potential energy.

In terms of these position coordinates, the radial component of the kinetic energy is $\frac{1}{2}m\dot{r}^2$ and the tangential component is $\frac{1}{2}m[(a+r)\dot{\theta}]^2$. Hence, the total kinetic energy is given by

$$T = \frac{1}{2}m[\dot{r}^2 + (a+r)^2\dot{\theta}^2].$$

Note that T is dependent on the velocities $\dot{r}$, $\dot{\theta}$ as well as the position r.

The potential energy has two components – one due to the spring and the other due to the gravitational potential. The elongation of the spring from unstretched condition is

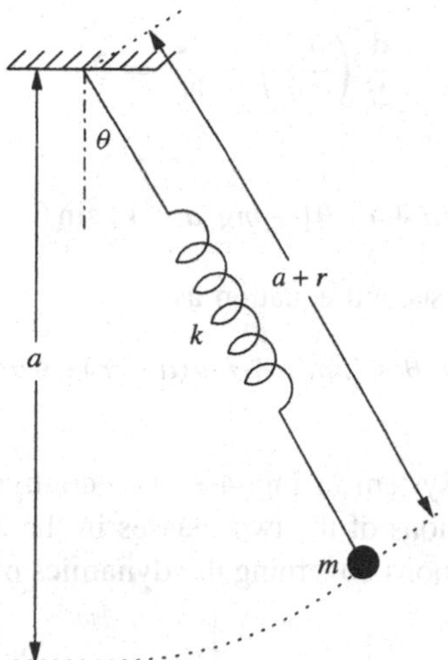

Figure 4.7 System pertaining to the example.

$(r + mg/k)$. Hence, the potential energy stored in the spring is $\frac{1}{2}k(r + mg/k)^2$. When the mass is in equilibrium vertically below the suspension, the gravitational potential is zero. At other positions the height of the mass from the zero level is $[a - (a + r)\cos\theta]$. Hence, the gravitational potential is $mg[a - (a + r)\cos\theta]$. Adding the two, we get the total potential as

$$V = \frac{1}{2}k(r + mg/k)^2 - mg(a + r)\cos\theta + mga.$$

This gives

$$\mathcal{L} = \frac{1}{2}m\dot{r}^2 + \frac{1}{2}m(a + r)^2\dot{\theta}^2 - \frac{1}{2}k(r + mg/k)^2 + mg(a + r)\cos\theta - mga$$

and

$$\frac{\partial\mathcal{L}}{\partial\dot{r}} = m\dot{r},$$

$$\frac{\partial\mathcal{L}}{\partial\dot{\theta}} = m(a + r)^2\dot{\theta},$$

$$\frac{\partial\mathcal{L}}{\partial r} = m\dot{\theta}^2(a + r) - k(r + mg/k) + mg\cos\theta,$$

$$\frac{\partial\mathcal{L}}{\partial\theta} = -mg(a + r)\sin\theta.$$

Hence, the Lagrangian equation

$$\frac{\mathrm{d}}{\mathrm{d}t}\left(\frac{\partial\mathcal{L}}{\partial\dot{r}}\right) - \frac{\partial\mathcal{L}}{\partial r} = 0$$

gives

$$m\ddot{r} - m\dot{\theta}^2(a + r) + k(r + mg/k) - mg\cos\theta = 0,$$

and the other equation

$$\frac{\mathrm{d}}{\mathrm{d}t}\left(\frac{\partial \mathcal{L}}{\partial \dot{\theta}}\right) - \frac{\partial \mathcal{L}}{\partial \theta} = 0$$

gives

$$m\frac{\mathrm{d}}{\mathrm{d}t}[(a+r)^2\dot{\theta}] + mg(a+r)\sin\theta = 0.$$

Upon simplification we get the second equation as

$$m(a+r)\ddot{\theta} + 2m\dot{\theta}\dot{r} + mg(a+r)\sin\theta = 0.$$ ◀

► **Example 4.7** We take the system in Fig. 4.8. The configuration of the system can be completely defined by the positions of the two masses in the horizontal direction, given by q_1 and q_2. The differential equations governing the dynamics of the system can be expressed in terms of these two variables.

In this particular case,

$$T = \frac{1}{2}m_1\dot{q}_1^2 + \frac{1}{2}m_2\dot{q}_2^2, \tag{4.19}$$

$$V = \frac{1}{2}k_1q_1^2 + \frac{1}{2}k_2(q_1-q_2)^2 + \frac{1}{2}k_3q_2^2. \tag{4.20}$$

We form the partial derivatives

$$\frac{\partial \mathcal{L}}{\partial \dot{q}_1} = m_1\dot{q}_1.$$

This is the momentum of the mass m_1.

$$\frac{\partial \mathcal{L}}{\partial q_1} = -k_1q_1 - k_2(q_1-q_2).$$

This is the restoring force on the mass m_1. So the Lagrangian equation for the q_1 coordinate is

$$m_1\ddot{q}_1 + k_1q_1 + k_2(q_1-q_2) = 0.$$

Similarly, for the q_2 axis we get

$$m_2\ddot{q}_2 + k_3q_2 - k_2(q_1-q_2) = 0.$$

These are the two equations governing the system dynamics. ◀

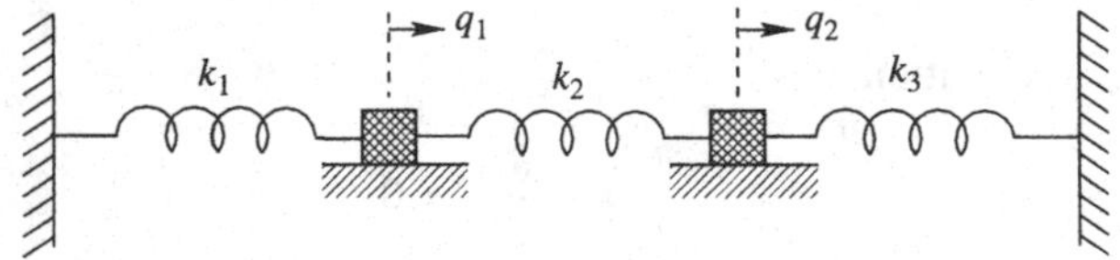

Figure 4.8 A mechanical system with two degrees of freedom.

► **Example 4.8** For the spring-connected triple pendulum system shown in Fig. 4.9, the minimum set of coordinates that uniquely define the positional status of this system are the three angles θ_1, θ_2 and θ_3. In terms of these generalized coordinates, the kinetic energy of the system in Fig. 4.9 is

$$T = \frac{1}{2}ml^2(\dot{\theta}_1^2 + \dot{\theta}_2^2 + \dot{\theta}_3^2).$$

The potential energy consists of two parts: the energy due to the gravitational force and the strain energy of the springs. The energy due to gravity is

$$V_g = mgl(1 - \cos\theta_1) + mgl(1 - \cos\theta_2) + mgl(1 - \cos\theta_3).$$

If the angles are small, this can be approximated to

$$V_g \approx \frac{1}{2}mgl(\theta_1^2 + \theta_2^2 + \theta_3^2).$$

The elongations of the springs are given by

$$h(\sin\theta_2 - \sin\theta_1) \approx h(\theta_2 - \theta_1)$$

and

$$h(\sin\theta_3 - \sin\theta_2) \approx h(\theta_3 - \theta_2)$$

respectively.

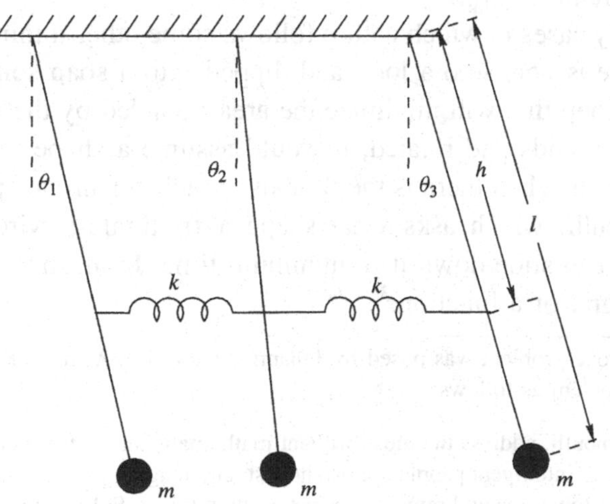

Figure 4.9 System pertaining to the example.

Hence, the energy stored in the springs is

$$V_s = \frac{1}{2}kh^2[(\theta_2 - \theta_1)^2 + (\theta_3 - \theta_2)^2].$$

This gives the total potential energy as

$$V = \frac{1}{2}mgl(\theta_1^2 + \theta_2^2 + \theta_3^2) + \frac{1}{2}kh^2[(\theta_2 - \theta_1)^2 + (\theta_3 - \theta_2)^2].$$

There are three generalized coordinates and hence the system will be described by three Lagrangian equations. For θ_1, θ_2 and θ_3, the Lagrangian equation gives respectively

$$l^2m\ddot{\theta}_1 + mgl\theta_1 + kh^2(\theta_1 - \theta_2) = 0,$$
$$l^2m\ddot{\theta}_2 + mgl\theta_2 + kh^2(\theta_2 - \theta_1) + kh^2(\theta_2 - \theta_3) = 0,$$
$$l^2m\ddot{\theta}_3 + mgl\theta_3 + kh^2(\theta_3 - \theta_2) = 0.$$ ◀

4.3 The Principle of Least Action

For conservative dynamical systems there is a very simple rule in classical mechanics, called the principle of least action, from which the path can be calculated. It says that if the system moved from $\mathbf{q} = \mathbf{x}_1$ at time t_1 to $\mathbf{q} = \mathbf{x}_2$ at time t_2, the path in between would be the one for which the integral of the Lagrangian function is a minimum. In terms of electrical circuits, it can be translated to say that the system always changes in a fashion that minimizes the integral of the difference between the energy stored in the inductors and the energy stored in the capacitors. It is quite surprising that the rule really works. To show that it does, let us calculate the path from the above premise.

There are many cases in which nature follows some other minimization rule. For example, if a wire is bent into a loop and dipped into a soap solution, a film will form to span the loop that will minimize the area bounded by the wire. If a flexible wire is held at two ends and rotated, it would assume a shape that minimizes the surface of revolution. Then there is the famous "brachistochrone" problem invented by Johann Bernoulli, which asks what shape a frictionless wire should have in order that a bead can slide down it in minimum time. In all these cases one has to minimize an integral of a function.[3]

[3]The brachistochrone problem was posed by Johann Bernoulli in *Acta Eruditorum* in June 1696. He introduced the problem as follows:

> "I, Johann Bernoulli, address the most brilliant mathematicians in the world. Nothing is more attractive to intelligent people than an honest, challenging problem, whose possible solution will bestow fame and remain as a lasting monument. Following the example set

Mathematically, the rule can be stated to say that a term (call it S) would be minimized over the path, where

$$S = \int_{t_1}^{t_2} f \, \mathrm{d}t. \tag{4.21}$$

In the case of dynamical systems, $f =$ the Lagrangian function $\mathcal{L}$. We first work out the solution for one-dimensional systems for which $\mathcal{L}$ is a function of $q, \dot{q}$ and t. And it is the integral

$$S = \int_{t_1}^{t_2} \mathcal{L}(q, \dot{q}, t) \, \mathrm{d}t \tag{4.22}$$

that we have to minimize. The problem is not quite the same as the minimization problems that one handles in elementary calculus. There it is a function that one has to minimize, while here it is an integral. Still, we shall use the same argument, which is as follows.

If we have a function q of an independent variable t, the minima has the particular property that if t is varied slightly, the variation in q is negligible (Fig. 4.10). In other words, $\frac{\mathrm{d}q}{\mathrm{d}t} = 0$. This property is not shared by the other points where, in general, $\Delta q \propto \Delta t$.

How do we apply this argument to paths? Suppose there is a path $q(t)$ that minimizes S. If we can vary it somehow by a slight amount, the resulting variation in S should be negligible. To vary the path, we arbitrarily draw a function $\eta(t)$ and obtain a varied path as

$$\bar{q}(t) = q(t) + \alpha \cdot \eta(t), \tag{4.23}$$

where α is a variable quantity. A small α will make $\bar{q}(t)$ deviate slightly from $q(t)$ and a large α will cause a large variation. Thus, α becomes the quantity with which we vary S. To obtain the varied path from this formulation, $\eta(t)$ should vanish at the two ends, that is,

$$\eta(t_1) = \eta(t_2) = 0. \tag{4.24}$$

---

by Pascal, Fermat, etc., I hope to gain the gratitude of the whole scientific community by placing before the finest mathematicians of our time a problem which will test their methods and the strength of their intellect. If someone communicates to me the solution of the proposed problem, I shall publicly declare him worthy of praise."

The problem he posed was the following. "Given two points A and B in a vertical plane, what is the curve traced out by a point acted on only by gravity, which starts at A and reaches B in the shortest time."

Five solutions were received before the deadline. Apart from Johann Bernoulli himself, the problem was successfully solved by Johann's brother Jacob Bernoulli, Leibniz, L'Hospital and Newton.

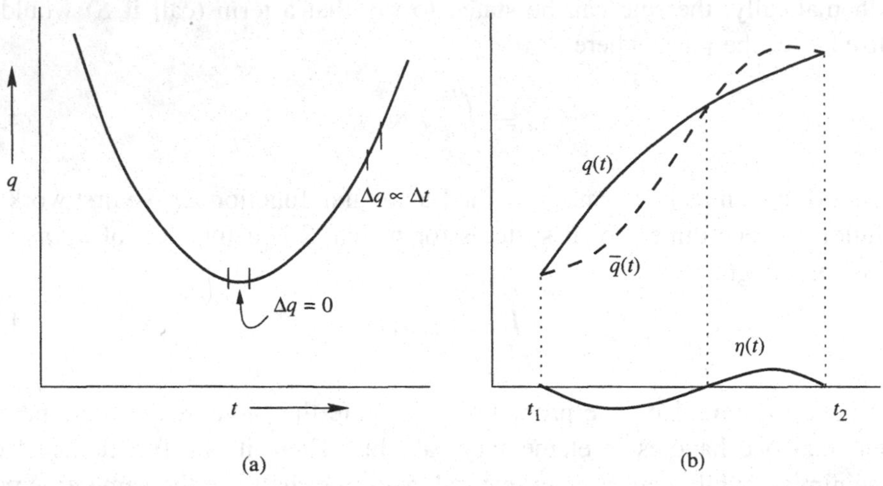

Figure 4.10 (a) The minimal point in a family of points (a function) is obtained by varying t and observing the resulting change in q. (b) The minimal function in a family of functions obtained by varying α and observing the resulting variation in S.

We will also need the derivative of (4.23):

$$\dot{\bar{q}}(t) = \dot{q}(t) + \alpha \cdot \dot{\eta}(t). \tag{4.25}$$

We have thus varied the path by α and obtained a new path which has a new S given by

$$S = \int_{t_1}^{t_2} \mathcal{L}(\bar{q}, \dot{\bar{q}}, t)\, \mathrm{d}t. \tag{4.26}$$

Substituting (4.23) and (4.25), we get

$$S = \int_{t_1}^{t_2} \mathcal{L}(q + \alpha\eta, \dot{q} + \alpha\dot{\eta}, t)\, \mathrm{d}t. \tag{4.27}$$

This is a function of α and the condition for minimization of S is $\frac{\partial S}{\partial \alpha} = 0$ at $\alpha = 0$. Now

$$\frac{\partial S}{\partial \alpha} = \int_{t_1}^{t_2} \frac{\partial}{\partial \alpha} \mathcal{L}(\bar{q}, \dot{\bar{q}}, t)\, \mathrm{d}t. \tag{4.28}$$

Since $\mathcal{L}$ is a function of several variables, by chain rule we get

$$\frac{\partial}{\partial \alpha} \mathcal{L}(\bar{q}, \dot{\bar{q}}, t) = \frac{\partial \mathcal{L}}{\partial \bar{q}} \frac{\partial \bar{q}}{\partial \alpha} + \frac{\partial \mathcal{L}}{\partial \dot{\bar{q}}} \frac{\partial \dot{\bar{q}}}{\partial \alpha} + \frac{\partial \mathcal{L}}{\partial t} \frac{\partial t}{\partial \alpha}. \tag{4.29}$$

Since t is independent of α, $\frac{\partial t}{\partial \alpha} = 0$. The other two terms can be obtained by differentiating (4.23) and (4.25) with respect to α. Thus, $\frac{\partial \bar{q}}{\partial \alpha} = \eta(t)$ and $\frac{\partial \dot{\bar{q}}}{\partial \alpha} = \dot{\eta}(t)$. Substituting, we get

$$\frac{\partial}{\partial \alpha}\mathcal{L}(\bar{q}, \dot{\bar{q}}, t) = \frac{\partial \mathcal{L}}{\partial \bar{q}}\eta(t) + \frac{\partial \mathcal{L}}{\partial \dot{\bar{q}}}\dot{\eta}(t). \tag{4.30}$$

Now recall the condition for minimization of S, which is $\frac{\partial S}{\partial \alpha} = 0$ at $\alpha = 0$. At that value of α, $\bar{q} = q$ and $\dot{\bar{q}} = \dot{q}$.

Substituting these into (4.28), we get

$$\frac{\partial S}{\partial \alpha} = \int_{t_1}^{t_2}\left[\frac{\partial \mathcal{L}}{\partial q}\eta(t) + \frac{\partial \mathcal{L}}{\partial \dot{q}}\dot{\eta}(t)\right]\mathrm{d}t = 0. \tag{4.31}$$

Integrating the second term by parts, we get

$$\int_{t_1}^{t_2}\frac{\partial \mathcal{L}}{\partial q}\eta(t)\,\mathrm{d}t + \frac{\partial \mathcal{L}}{\partial \dot{q}}\eta(t)\bigg|_{t_1}^{t_2} - \int_{t_1}^{t_2}\frac{\mathrm{d}}{\mathrm{d}t}\frac{\partial \mathcal{L}}{\partial \dot{q}}\eta(t)\,\mathrm{d}t = 0. \tag{4.32}$$

By (4.24) the second term would always vanish. Thus, the condition for minimum S becomes

$$\int_{t_1}^{t_2}\eta(t)\left[\frac{\partial \mathcal{L}}{\partial q} - \frac{\mathrm{d}}{\mathrm{d}t}\left(\frac{\partial \mathcal{L}}{\partial \dot{q}}\right)\right]\mathrm{d}t = 0. \tag{4.33}$$

Since $\eta(t)$ can be any arbitrary function and the integral must vanish for all ηs, the term in bracket must be zero. Hence,

$$\frac{\mathrm{d}}{\mathrm{d}t}\left(\frac{\partial \mathcal{L}}{\partial \dot{q}}\right) - \frac{\partial \mathcal{L}}{\partial q} = 0. \tag{4.34}$$

This is the Lagrangian equation for one generalized coordinate.[4] Thus, we see that the principle of least action yields the same result as Newtonian or Lagrangian mechanics.

---

[4]The general solution to the problem (4.21) yields

$$\frac{\mathrm{d}}{\mathrm{d}t}\left(\frac{\partial f}{\partial \dot{q}}\right) - \frac{\partial f}{\partial q} = 0, \tag{4.35}$$

which is known as the *Euler's equation*. One may notice that the procedure does not ensure that S is a minimum. It could be a maximum or a point of inflection. What the procedure ensures is that the function obtained is *stationary* against variations. The type of extremum is given by the particular problem in hand and is obtained from the analysis of the second derivative.

Systems with two degrees of freedom

Now let us take a system with two degrees of freedom. What does the principle of least action say for this case? Since the system has two degrees of freedom q_1 and q_2, (4.22) would have the form

$$S = \int_{t_1}^{t_2} \mathcal{L}(q_1, \dot{q}_1, q_2, \dot{q}_2, t)\,\mathrm{d}t. \tag{4.36}$$

In this case, we would define two arbitrary functions $\eta_1(t)$ and $\eta_2(t)$ with the boundary conditions

$$\eta_1(t_1) = \eta_1(t_2) = \eta_2(t_1) = \eta_2(t_2) = 0.$$

Then we obtain varied functions $\bar{q_1}$ and $\bar{q_2}$ with the help of a variable α as

$$\bar{q_1}(t) = q_1(t) + \alpha\eta_1(t),$$
$$\bar{q_2}(t) = q_2(t) + \alpha\eta_2(t).$$

The condition for minimum S in this case is the same, that is, $\frac{\partial S}{\partial \alpha} = 0$ at $\alpha = 0$. Following the same procedure, we get the condition as

$$\frac{\partial S}{\partial \alpha} = \int_{t_1}^{t_2} \left[\frac{\partial \mathcal{L}}{\partial q_1}\eta_1(t) + \frac{\partial \mathcal{L}}{\partial \dot{q}_1}\dot{\eta}_1(t) + \frac{\partial \mathcal{L}}{\partial q_2}\eta_2(t) + \frac{\partial \mathcal{L}}{\partial \dot{q}_2}\dot{\eta}_2(t)\right]\mathrm{d}t = 0. \tag{4.37}$$

The terms involving $\dot{\eta}_1$ and $\dot{\eta}_2$ are then integrated by parts that yield (after putting the boundary conditions of η_1 and η_2)

$$\int_{t_1}^{t_2} \left\{\eta_1(t)\left[\frac{\partial \mathcal{L}}{\partial q_1} - \frac{\mathrm{d}}{\mathrm{d}t}\left(\frac{\partial \mathcal{L}}{\partial \dot{q}_1}\right)\right] + \eta_2(t)\left[\frac{\partial \mathcal{L}}{\partial q_2} - \frac{\mathrm{d}}{\mathrm{d}t}\left(\frac{\partial \mathcal{L}}{\partial \dot{q}_2}\right)\right]\right\}\mathrm{d}t = 0. \tag{4.38}$$

Since this must hold for all arbitrary choices of the functions $\eta_1(t)$ and $\eta_2(t)$, we get two Euler's equations:

$$\left[\frac{\mathrm{d}}{\mathrm{d}t}\left(\frac{\partial \mathcal{L}}{\partial \dot{q}_1}\right) - \frac{\partial \mathcal{L}}{\partial q_1}\right] = 0, \tag{4.39}$$

$$\left[\frac{\mathrm{d}}{\mathrm{d}t}\left(\frac{\partial \mathcal{L}}{\partial \dot{q}_2}\right) - \frac{\partial \mathcal{L}}{\partial q_2}\right] = 0. \tag{4.40}$$

Thus, the law of least action yields two Lagrangian equations for q_1 and q_2. Generalizing the line of argument, one can see that for a system with n configuration coordinates, the same least action principle is equivalent to a set of n Lagrangian equations, one for each of the independent degrees of freedom.

4.4 Lagrangian Method Applied to Electrical Circuits

Though the Lagrangian formalism was developed to remove the hurdles in obtaining the differential equations for mechanical systems, the same method can also be applied to obtain the differential equations for electrical circuits. A particular advantage of using the same technique for electrical as well as mechanical systems is that it offers a unified framework for electromechanical systems, that is, systems with both electrical and mechanical components.

We note, first, that the generalized coordinates in a mechanical system are the *position* variables that are consistent with the constraints. The constraints in an electrical circuit are given by the way the circuit components are connected, and the electrical equivalent of position variable is the charge. Therefore, our choice of variables should be *charge*, and these have to be defined depending on the circuit connection. We already know that a simple way of doing so is to define the variables as the charges flowing in the meshes. This ensures that the charges flowing in all the branches are defined in terms of the variables chosen. Once the configuration coordinates q_i in an electrical circuit are defined in the above manner, the $\dot{q}_i$ become simply the mesh currents.

The equivalence between the formulations in a mechanical system and an electrical system is illustrated by the circuit shown in Fig. 4.11, whose dynamical equations are the same as that of the mechanical system in Fig. 4.8. In that sense, we call it the electrical equivalent of the mechanical system.

To show that it indeed has the same dynamics, recall the equivalences between mechanical and electrical systems shown in Chapter 1. An inductor is an inertial element like a mass, and a capacitor is a compliant element like a spring. The position coordinate is represented by the charges, and in this circuit, we define the charges flowing in the two loops as the generalized coordinates. In terms of these coordinates, the kinetic energy and the potential energy – and hence the Lagrangian – take exactly the same form as in the mechanical system. This yields exactly the same differential equations. The initial condition may be some energy initially stored in

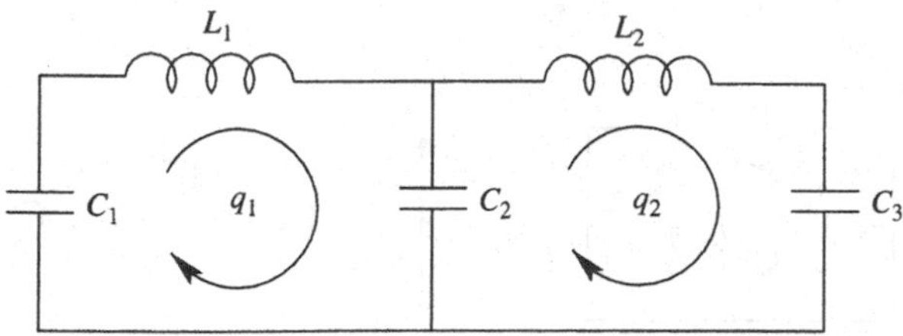

Figure 4.11 An electrical system that is dynamically equivalent to the mechanical system in Fig. 4.8.

the capacitors, which will be equivalent to some initial displacement of the masses in the mechanical system. The dynamics of the two systems would be identical if the corresponding parameters are the same.

It may be noted that the choice of the generalized coordinates is not unique. In fact the charges flowing in any set of branches would suffice so long as the charges in all the branches are given in terms of the chosen set. For example, one can choose the charge flowing in the storage elements in the system plus that in the minimum number of other branches so that the charges in all the branches can be described in terms of the chosen variables. The charges in the meshes form only one such convenient choice that is ensured to be sufficient.

4.5 Systems with External Forces or Electromotive Forces

So far we have considered only conservative systems with no externally applied force or voltage source. If there are external forces or voltage sources $F_i(t)$ present in the system as in Fig. 4.12, the same general framework can be used to formulate the differential equations. The forces F_i may be constant, or may vary as functions of time (as in a sinusoidal voltage source). In this case, we only need to incorporate the additional force in the potential function. This can be done by adding a term to V. This new term is external force times the generalized coordinate along which the external force acts, that is,

$$V \text{ due to external force} = -F_i q_i.$$

This term, when differentiated with respect to that generalized coordinate, will yield the external force. Therefore, the generalized force is still obtainable from the potential function as

$$Q_i = -\frac{\partial V}{\partial q_i}.$$

Hence, the system is still conservative, and the same Lagrangian equation can be applied.

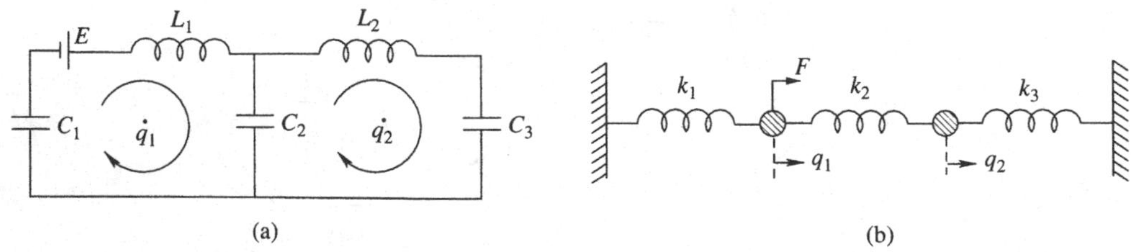

Figure 4.12 A circuit with a voltage source and its mechanical equivalent.

Note the negative sign in this expression, which is necessary to set the signs right (because $\mathcal{L} = T - V$).

► **Example 4.9** The system in Fig. 4.12 is different from that in Fig. 4.11 in the addition of the voltage source in the first loop. To account for the source, the potential function would be written as

$$V = \frac{1}{2C_1}q_1^2 + \frac{1}{2C_3}q_2^2 + \frac{1}{2C_2}(q_1 - q_2)^2 - Eq_1.$$

Note the last term coming into account for the voltage source; the voltage times the coordinate q_1 along which it acts. Thus,

$$\frac{\partial \mathcal{L}}{\partial q_1} = -\frac{q_1}{C_1} - \frac{1}{C_2}(q_1 - q_2) + E.$$

The first Lagrangian equation will be modified to

$$L_1\ddot{q}_1 + \frac{q_1}{C_1} + \frac{1}{C_2}(q_1 - q_2) - E = 0.$$

There will be no change in the second equation. ◄

► **Example 4.10** An inverted pendulum with rigid massless rod (length l) is placed on a cart as shown in Fig. 4.13. The mass of the cart is m_1 and that of the bob is m_2. An external force $F(t)$ is applied on the cart. Derive the dynamical equations of the system.

Solution: Let the generalized coordinates be x and θ as shown in the figure. The kinetic energy of the cart is $\frac{1}{2}m_1\dot{x}^2$. The kinetic energy of the bob has two components. The horizontal component is $\frac{1}{2}m_2(l\dot{\theta}\cos\theta + \dot{x})^2$ and the vertical component is $\frac{1}{2}m_2(l\dot{\theta}\sin\theta)^2$. The potential also has two parts. One is due to the height of the bob from the level of contact between the rod and the cart. Its value is $m_2 gl\cos\theta$. The other is due to the applied force and its value is $-Fx$. Thus, the Lagrangian becomes

$$\mathcal{L} = \frac{1}{2}m_1\dot{x}^2 + \frac{1}{2}m_2\dot{x}^2 + \frac{1}{2}m_2 l^2\dot{\theta}^2 + m_2 l\dot{\theta}\dot{x}\cos\theta - m_2 gl\cos\theta + Fx.$$

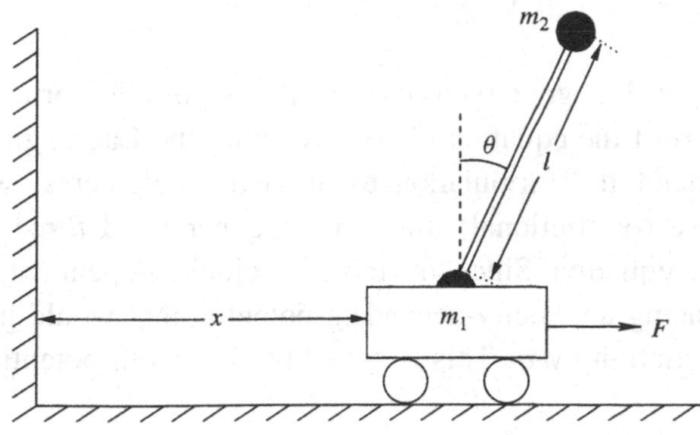

Figure 4.13 Inverted pendulum on cart pertaining to the example.

We obtain the partial derivatives as

$$\frac{\partial \mathcal{L}}{\partial \dot{x}} = m_1\dot{x} + m_2\dot{x} + m_2 l\dot{\theta}\cos\theta,$$
$$\frac{\partial \mathcal{L}}{\partial \dot{\theta}} = m_2 l^2\dot{\theta} + m_2 l\dot{x}\cos\theta,$$
$$\frac{\partial \mathcal{L}}{\partial x} = F,$$
$$\frac{\partial \mathcal{L}}{\partial \theta} = -m_2 l\dot{\theta}\dot{x}\sin\theta + m_2 gl\sin\theta.$$

Putting these into the Lagrangian equations, we get the equations for x and θ as

$$m_1\ddot{x} + m_2\ddot{x} + m_2 l\cos\theta\ddot{\theta} - m_2 l\dot{\theta}^2\sin\theta - F = 0$$

and

$$l\ddot{\theta} + \ddot{x}\cos\theta - g\sin\theta = 0.$$ ◄

4.6 Systems with Resistance or Friction

So far, we have considered only conservative systems. In such systems, the generalized forces are derivable from a scalar potential. The problem with frictional elements is that the force (or emf) across such an element is proportional to the velocity (or current). Naturally, the force cannot be derived from a potential that is a function of the position coordinates.

We would, however, like to retain the advantages of the Newton–Lagrange formalism. In doing so, let us recall the primary form of Lagrangian equation that we derived in (4.12):

$$\frac{\mathrm{d}}{\mathrm{d}t}\left(\frac{\partial T}{\partial \dot{q}_i}\right) - \frac{\partial T}{\partial q_i} - Q_i = 0.$$

Starting from here, we had gone on to include all the possible forms of *conservative forces* and had derived the equation (4.15) involving the Lagrangian function.

In order to extend the formulation to dissipative elements, we will have to include the resistive (or frictional) force in the generalized force term Q_i in this original Lagrangian equation. Since this force is velocity dependent, we can achieve this feat by introducing a velocity-dependent potential that would give the resistive force upon differentiation by $\dot{q}$. This is called the Rayleigh potential, given by

$$\Re = \sum \frac{1}{2} R_i\dot{q}_i^2, \tag{4.41}$$

where R_i are the resistances along the q_i coordinates. Now the generalized force term Q_i would consist of the conservative component and the dissipative component as

$$Q_i = -\frac{\partial V}{\partial q_i} - \frac{\partial \mathfrak{R}}{\partial \dot{q}_i}.$$

Thus, the equation (4.12) would be modified as

$$\frac{\mathrm{d}}{\mathrm{d}t}\left(\frac{\partial T}{\partial \dot{q}_i}\right) - \frac{\partial T}{\partial q_i} + \frac{\partial V}{\partial q_i} + \frac{\partial \mathfrak{R}}{\partial \dot{q}_i} = 0$$

or

$$\frac{\mathrm{d}}{\mathrm{d}t}\left(\frac{\partial (T-V)}{\partial \dot{q}_i}\right) - \frac{\partial (T-V)}{\partial q_i} + \frac{\partial \mathfrak{R}}{\partial \dot{q}_i} = 0$$

or

$$\frac{\mathrm{d}}{\mathrm{d}t}\left(\frac{\partial \mathcal{L}}{\partial \dot{q}_i}\right) - \frac{\partial \mathcal{L}}{\partial q_i} + \frac{\partial \mathfrak{R}}{\partial \dot{q}_i} = 0. \tag{4.42}$$

Notice that the term $\frac{\partial \mathfrak{R}}{\partial \dot{q}_i}$, even though dependent on $\dot{q}_i$, should not be within the time derivative term $\frac{\mathrm{d}}{\mathrm{d}t}\left(\frac{\partial \mathcal{L}}{\partial \dot{q}_i}\right)$. It is added in the generalized force term for the formal completion of the Lagrange's formula for dissipative systems.

► **Example 4.11** Let us consider the electrical circuit in Fig. 4.14(a) and the mechanical equivalent system in Fig. 4.14(b). The voltage across the resistances (in the mechanical system, the forces exerted by frictional elements) are $R_1\dot{q}_1$ and $R_2\dot{q}_2$ respectively. These can be taken into account by defining the Rayleigh potential as

$$\mathfrak{R} = \frac{1}{2}R_1\dot{q}_1^2 + \frac{1}{2}R_2\dot{q}_2^2, \tag{4.43}$$

so that

$$\frac{\partial \mathfrak{R}}{\partial \dot{q}_1} = R_1\dot{q}_1,$$

$$\frac{\partial \mathfrak{R}}{\partial \dot{q}_2} = R_2\dot{q}_2.$$

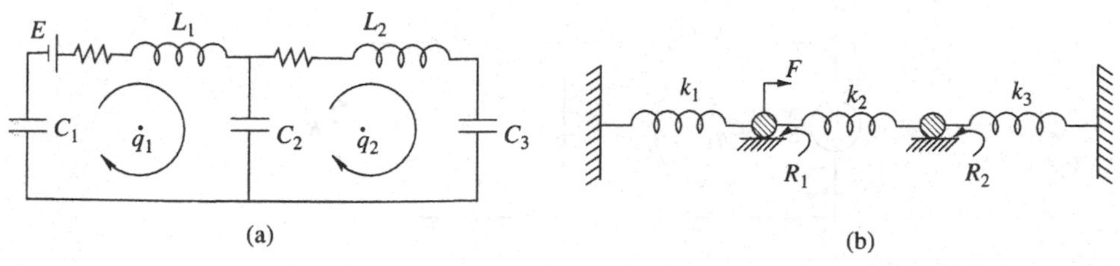

Figure 4.14 A circuit with a resistive element and its mechanical equivalent.

Recall from Examples 4.7 and 4.9 that

$$
\begin{aligned}
T &= \frac{1}{2}L_1\dot{q}_1^2 + \frac{1}{2}L_2\dot{q}_2^2 \quad \text{in electrical domain,} \\
&= \frac{1}{2}m_1\dot{q}_1^2 + \frac{1}{2}m_2\dot{q}_2^2 \quad \text{in mechanical domain.} \\
V &= \frac{1}{2C_1}q_1^2 + \frac{1}{2C_2}q_2^2 + \frac{1}{2C_3}(q_1 - q_2)^2 - Eq_1 \quad \text{in electrical domain,} \\
&= \frac{1}{2}k_1q_1^2 + \frac{1}{2}k(q_1 - q_2)^2 + \frac{1}{2}k_2q_2^2 - Fq_1 \quad \text{in mechanical domain.}
\end{aligned}
$$

Applying (4.42), the Lagrangian equation for the q_1 coordinate is

$$L_1\ddot{q}_1 + \frac{q_1}{C_1} + \frac{1}{C_2}(q_1 - q_2) - E + R_1\dot{q}_1 = 0.$$

Similarly, for the q_2 axis we get

$$L_2\ddot{q}_2 + \frac{q_2}{C_3} - \frac{1}{C_2}(q_1 - q_2) = 0.$$

These are the two equations governing the dynamics of the electrical circuit.

It is easy to see that the equations for the mechanical system have similar form, only L is replaced by m and $1/C$ is replaced by k. ◀

▶ **Example 4.12** In this circuit, there are three meshes, and we define the coordinates as the charges flowing in the meshes as shown in Fig. 4.15. In terms of these variables,

$$
\begin{aligned}
T &= \frac{1}{2}L(\dot{q}_1 - \dot{q}_3)^2, \\
V &= \frac{1}{2C_1}(q_1 - q_2)^2 + \frac{1}{2C_2}q_3^2 - Eq_1,
\end{aligned}
$$

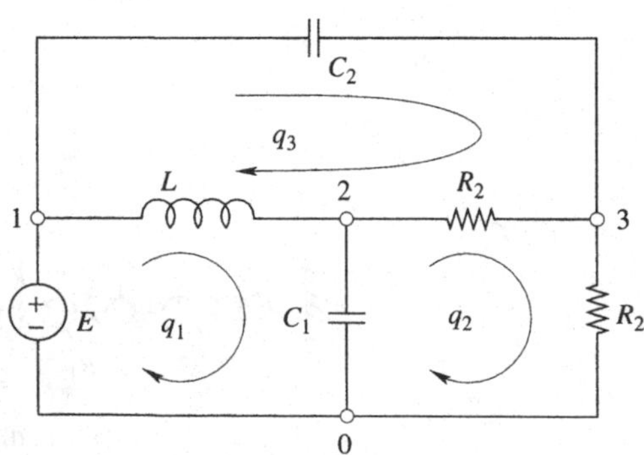

Figure 4.15 The circuit pertaining to Example 4.12.

$$\mathcal{L} = \frac{1}{2}L(\dot{q}_1 - \dot{q}_3)^2 - \frac{1}{2C_1}(q_1 - q_2)^2 - \frac{1}{2C_2}q_3^2 + Eq_1,$$

$$\mathfrak{R} = \frac{1}{2}R_1(\dot{q}_2 - \dot{q}_3)^2 + \frac{1}{2}R_2\dot{q}_2^2.$$

From these, we obtain

$$\frac{\partial \mathcal{L}}{\partial \dot{q}_1} = L(\dot{q}_1 - \dot{q}_3),$$

$$\frac{\partial \mathcal{L}}{\partial q_1} = -\frac{1}{C_1}(q_1 - q_2) + E,$$

$$\frac{\partial \mathfrak{R}}{\partial \dot{q}_1} = 0.$$

Thus, the Lagrangian equation in the q_1 coordinate is

$$L(\ddot{q}_1 - \ddot{q}_3) + \frac{1}{C_1}(q_1 - q_2) - E = 0. \tag{4.44}$$

In the q_2 coordinate,

$$\frac{\partial \mathcal{L}}{\partial \dot{q}_2} = 0,$$

$$\frac{\partial \mathcal{L}}{\partial q_2} = \frac{1}{C_1}(q_1 - q_2),$$

$$\frac{\partial \mathfrak{R}}{\partial \dot{q}_2} = R_1(\dot{q}_2 - \dot{q}_3) + R_2\dot{q}_2,$$

and thus the differential equation is

$$-\frac{1}{C_1}(q_1 - q_2) + R_1(\dot{q}_2 - \dot{q}_3) + R_2\dot{q}_2 = 0. \tag{4.45}$$

Similarly, in the q_3 coordinate,

$$\frac{\partial \mathcal{L}}{\partial \dot{q}_3} = -L(\dot{q}_1 - \dot{q}_3),$$

$$\frac{\partial \mathcal{L}}{\partial q_3} = \frac{1}{C_2}q_3,$$

$$\frac{\partial \mathfrak{R}}{\partial \dot{q}_3} = -R_1(\dot{q}_2 - \dot{q}_3),$$

and thus the differential equation is

$$L(\ddot{q}_1 - \ddot{q}_3) + \frac{1}{C_2}q_3 + R_1(\dot{q}_2 - \dot{q}_3) = 0. \tag{4.46}$$

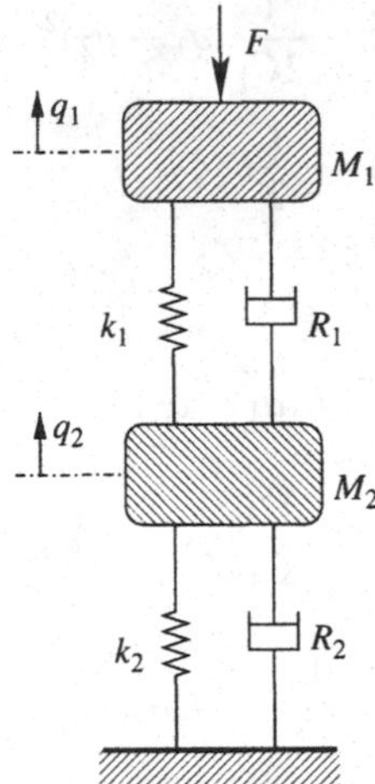

Figure 4.16 The mechanical system pertaining to Example 4.13.

The differential equations of the system are given by (4.44), (4.45) and (4.46). A closer scrutiny reveals that these are nothing but the KVL equations in the three meshes, expressed in terms of the chosen coordinates. ◀

► **Example 4.13** Take the mechanical system in Fig. 4.16. Let the coordinates q_1 and q_2 have value zero at the unstretched positions of the two springs. In terms of these coordinates,

$$T = \frac{1}{2}M_1\dot{q}_1^2 + \frac{1}{2}M_2\dot{q}_2^2,$$

$$V = \frac{1}{2}k_1(q_1 - q_2)^2 + \frac{1}{2}k_2 q_2^2 + Fq_1,$$

$$\mathfrak{R} = \frac{1}{2}R_1(\dot{q}_1 - \dot{q}_2)^2 + \frac{1}{2}R_2\dot{q}_2^2.$$

From here it is trivial to obtain the differential equations. ◀

4.7 Accounting for Current Sources

If a current source is present in a circuit, it essentially defines the charge flow in that particular branch. Therefore, current sources are automatically accounted when these branch currents are expressed in terms of the source current function. Let us illustrate it with an example.

► **Example 4.14** In the circuit of Fig. 4.17, the loop current $\dot{q}_1$ is equal to the externally impressed current I. Therefore, the corresponding charge flow is

$$\int_0^t I(\epsilon)\,\mathrm{d}\epsilon.$$

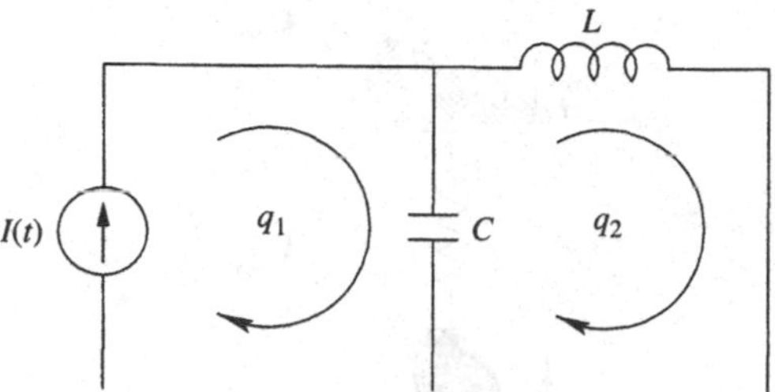

Figure 4.17 The circuit pertaining to Example 4.14.

The energy functions can then be written as

$$T = \frac{1}{2}L\dot{q}_2^2,$$

$$V = \frac{1}{2C}\left(\int_0^t I(\epsilon)\,\mathrm{d}\epsilon - q_2\right)^2,$$

$$\mathcal{L} = \frac{1}{2}L\dot{q}_2^2 - \frac{1}{2C}\left(\int_0^t I(\epsilon)\,\mathrm{d}\epsilon - q_2\right)^2.$$

Notice that this Lagrangian yields no equation along the q_1 direction because it is no longer an independent coordinate. The equation along the q_2 coordinate is given by

$$L\ddot{q}_2 - \frac{1}{C}\int_0^t I(\epsilon)\,\mathrm{d}\epsilon + \frac{1}{C}q_2 = 0.$$

It may be noted that here the source function appears in an integral. This is natural because the externally impressed function is a current, and the chosen variables demand the charge in the capacitor to be expressed in terms of the source function. If a different variable is chosen, it may demand a different representation. For example, in this system, if the charge through the capacitor is chosen as the variable, the current source will be required to be included in the kinetic energy function. This will make the *derivative* of the current source function to appear in the differential equation. We leave this exercise to the reader.

The source of flow in a mechanical system is a cam, and so its representation follows the same procedure. The system shown in Fig. 4.18 is equivalent to the circuit considered in this problem. Using the position variables as shown, this system yields the same set of equations. ◀

4.8 Modelling Mutual Inductances

In case the inductances in a system have magnetic coupling, leading to mutual inductance effect, it can be easily accounted for in terms of a change in the kinetic

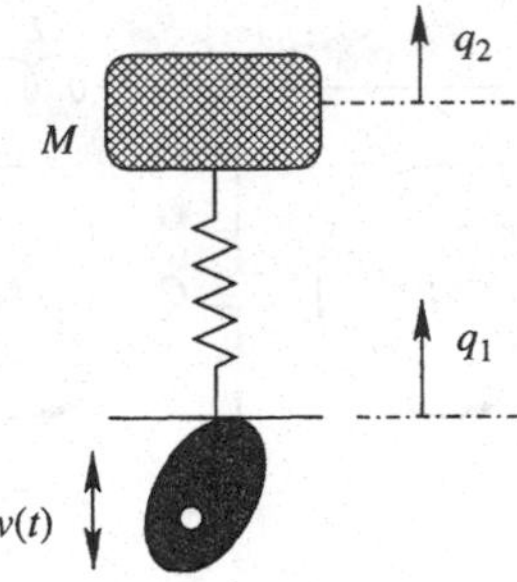

Figure 4.18 The mechanical system equivalent to the circuit of Fig. 4.17.

energy. We know that if there are two coils with self inductances L_1 and L_2, and mutual inductance M, and if currents i_1 and i_2 flow in them, then the total energy stored is

$$T = \frac{1}{2}L_1 i_1^2 + \frac{1}{2}L_2 i_2^2 \pm M i_1 i_2.$$

The sign of the mutual inductance term depends on whether the magnetic coupling is additive or subtractive. If the coils are wound in the same sense, the magnetic fields reinforce each other, resulting in a higher amount of energy storage. In that case, we take the positive sign. If, on the other hand, the coils are wound in the opposite sense, the total magnetic field reduces and we have to take the negative sign.

Normally this property is denoted by a dot placed on one of the sides of the inductors. By the dot convention, the magnetomotive forces due to the two currents are additive if both the currents enter (or leave) the respective coils through the ends marked by the dot.

▶ **Example 4.15** In the circuit of Fig. 4.19, there are two loops, and so the coordinates are chosen as q_1 and q_2. In terms of these coordinates,

$$T = \frac{1}{2}L_1\dot{q}_1^2 + \frac{1}{2}L_2(\dot{q}_1 - \dot{q}_2)^2 - M\dot{q}_1(\dot{q}_1 - \dot{q}_2),$$

$$V = \frac{1}{2C}q_2^2 - Eq_1,$$

$$\mathcal{L} = \frac{1}{2}L_1\dot{q}_1^2 + \frac{1}{2}L_2(\dot{q}_1 - \dot{q}_2)^2 - M\dot{q}_1(\dot{q}_1 - \dot{q}_2) - \frac{1}{2C}q_2^2 + Eq_1,$$

$$\mathfrak{R} = \frac{1}{2}R_1\dot{q}_2^2.$$

These give

$$\frac{\partial \mathcal{L}}{\partial \dot{q}_1} = L_1\dot{q}_1 + L_2(\dot{q}_1 - \dot{q}_2) - 2M\dot{q}_1 + M\dot{q}_2,$$

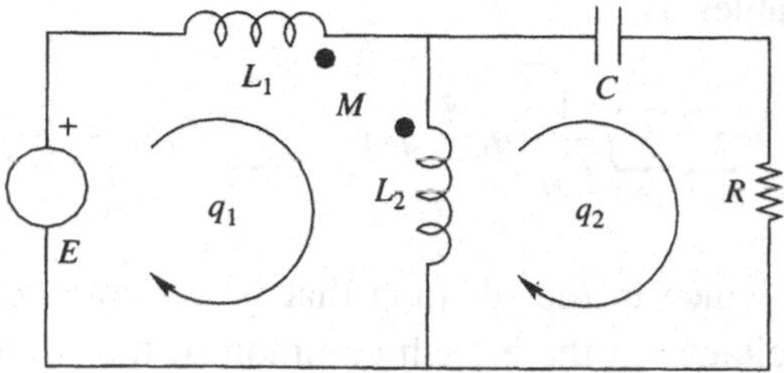

Figure 4.19 The circuit pertaining to Example 4.15.

$$\frac{\partial \mathcal{L}}{\partial q_1} = E,$$

$$\frac{\partial \mathfrak{R}}{\partial \dot{q}_1} = 0$$

and

$$\frac{\partial \mathcal{L}}{\partial \dot{q}_2} = -L_2(\dot{q}_1 - \dot{q}_2) + M\dot{q}_1,$$

$$\frac{\partial \mathcal{L}}{\partial q_2} = -\frac{q_2}{C},$$

$$\frac{\partial \mathfrak{R}}{\partial \dot{q}_2} = R\dot{q}_2.$$

Thus, the two Lagrangian equations become

$$L_1\ddot{q}_1 + L_2(\ddot{q}_1 - \ddot{q}_2) - 2M\ddot{q}_1 + M\ddot{q}_2 - E = 0$$

and

$$-L_2(\ddot{q}_1 - \ddot{q}_2) + M\ddot{q}_1 + \frac{q_2}{C} + R\dot{q}_2 = 0.$$ ◄

4.9 A General Methodology for Electrical Networks

In case of electrical circuits, we have already seen that the mesh currents form a minimum set of independent variables. The same approach can be followed to identify the generalized coordinates and to formulate the differential equations in terms of these coordinates. The only difference is that in the Lagrangian methodology, the generalized coordinates should be position variables that are equivalent to *charges* in the electrical domain. Therefore, the independent generalized coordinates become the charges flowing in the loops.

If there is an electrical network with n "windows" or loops, assign a generalized coordinate q_i to each loop. Thus, the loop currents will be $\dot{q}_i$. Write V, T, and $\mathfrak{R}$

in terms of these variables as

$$V = \frac{1}{2}\sum_{i=1}^{n}\frac{1}{C_i}q_i^2 + \frac{1}{2}\sum_{i=1}^{n}\sum_{j=1}^{i-1}\frac{1}{C_{ij}}(q_i - q_j)^2 - \sum_{i=1}^{n}E_i q_i - \sum_{i=1}^{n}\sum_{j=1}^{i-1}E_{ij}(q_i - q_j), \tag{4.47}$$

where C_i is the capacitance in the ith loop that is not common with other loops; $C_{ij} = C_{ji}$ is the capacitance in the branch common to the ith and jth loops; E_i is the impressed voltage (positive in the positive direction of $\dot{q}_i$) in the ith loop and E_{ij} is the voltage source in the branch common to the ith and jth loops. Expressed in a similar manner, we have

$$T = \frac{1}{2}\sum_{i=1}^{n}L_i\dot{q}_i^2 + \frac{1}{2}\sum_{i=1}^{n}\sum_{j=1}^{i-1}L_{ij}(\dot{q}_i - \dot{q}_j)^2 + \frac{1}{2}\sum_{i\neq j}M_{ij}\dot{q}_i\dot{q}_j \tag{4.48}$$

and

$$\Re = \frac{1}{2}\sum_{i=1}^{n}R_i\dot{q}_i^2 + \frac{1}{2}\sum_{i=1}^{n}\sum_{j=1}^{i-1}R_{ij}(\dot{q}_i - \dot{q}_j)^2, \tag{4.49}$$

where R_i and L_i are the resistance and the coefficient of self inductance in the ith loop not common with any other loop; R_{ij} and L_{ij} are the resistance and the coefficient of self inductance of the branch common to ith and jth loops. M_{ij} are the mutual inductances between the ith and the jth loops. According to the dot convention, M_{ij} will be positive if both the currents enter the dots or come out of the dots (fluxes help each other and hence the stored energy increases), and will be negative if one enters and the other leaves the dots.

The above methodology of obtaining the differential equations follows the mesh current method. It has been shown in Chapter 3 that the mesh current method cannot identify the minimum number of independent variables in case the system has dependent storage elements. This weakness is shared by the Lagrangian method, which yields correct equations in all circumstances, but these may not be the minimal set required to describe the dynamics of a system. This method is therefore apt for systems where there is no closed path containing only capacitances and voltage sources, or a cutset containing only inductances.

4.10 Modelling Coulomb Friction

Coulomb (or dry) friction in mechanical systems poses a problem in modelling because of its nonlinear nature (see Section 1.1.3). Nevertheless, we can approach

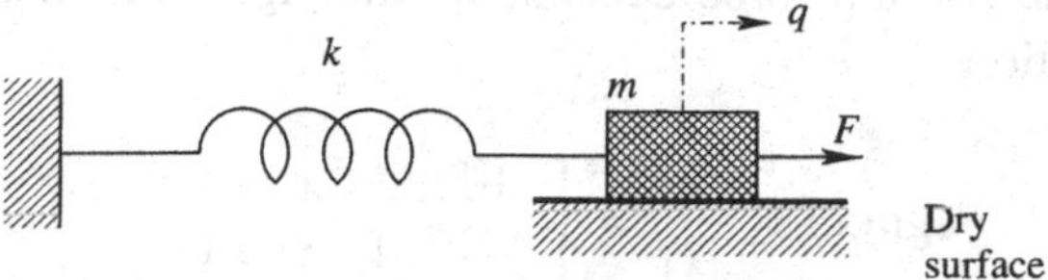

Figure 4.20 The mass-spring system with Coulomb damping.

the problem first by assuming it to be represented by a resistance element R, and then substituting the nonlinear functional form of R.

To illustrate, let us take the simple mass-spring system shown in Fig. 4.20. We derive the Lagrangian equation in the usual way:

$$\mathcal{L} = \frac{1}{2}m\dot{q}^2 - \frac{1}{2}kq^2 + Fq,$$

$$\Re = \frac{1}{2}R\dot{q}^2,$$

which give the Lagrangian equation as

$$m\ddot{q} + kq - F + R\dot{q} = 0.$$

Now we note that R can be approximated by a piecewise constant functional form shown in Fig. 4.21. If μ is the coefficient of kinetic friction and $N = mg$ the normal reaction, then the Coulomb friction force is $+\mu N$ for positive values of $\dot{q}$ and $-\mu N$ for negative values of $\dot{q}$ (assuming the force acts in the direction opposite to $\dot{q}$).

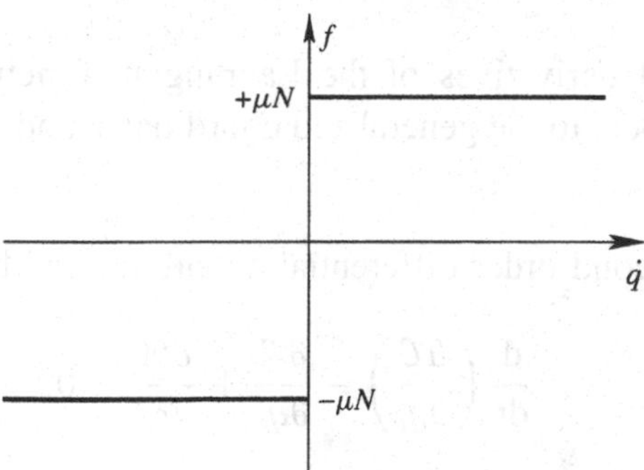

Figure 4.21 The kinetic friction force in Coulomb damping.

Since the sign of the friction force depends on the sign of $\dot{q}$, it can be represented by the *signum* function

$$\operatorname{sgn}(\dot{x}) = \frac{\dot{x}}{|\dot{x}|} = \begin{cases} +1 & \text{if } \dot{x} > 0 \\ -1 & \text{if } \dot{x} < 0 \end{cases} .$$

Therefore, the Lagrangian equation becomes

$$m\ddot{q} + kq - F + \mu mg\frac{\dot{x}}{|\dot{x}|} = 0.$$

4.11 Chapter Summary

The Lagrangian formalism offers a powerful method of obtaining the second-order differential equations for any mechanical, electrical or electromechanical system. The steps to apply this methodology are:

- Identify the configuration coordinates – the minimum number of variables that specify the positional status of a system.
- Express the potential energy and the kinetic energy as functions of these coordinates and their time derivatives. The potential energy term includes energy stored in capacitors and springs, potential energy due to gravity, and the energy due to any externally applied force or emf (force × displacement).
- Express the Lagrangian function as $\mathcal{L} = T - V$.
- For non-conservative systems, define a Rayleigh potential as $\sum \frac{1}{2} R_i \dot{q}_i^2$, where $\dot{q}_i$ is the relative velocity at the ith friction element or the current through the ith resistance.
- Obtain the partial derivatives of the Lagrangian function and the Rayleigh function with respect to the generalized coordinates and the generalized velocities.
- The system of second-order differential equations are then given by

$$\frac{\mathrm{d}}{\mathrm{d}t}\left(\frac{\partial \mathcal{L}}{\partial \dot{q}_i}\right) - \frac{\partial \mathcal{L}}{\partial q_i} + \frac{\partial \mathfrak{R}}{\partial \dot{q}_i} = 0.$$

Further Reading

F. Gantmacher, *Lectures in Analytical Mechanics*, Mir Publishers, Moscow, Russia, 1975.

H. Goldstein, *Classical Mechanics*, Addison Wesley, USA, 1980.

Problems

1. Obtain the differential equations for the mechanical systems given below.

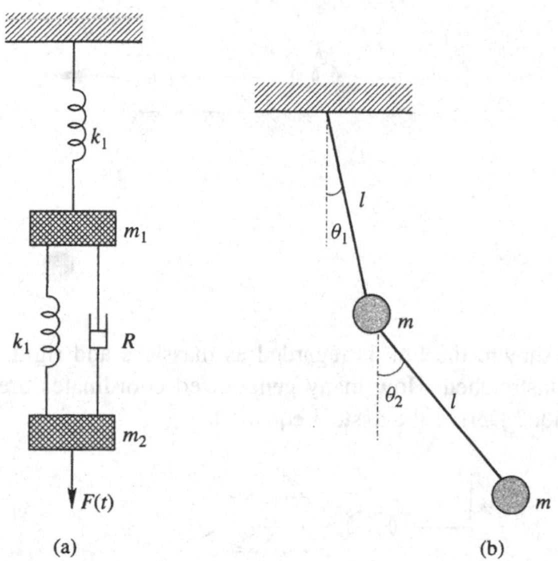

(a) (b)

2. Obtain the differential equations for the system shown in the figure. Note that the force F is applied only on the mass M_2.

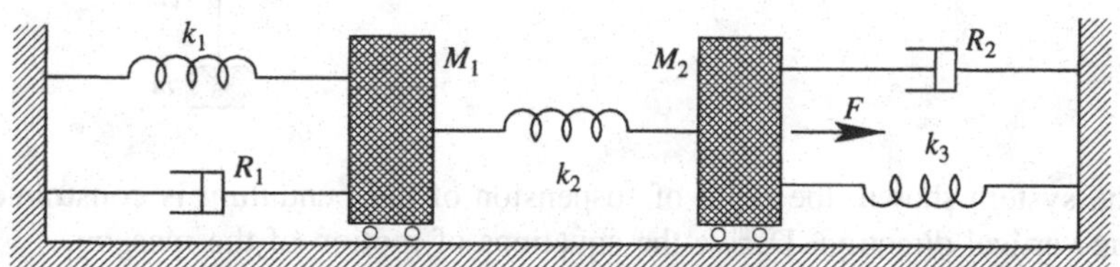

3. Obtain the differential equations for the electrical circuit with coupled inductances.

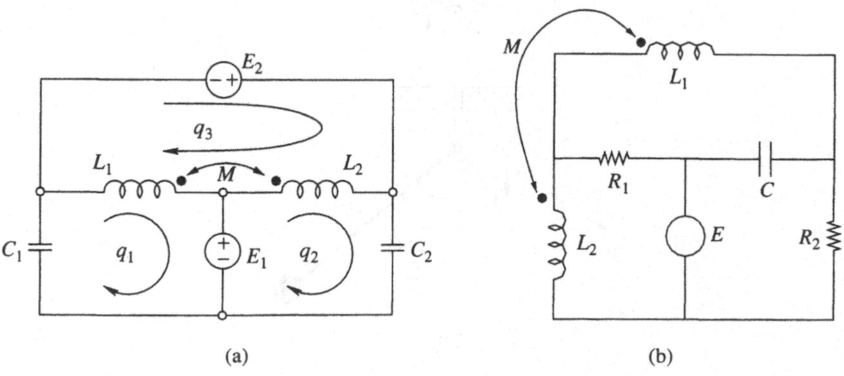

(a) (b)

4. In the figure shown there is a channel cut in the floor, through which the suspension wire can pass, allowing the pendulum to oscillate in a plane. Obtain the Lagrangian equations.

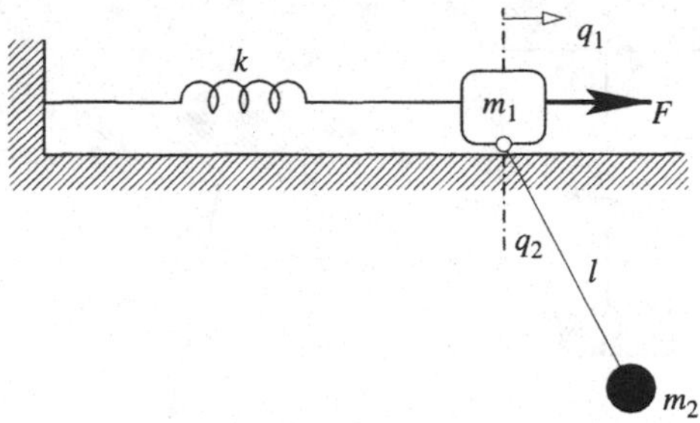

5. In the system shown, the link is regarded as massless and rigid. It is horizontal when the spring is unstretched. How many generalized coordinates are necessary to specify the configuration? Derive the system equations.

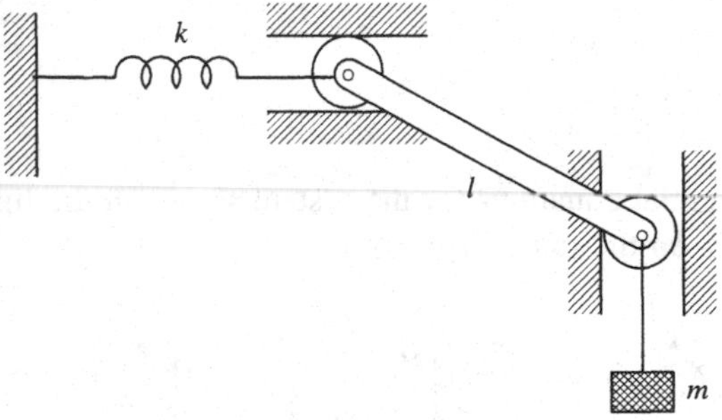

6. In the system shown, the point of suspension of the pendulum is constrained to move in the vertical direction. Derive the equations of motion of the mass m.

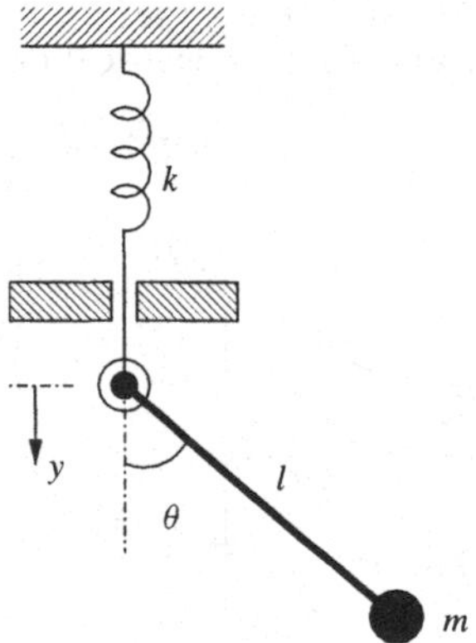

7. For the forcing function $F = A \sin \omega t$, derive the equation of motion for the two masses.

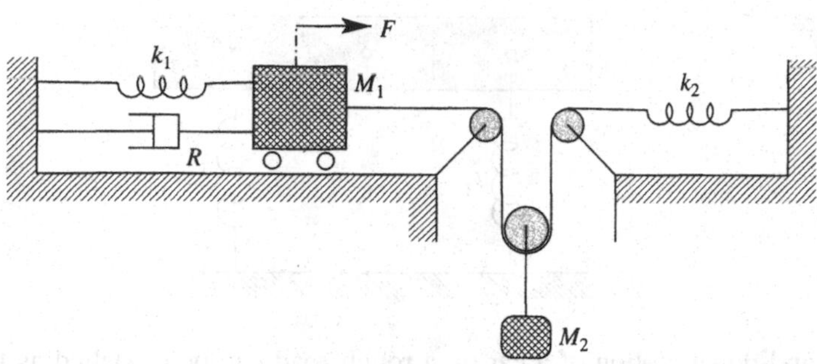

8. In the spring pendulum shown, a rigid rod of mass m and length l is attached to the end of a spring of constant k and unstretched length l_0. Obtain the dynamical equations.

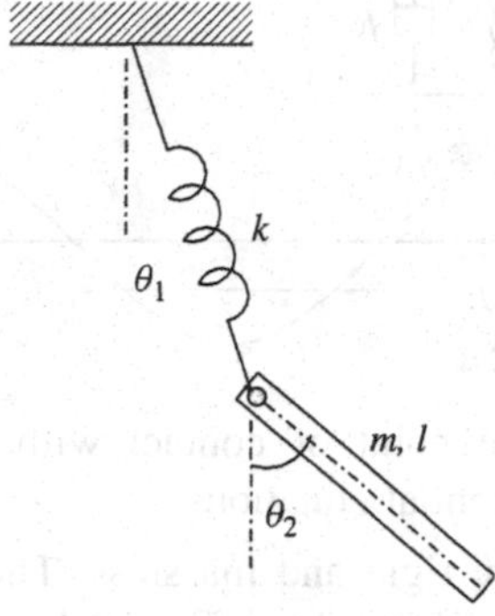

9. A rigid uniform bar is supported by two translational springs and a torsional spring. Derive the equations of motion.

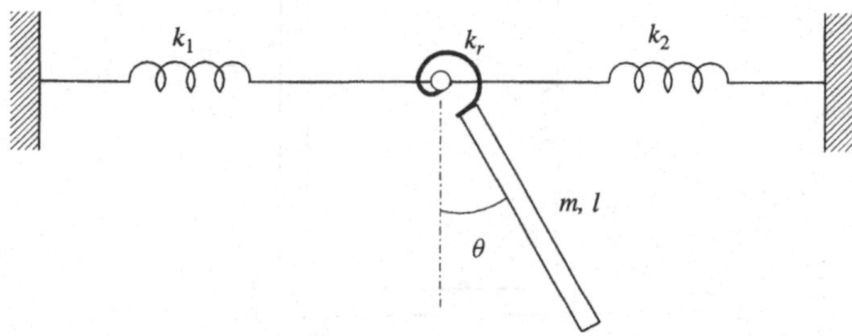

10. Derive the equations of motion of a bar of mass M and length l, supported by two different springs.

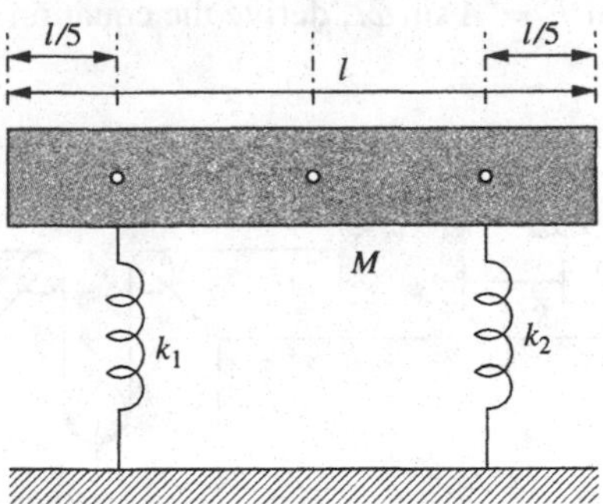

11. The up-and-down motion of a car on a rough road can be modelled as follows:

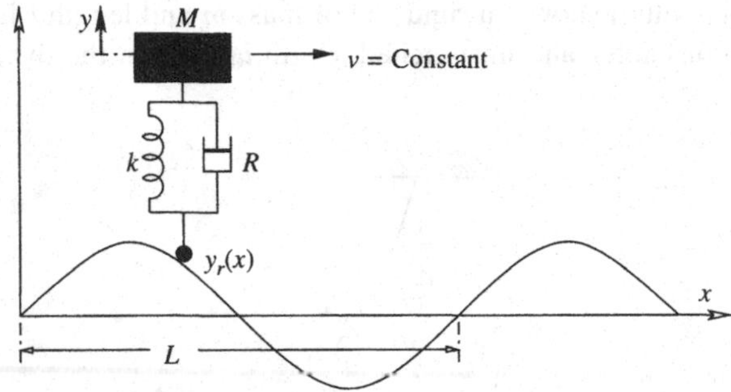

The vertical motion of the point of contact with the ground is given by $y_r(x) = A \sin \frac{2\pi}{L} x$. Derive the differential equations.

12. The links in this system are rigid and massless. The position at the upper end of the spring is varied as $A \cos \omega t$. Obtain the differential equations for the motion of the mass.

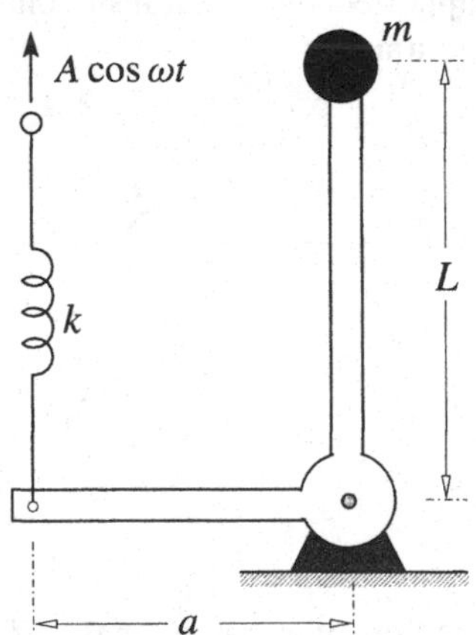

13. For the electrical circuits given in Problem 1 of Chapter 3, obtain the differential equations by the Lagrangian method.

5

Obtaining First-order Equations

It was seen in the last chapter that the Lagrangian methodology yields the system equations in the second order. The next step would be to solve the equations to obtain the course of evolution of the system starting from any given initial condition.

It is known that differential equations are much easier to solve if they are expressed in the first order. Especially if the solution is to be obtained through numerical methods in a computer, it is desirable to express the equations in the first-order form. Standard numerical routines like the Runge-Kutta method can then be applied to obtain the solution.

5.1 First-order Equations from the Lagrangian Method

A second-order differential equation can be expressed in the form of two first-order equations by defining an additional variable. It will be convenient to define the additional variables as the *generalized momenta*, given by

$$p_i = \frac{\partial \mathcal{L}}{\partial \dot{q}_i}. \tag{5.1}$$

Owing to the associatedness of p_i with q_i, these new sets of variables are also called *conjugate momenta*. The advantage of defining the additional variables this way is that the Lagrangian takes a simple form in the generalized momenta. Since

$$\frac{\mathrm{d}}{\mathrm{d}t}\left(\frac{\partial \mathcal{L}}{\partial \dot{q}_i}\right) = \dot{p}_i,$$

the Lagrangian equation

$$\frac{\mathrm{d}}{\mathrm{d}t}\left(\frac{\partial \mathcal{L}}{\partial \dot{q}_i}\right) - \frac{\partial \mathcal{L}}{\partial q_i} + \frac{\partial \mathfrak{R}}{\partial \dot{q}_i} = 0$$

Dynamics for Engineers S. Banerjee

becomes simply

$$\dot{p}_i - \frac{\partial \mathcal{L}}{\partial q_i} + \frac{\partial \mathfrak{R}}{\partial \dot{q}_i} = 0. \tag{5.2}$$

These are first-order equations. The desirable form of the first-order equations is such that the derivative quantities ($\dot{q}_i$ and $\dot{p}_i$) may be expressed as functions of the fundamental variables. The equations for $\dot{q}_i$ are obtained from (5.1) and those for $\dot{p}_i$ are obtained from (5.2). In some cases, substitution of the expressions for $\dot{q}_i$ into (5.2) would be necessary to eliminate derivative quantities from the right-hand side.

► **Example 5.1** Find the generalized momenta and the first-order equation for the spring-pendulum system of Example 4.6.
Solution: Recall that the Lagrangian was given by

$$\mathcal{L} = \frac{1}{2}m\dot{r}^2 + \frac{1}{2}m(a+r)^2\dot{\theta}^2 - \frac{1}{2}k(r + mg/k)^2 + mg(a+r)\cos\theta - mga.$$

Here the generalized coordinates are $q_1 = r$ and $q_2 = \theta$. Hence, the generalized momenta are

$$p_1 = \frac{\partial \mathcal{L}}{\partial \dot{q}_1} = m\dot{q}_1$$

and

$$p_2 = \frac{\partial \mathcal{L}}{\partial \dot{q}_2} = m(a+q_1)^2\dot{q}_2.$$

Putting the derivative quantities in the left-hand side, we get two first-order equations as

$$\dot{q}_1 = \frac{p_1}{m}, \tag{5.3}$$

$$\dot{q}_2 = \frac{p_2}{m(a+q_1)^2}. \tag{5.4}$$

The differential equations in the conjugate momenta are obtained from (5.2), with the Rayleigh term set to zero. The first equation is obtained from

$$\dot{p}_1 - \frac{\partial \mathcal{L}}{\partial q_1} = 0,$$

as

$$\dot{p}_1 - m\dot{q}_2^2(a+q_1) + k(q_1 + mg/k) - mg\cos q_2 = 0.$$

Substituting $\dot{q}_2$, we get

$$\dot{p}_1 = \frac{p_2^2}{m(a+q_1)^3} - k(q_1 + mg/k) + mg\cos q_2. \tag{5.5}$$

The equation in the second coordinate is obtained from

$$\dot{p}_2 - \frac{\partial \mathcal{L}}{\partial q_2} = 0,$$

as

$$\dot{p}_2 = -mg(a + q_1)\sin q_2. \tag{5.6}$$

We thus get four first-order equations (5.3), (5.4), (5.5) and (5.6). ◄

► **Example 5.2** In the circuit in Fig. 5.1,

$$T = \frac{1}{2}L_1(\dot{q}_1 - \dot{q}_2)^2 + \frac{1}{2}L_2\dot{q}_2^2,$$

$$V = \frac{1}{2C}q_1^2 - q_1 E,$$

$$\mathfrak{R} = \frac{1}{2}R\dot{q}_2^2.$$

Hence,

$$\mathcal{L} = \frac{1}{2}L_1(\dot{q}_1 - \dot{q}_2)^2 + \frac{1}{2}L_2\dot{q}_2^2 - \frac{1}{2C}q_1^2 + q_1 E.$$

The Lagrangian equations become

$$L_1(\ddot{q}_1 - \ddot{q}_2) + \frac{q_1}{C} - E = 0,$$

$$-L_1(\ddot{q}_1 - \ddot{q}_2) + L_2\ddot{q}_2 + R\dot{q}_2 = 0.$$

We now define the conjugate momenta as

$$p_1 = \frac{\partial \mathcal{L}}{\partial \dot{q}_1} = L_1(\dot{q}_1 - \dot{q}_2)$$

and

$$p_2 = \frac{\partial \mathcal{L}}{\partial \dot{q}_2} = -L_1(\dot{q}_1 - \dot{q}_2) + L_2\dot{q}_2.$$

These provide the equations for $\dot{q}_1$ and $\dot{q}_2$ as

$$\dot{q}_1 = \left(\frac{1}{L_1} + \frac{1}{L_2}\right)p_1 + \frac{1}{L_2}p_2,$$

$$\dot{q}_2 = \frac{1}{L_2}p_1 + \frac{1}{L_2}p_2.$$

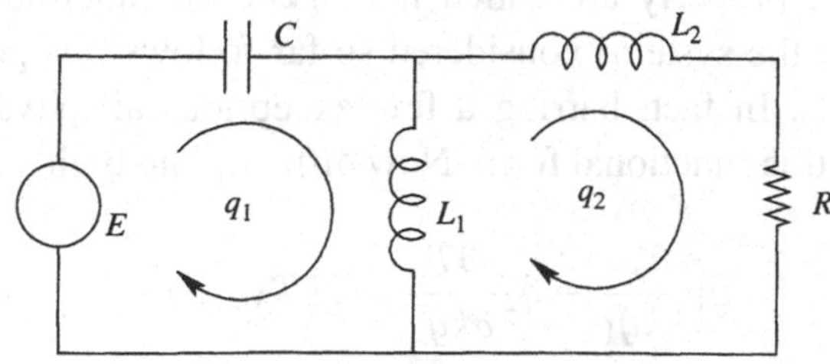

Figure 5.1 System pertaining to Example 5.2.

In terms of p_1 and p_2, the Lagrangian equations become

$$\dot{p}_1 = -\frac{q_1}{C} + E,$$

$$\dot{p}_2 = -R\dot{q}_2 = -\frac{R}{L_2}(p_1 + p_2).$$

These are the four dynamical equations in the first order. ◀

5.2 The Hamiltonian Formalism

If we have to express the system equations finally in the first-order form, why don't we *derive* them in the first order? In this section, we shall show that this can be done, and the Hamiltonian method provides the way.

For this, instead of the Lagrangian function $\mathcal{L} = T - V$, we shall use the total energy function and denote it by $\mathcal{H}$. Hence[1]

$$\mathcal{H} = T + V.$$

We then make use of the functional forms of T and V. We note that the potential V is dependent only on the generalized coordinates and not on the generalized velocities. Hence,

$$\frac{\partial V}{\partial \dot{q}_i} = 0,$$

and therefore,

$$\frac{\partial \mathcal{L}}{\partial \dot{q}_i} = \frac{\partial (T - V)}{\partial \dot{q}_i} = \frac{\partial T}{\partial \dot{q}_i}. \tag{5.7}$$

At the next step, we note that the kinetic energy is a homogeneous function of degree 2 in the generalized velocities. To illustrate what this means, consider a two-dimensional system, where T is a function of $\dot{q}_1$ and $\dot{q}_2$. The functional form of T would be such that if we multiply a constant k with $\dot{q}_1$ and $\dot{q}_2$, then

$$T(k\dot{q}_1, k\dot{q}_2) = k^2 T(\dot{q}_1, \dot{q}_2).$$

Functions with such property are called homogeneous function of degree 2. The kinetic energy of all the systems considered so far follows this property (the reader may check this fact). In fact, barring a few exceptions all physical systems have kinetic energy with this functional form. Now differentiate both sides of the equation by k, to get

$$\dot{q}_1 \frac{\partial T}{\partial k\dot{q}_1} + \dot{q}_2 \frac{\partial T}{\partial k\dot{q}_2} = 2kT(\dot{q}_1, \dot{q}_2).$$

[1]Note the calligraphic symbol $\mathcal{H}$. We reserve the symbol H for something else.

Since the choice of k is arbitrary, this equation would be valid for $k = 1$ also. Hence,

$$\dot{q}_1 \frac{\partial T}{\partial \dot{q}_1} + \dot{q}_2 \frac{\partial T}{\partial \dot{q}_2} = 2T(\dot{q}_1, \dot{q}_2).$$

The same result would hold for a homogeneous function in more than two variables (the case for higher dimensional systems). Thus, in general, we have

$$\sum_i \dot{q}_i \frac{\partial T}{\partial \dot{q}_i} = 2T.$$

This result also goes by the name of the great mathematician Euler, and is called the *Euler theorem* for homogeneous functions. From it, by (5.7) we have

$$\sum_i \dot{q}_i \frac{\partial \mathcal{L}}{\partial \dot{q}_i} = 2T.$$

With the help of this equation, the total energy function $\mathcal{H}$ can be written as

$$\mathcal{H} = T + V = 2T - (T - V) = 2T - \mathcal{L}$$

or

$$\mathcal{H} = \sum_i \dot{q}_i \frac{\partial \mathcal{L}}{\partial \dot{q}_i} - \mathcal{L} = \sum_i \dot{q}_i p_i - \mathcal{L}. \tag{5.8}$$

The function in the right-hand side is called the *Hamiltonian function* (denoted by H) in classical mechanics. We thus find that the Hamiltonian function is nothing but the total energy (i.e. $H = \mathcal{H}$) for systems in which the above functional forms of T and V hold. Since most of the systems an engineer has to deal with obey these functional forms and since total energy is a much clearer notion for an engineer, we have preferred to begin from that premise to arrive at the Hamiltonian function.

In the Hamiltonian formalism, the fundamental variables are q_i, p_i and t. It is therefore necessary to express the H function in terms of the generalized positions, momenta and time. Since in our earlier discussions we have expressed T as a function of $\dot{q}_i$, it will be necessary to substitute $\dot{q}_i$ in terms of p_i. This can be easily obtained by using the definition of generalized momenta as (5.1).

After $H(p_i, q_i, t)$ is obtained, the equations of motion in terms of H can be obtained by taking the differential of H as

$$\mathrm{d}H = \sum_i \frac{\partial H}{\partial p_i} \mathrm{d}p_i + \sum_i \frac{\partial H}{\partial q_i} \mathrm{d}q_i + \frac{\partial H}{\partial t} \mathrm{d}t. \tag{5.9}$$

Since $\mathcal{L}$ is a function of q_i, $\dot{q}_i$ and t, using (5.8), we can also write $\mathrm{d}H$ as

$$\begin{aligned}\mathrm{d}H &= \sum_i \dot{q}_i\,\mathrm{d}p_i + \sum_i p_i\,\mathrm{d}\dot{q}_i - \sum_i \frac{\partial \mathcal{L}}{\partial q_i}\,\mathrm{d}q_i - \sum_i \frac{\partial \mathcal{L}}{\partial \dot{q}_i}\,\mathrm{d}\dot{q}_i - \frac{\partial \mathcal{L}}{\partial t}\,\mathrm{d}t,\\ &= \sum_i \dot{q}_i\,\mathrm{d}p_i + \sum_i \mathrm{d}\dot{q}_i\left(p_i - \frac{\partial \mathcal{L}}{\partial \dot{q}_i}\right) - \sum_i \frac{\partial \mathcal{L}}{\partial q_i}\,\mathrm{d}q_i - \frac{\partial \mathcal{L}}{\partial t}\,\mathrm{d}t.\end{aligned}$$

The second term is zero because of (5.1). Hence,

$$\mathrm{d}H = \sum_i \dot{q}_i\,\mathrm{d}p_i - \sum_i \frac{\partial \mathcal{L}}{\partial q_i}\,\mathrm{d}q_i - \frac{\partial \mathcal{L}}{\partial t}\,\mathrm{d}t.$$

By (5.2), we have

$$\frac{\partial \mathcal{L}}{\partial q_i} = \dot{p}_i + \frac{\partial \mathfrak{R}}{\partial \dot{q}_i}.$$

Substituting, we get

$$\mathrm{d}H = \sum_i \dot{q}_i\,\mathrm{d}p_i + \sum_i \left(-\dot{p}_i - \frac{\partial \mathfrak{R}}{\partial \dot{q}_i}\right)\mathrm{d}q_i - \frac{\partial \mathcal{L}}{\partial t}\,\mathrm{d}t. \tag{5.10}$$

Comparing (5.9) and (5.10), we have the relations

$$\frac{\partial H}{\partial p_i} = \dot{q}_i, \tag{5.11}$$

$$\frac{\partial H}{\partial q_i} = -\dot{p}_i - \frac{\partial \mathfrak{R}}{\partial \dot{q}_i}, \tag{5.12}$$

$$\frac{\partial H}{\partial t} = -\frac{\partial \mathcal{L}}{\partial t}. \tag{5.13}$$

The first two equations, expressed in the first-order form, are called the Hamiltonian equations of motion. These provide an easy way of deriving the first-order differential equations of any system directly. The only drawback is the existence of the dissipative term as a function of the generalized velocities that are not the fundamental variables in the Hamiltonian representation. This, however, is no great deterrent since $\dot{q}_i$ can always be substituted in terms of the fundamental variables by using (5.11). Note that equation (5.11) gives basically the same functional relationship between $\dot{q}_i$ and p_i as equation (5.1).

It may also be noted that (5.13) is not a differential equation and hence is not needed to represent the dynamics. It says simply that if the Lagrangian does not have an explicit dependence on time (i.e. if t does not appear in its functional form), then H also would not be explicitly dependent on time. This is obvious, since both are composed of the functions T and V.

In the absence of dissipative forces, the first two equations take quite a symmetrical form:

$$\dot{q}_i = \frac{\partial H}{\partial p_i}, \tag{5.14}$$

$$\dot{p}_i = -\frac{\partial H}{\partial q_i}. \tag{5.15}$$

Such symmetry in the form of dynamical equations has prompted generations of scientists to probe these systems in detail, and through this a body of knowledge has emerged. Dynamical systems where these equations hold are called *Hamiltonian systems*. We are, however, not much interested in such systems that are almost non-existent in engineering, and will use differential equations of the form

$$\dot{q}_i = \frac{\partial H}{\partial p_i}, \tag{5.16}$$

$$\dot{p}_i = -\frac{\partial H}{\partial q_i} - \frac{\partial \mathfrak{R}}{\partial \dot{q}_i}. \tag{5.17}$$

► **Example 5.3** Let us take the system in the Example 5.2 and derive the system equations by the Hamiltonian method.

$$\begin{aligned} H &= T + V, \\ &= \frac{1}{2}L_1(\dot{q}_1 - \dot{q}_2)^2 + \frac{1}{2}L_2\dot{q}_2^2 + \frac{1}{2C}q_1^2 - q_1 E, \\ &= \frac{1}{2}L_1\left(\frac{p_1}{L_1}\right)^2 + \frac{1}{2}L_2\left(\frac{p_1 + p_2}{L_2}\right)^2 + \frac{1}{2C}q_1^2 - q_1 E, \\ &= \frac{1}{2L_1}p_1^2 + \frac{1}{2L_2}(p_1 + p_2)^2 + \frac{1}{2C}q_1^2 - q_1 E. \end{aligned}$$

Hence, the Hamiltonian equations are

$$\begin{aligned} \dot{q}_1 &= \frac{\partial H}{\partial p_1} = \frac{p_1}{L_1} + \frac{p_1 + p_2}{L_2}, \\ \dot{q}_2 &= \frac{\partial H}{\partial p_2} = \frac{p_1 + p_2}{L_2}, \\ \dot{p}_1 &= -\frac{\partial H}{\partial q_1} - \frac{\partial \mathfrak{R}}{\partial \dot{q}_1} = -\frac{q_1}{C} + E, \\ \dot{p}_2 &= -\frac{\partial H}{\partial q_1} - \frac{\partial \mathfrak{R}}{\partial \dot{q}_2} = -R\dot{q}_2 = -\frac{R}{L_2}(p_1 + p_2). \end{aligned}$$

Note that these equations are the same as those derived in Example 5.2. ◄

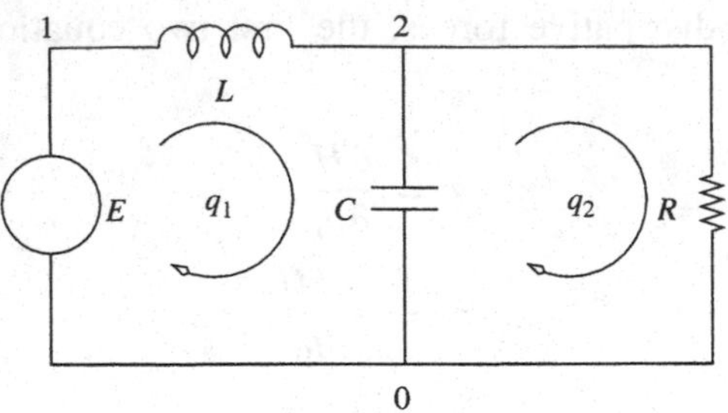

Figure 5.2 The circuit pertaining to Example 5.4.

► **Example 5.4** The simple circuit of Fig. 5.2 was taken up in Chapter 3. If we want to obtain the first-order equations by the Hamiltonian method, we note

$$\mathcal{L} = \frac{1}{2}L\dot{q}_1^2 - \frac{1}{2C}(q_1 - q_2)^2 + Eq_1$$

$$\text{and} \quad \mathfrak{R} = \frac{1}{2}R\dot{q}_2^2.$$

Therefore,

$$p_1 = \frac{\partial \mathcal{L}}{\partial \dot{q}_1} = L\dot{q}_1$$

$$\text{or} \quad \dot{q}_1 = p_1/L,$$

$$\text{and} \quad p_2 = \frac{\partial \mathcal{L}}{\partial \dot{q}_2} = 0.$$

The Hamiltonian function can then be expressed as

$$\begin{aligned} H &= \frac{1}{2}L\dot{q}_1^2 + \frac{1}{2C}(q_1 - q_2)^2 - Eq_1, \\ &= \frac{1}{2L}p_1^2 + \frac{1}{2C}(q_1 - q_2)^2 - Eq_1. \end{aligned}$$

In terms of this Hamiltonian function,

$$\dot{q}_1 = \frac{\partial H}{\partial p_1} = p_1/L, \tag{5.18}$$

and since $p_2 = 0$, $\partial H/\partial p_2$ cannot be evaluated. The momentum equations are

$$\begin{aligned} \dot{p}_1 &= -\frac{\partial H}{\partial q_1} - \frac{\partial \mathfrak{R}}{\partial \dot{q}_1} = -\frac{1}{C}(q_1 - q_2) + E, \\ \dot{p}_2 &= -\frac{\partial H}{\partial q_2} - \frac{\partial \mathfrak{R}}{\partial \dot{q}_2} = \frac{1}{C}(q_1 - q_2) - R\dot{q}_2. \end{aligned} \tag{5.19}$$

Since $p_2 = 0$, $\dot{p}_2 = 0$. Therefore,

$$\dot{q}_2 = \frac{1}{RC}(q_1 - q_2). \tag{5.20}$$

Therefore, (5.18), (5.19) and (5.20) are the three first-order equations. ◄

It may be noted that though the obtained equations are correct (in the sense that they suffice in defining the time evolution), these are not the minimum set of equations required to define the dynamics of the systems. In Example 5.3, there are three independent storage elements, and so a minimum of three first-order differential equations should suffice in defining the dynamics. In Example 5.4, there are two storage elements, and hence there should be two independent differential equations. But the equations obtained are one more than the minimum necessary number.

The problem lies in our assumption of the position coordinates as the charges flowing in the loops irrespective of what constitute the elements of the loop. One possible way out is to consider the charges flowing in the storage elements as the variables.

► **Example 5.5** In this example, we work out the differential equations of the system in the last example in terms of the currents flowing through the inductor and the capacitor as shown in Fig. 5.3.

In this case,

$$\mathcal{L} = \frac{1}{2}L\dot{q}_1^2 - \frac{1}{2C}q_2^2 + Eq_1,$$

$$\Re = \frac{1}{2}R(\dot{q}_1 - \dot{q}_2)^2.$$

Therefore,

$$p_1 = \frac{\partial \mathcal{L}}{\partial \dot{q}_1} = L\dot{q}_1$$

$$\text{or} \quad \dot{q}_1 = p_1/L,$$

$$\text{and} \quad p_2 = \frac{\partial \mathcal{L}}{\partial \dot{q}_2} = 0.$$

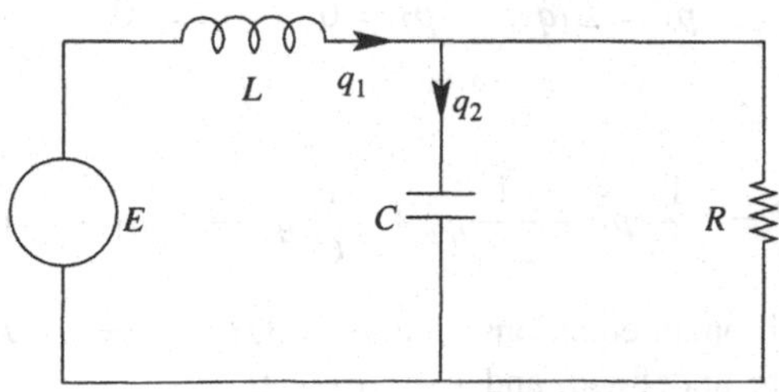

Figure 5.3 The circuit of Example 5.4 with variables redefined.

The Hamiltonian function can then be expressed as

$$H = \frac{1}{2L} p_1^2 + \frac{1}{2C} q_2^2 - E q_1.$$

In terms of this Hamiltonian function,

$$\dot{q}_1 = \frac{\partial H}{\partial p_1} = p_1/L. \tag{5.21}$$

Since $p_2 = 0$, in this case also $\partial H/\partial p_2$ cannot be evaluated. The momentum equations are

$$\dot{p}_1 = -\frac{\partial H}{\partial q_1} - \frac{\partial \Re}{\partial \dot{q}_1} = E - R(\dot{q}_1 - \dot{q}_2), \tag{5.22}$$

$$\dot{p}_2 = -\frac{\partial H}{\partial q_2} - \frac{\partial \Re}{\partial \dot{q}_2} = -\frac{1}{C} q_2 + R(\dot{q}_1 - \dot{q}_2).$$

Since $p_2 = 0$, $\dot{p}_2 = 0$. Therefore, the last equation gives

$$\dot{q}_2 = -\frac{1}{RC} q_2 + \frac{1}{L} p_1. \tag{5.23}$$

Substituting (5.21) and (5.23), (5.22) takes the form

$$\dot{p}_1 = -\frac{1}{C} q_2 + E. \tag{5.24}$$

Therefore, (5.24) and (5.23) are the two first-order equations. ◀

► **Example 5.6** In this example, we consider the same system as in Example 4.12 and define the variables differently – the charges flowing in the storage elements forming the coordinates. Since these branches form a tree, these variables suffice in defining the charges in all other branches, as shown in Fig. 5.4. In this case

$$\mathcal{L} = \frac{1}{2} L \dot{q}_1^2 - \frac{1}{2C_1} q_2^2 - \frac{1}{2C_2} q_3^2 + E(q_1 + q_3),$$

$$\Re = \frac{1}{2} R_1 (\dot{q}_1 - \dot{q}_2)^2 + \frac{1}{2} R_2 (\dot{q}_1 - \dot{q}_2 + \dot{q}_3)^2.$$

This gives

$$p_1 = L_1 \dot{q}_1, \quad p_2 = 0, \quad p_3 = 0.$$

Thus, H can be written as

$$H = \frac{1}{2L_1} p_1^2 + \frac{1}{2C_1} q_2^2 + \frac{1}{2C_2} q_3^2 - E(q_1 + q_3).$$

The first set of Hamiltonian equations give $\dot{q}_1 = \partial H/\partial p_1 = p_1/L_1$. Since $p_2 = p_3 = 0$, $\partial H/\partial p_2$ and $\partial H/\partial p_3$ are undefined, and $\dot{p}_2 = \dot{p}_3 = 0$.

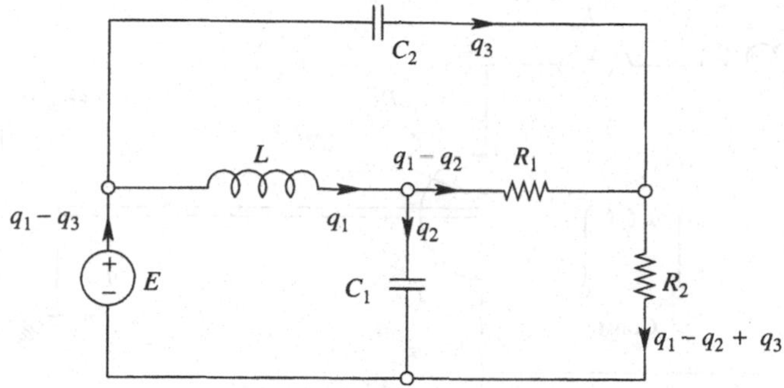

Figure 5.4 The circuit pertaining to Example 5.6, with charges flowing in the storage elements chosen as the coordinates.

The second set of Hamiltonian equations give

$$\dot{p}_1 = E - R_1(\dot{q}_1 - \dot{q}_2) - R_2(\dot{q}_1 - \dot{q}_2 + \dot{q}_3),$$

$$\dot{p}_2 = -\frac{q_2}{C_1} + R_1(\dot{q}_1 - \dot{q}_2) + R_2(\dot{q}_1 - \dot{q}_2 + \dot{q}_3) = 0,$$

$$\dot{p}_3 = -\frac{q_3}{C_2} + E - R_2(\dot{q}_1 - \dot{q}_2 + \dot{q}_3) = 0.$$

Algebraic manipulation of these three equations yield

$$\dot{p}_1 = E - \frac{q_2}{C_1},$$

$$\dot{q}_2 = -\frac{q_2}{R_1 C_1} - \frac{q_3}{R_1 C_2} + \frac{p_1}{L_1} + \frac{E}{R_1},$$

$$\dot{q}_3 = \frac{q_2}{R_2 C_1} - \frac{R_1 + R_2}{R_1 R_2}\left(\frac{q_2}{C_1} + \frac{q_3}{C_2} - E\right).$$

These three are the first-order differential equations of the system. ◀

A particular advantage of the Lagrangian–Hamiltonian formalism is that mechanical and electrical systems are treated in the same way. That makes it particularly suitable for handling electromechanical systems that have mechanical and electrical components. In such systems, one may come across situations where the potential energy term may become dependent on the generalized velocities, thus making the assumption $\frac{\partial V}{\partial \dot{q}_i} = 0$ invalid. In such cases, one has to be careful about the form of the Lagrangian equation to be used. We illustrate this with the example of a dc motor and mechanical load system.

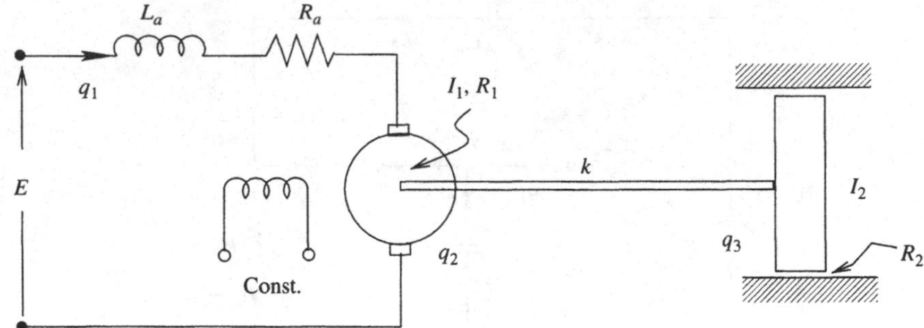

Figure 5.5 The separately excited dc motor and mechanical load system with a flexible shaft.

► **Example 5.7** The system under consideration is shown in Fig. 5.5. In this problem, the generalized coordinates are

q_1: the charge flowing in the armature circuit
q_2: the angle of the rotor
q_3: the angle of the load wheel.

The electrical circuit and the mechanical part can be treated as two sub-systems. These two sub-systems interact through the torque F exerted by the electrical side on the mechanical side and the back emf E_b exerted by the mechanical side on the electrical side. From the theory of electrical machines, we know that these are given by

$$E_b = K\phi\dot{q}_2$$

and

$$F = K\phi\dot{q}_1,$$

where K is a constant and ϕ is the field flux. Since the electrical and mechanical sub-systems are considered separately, the back emf and the shaft torque are given the status of external forces on the corresponding sub-systems. Therefore, the potentials caused by these force terms should not generate any momentum.

In the electrical sub-system,

$$T_e = \frac{1}{2}L_a\dot{q}_1^2,$$

$$V_e = -(E - E_b)q_1,$$

$$\Re_e = \frac{1}{2}R_a\dot{q}_1^2,$$

$$p_1 = \frac{\partial \mathcal{L}_e}{\partial \dot{q}_1} = \frac{\partial T_e}{\partial \dot{q}_1} = L_a\dot{q}_1.$$

Thus, we can write H_e, the Hamiltonian of the electrical sub-system in terms of q_1 and p_1 as

$$H_e = \frac{1}{2L_a}p_1^2 - Eq_1 + E_bq_1.$$

This gives the first-order equations as

$$\begin{aligned}\dot{q}_1 &= \frac{\partial H_e}{\partial p_1},\\ &= p_1/L_a, \qquad (5.25)\\ \dot{p}_1 &= -\frac{\partial H_e}{\partial q_1} - \frac{\partial \Re_e}{\partial \dot{q}_1},\\ &= E - E_b - R_a\dot{q}_1,\\ &= E - E_b - \frac{R_a}{L_a}p_1. \qquad (5.26)\end{aligned}$$

In the mechanical sub-system,

$$\begin{aligned}T_m &= \frac{1}{2}I_1\dot{q}_2^2 + \frac{1}{2}I_2\dot{q}_3^2,\\ V_m &= \frac{1}{2}k(q_2 - q_3)^2 - Fq_2,\\ \Re_m &= \frac{1}{2}R_1\dot{q}_2^2 + \frac{1}{2}R_2\dot{q}_3^2.\end{aligned}$$

The generalized momenta, p_2 and p_3, are given by

$$p_2 = \frac{\partial T_m}{\partial \dot{q}_2} = I_1\dot{q}_2,$$

and

$$p_3 = \frac{\partial T_m}{\partial \dot{q}_3} = I_2\dot{q}_3.$$

The Hamiltonian function of the mechanical system can be expressed as

$$H_m = \frac{1}{2I_1}p_2^2 + \frac{1}{2I_2}p_3^2 + \frac{1}{2}k(q_2 - q_3)^2 - Fq_2.$$

The first-order equations for the mechanical sub-system are obtained as

$$\begin{aligned}\dot{q}_2 &= \frac{\partial H_m}{\partial p_2},\\ &= p_2/I_1, \qquad (5.27)\\ \dot{q}_3 &= \frac{\partial H_m}{\partial p_3},\\ &= p_3/I_2, \qquad (5.28)\\ \dot{p}_2 &= -\frac{\partial H_m}{\partial q_2} - \frac{\partial \Re_m}{\partial \dot{q}_2},\\ &= -k(q_2 - q_3) + F - R_1\dot{q}_2,\\ &= -k(q_2 - q_3) + F - \frac{R_1}{I_1}p_2, \qquad (5.29)\end{aligned}$$

$$\begin{aligned}\dot{p}_3 &= -\frac{\partial H_m}{\partial q_3} - \frac{\partial \Re_m}{\partial \dot{q}_3},\\ &= k(q_2 - q_3) - R_2\dot{q}_3 = k(q_2 - q_3) - \frac{R_2}{I_2}p_3. \end{aligned} \tag{5.30}$$

The interaction between the mechanical and electrical sub-systems occurs through the back emf and the torque, which appear in equations (5.26) and (5.29) respectively. When expressed in terms of the generalized coordinates, these become

$$E_b = K\phi\frac{p_2}{I_1}$$

and

$$F = K\phi\frac{p_1}{L_a}.$$

Substituting these into (5.26) and (5.29), we get

$$\dot{p}_1 = E - K\phi\frac{p_2}{I_1} - \frac{R_a}{L_a}p_1, \tag{5.31}$$

$$\dot{p}_2 = -k(q_2 - q_3) + K\phi\frac{p_1}{L_a} - \frac{R_1}{I_1}p_2. \tag{5.32}$$

Equations (5.25), (5.27), (5.28), (5.30), (5.31), and (5.32) are the final set of differential equations.

Now let us demonstrate how to obtain the state equations by the Lagrangian methodology, without treating the electrical and mechanical sub-systems separately.

The total kinetic energy is

$$T = \frac{1}{2}L_a\dot{q}_1^2 + \frac{1}{2}I_1\dot{q}_2^2 + \frac{1}{2}I_2\dot{q}_3^2.$$

The total potential energy is

$$\begin{aligned}V &= -(E - E_b)q_1 + \frac{1}{2}k(q_2 - q_3)^2 - Fq_2,\\ &= -(E - K\phi\dot{q}_2)q_1 + \frac{1}{2}k(q_2 - q_3)^2 - K\phi\dot{q}_1q_2,\end{aligned}$$

and the total Rayleigh function is

$$\Re = \frac{1}{2}R_a\dot{q}_1^2 + \frac{1}{2}R_1\dot{q}_2^2 + \frac{1}{2}R_2\dot{q}_3^2.$$

Now notice that the potential energy has terms dependent on the generalized velocities. This violates the assumption

$$\frac{\partial \mathcal{L}}{\partial \dot{q}_i} = \frac{\partial T}{\partial \dot{q}_i}$$

used in deriving the Lagrangian equation (4.15). Therefore, in this case we have to use the form that is derived without using this assumption. This is

$$\frac{\mathrm{d}}{\mathrm{d}t}\left(\frac{\partial T}{\partial \dot{q}_i}\right) - \frac{\partial \mathcal{L}}{\partial q_i} + \frac{\partial \mathfrak{R}}{\partial \dot{q}_i} = 0. \tag{5.33}$$

Using this equation, we get the second-order equations as

$$L_a\ddot{q}_1 - (E - K\phi\dot{q}_2) + R_a\dot{q}_1 = 0,$$
$$I_1\ddot{q}_2 - K\phi\dot{q}_1 + k(q_2 - q_3) + R_1\dot{q}_2 = 0,$$
$$I_2\ddot{q}_3 - k(q_2 - q_3) + R_2\dot{q}_3 = 0.$$

In this case, the expression for the generalized momenta will be

$$p_i = \frac{\partial T}{\partial \dot{q}_i},$$

which gives

$$p_1 = L_a\dot{q}_1,$$
$$p_2 = I_1\dot{q}_2,$$
$$p_3 = I_2\dot{q}_3.$$

And the first-order equations are obtained from

$$\dot{p}_i = \frac{\partial \mathcal{L}}{\partial q_i} - \frac{\partial \mathfrak{R}}{\partial \dot{q}_i},$$

as

$$\begin{aligned}\dot{p}_1 &= E - K\phi\dot{q}_2 - R_a\dot{q}_1,\\ &= E - K\phi p_2/I_1 - R_a p_1/L_a,\end{aligned} \tag{5.34}$$

$$\begin{aligned}\dot{p}_2 &= K\phi\dot{q}_1 - k(q_2 - q_3) - R_1\dot{q}_2,\\ &= K\phi p_1/L_a - k(q_2 - q_3) - R_1 p_2/I_1,\end{aligned} \tag{5.35}$$

$$\begin{aligned}\dot{p}_3 &= k(q_2 - q_3) - R_2\dot{q}_3,\\ &= k(q_2 - q_3) - R_2 p_3/I_2.\end{aligned} \tag{5.36}$$

Series motor: If a series motor (with field resistance R_f and field inductance L_f) supplies the same load, no new state variable will be necessary. Since the field circuit is in series with the armature circuit, L_a will be replaced by $(L_a + L_f)$ and R_a by $(R_a + R_f)$ in the state variable equations.

However, there will be a major change in the torque and the emf equations as the field flux is dependent on the field current. In this case,

$$E_b = K\phi\dot{q}_2 = K'\dot{q}_1\dot{q}_2$$

and

$$F = K\phi\dot{q}_1 = K'\dot{q}_1^2.$$

Thus, (5.31) will be modified to

$$\dot{p}_1 = E - K'\dot{q}_1\dot{q}_2 - \frac{R_a + R_f}{L_a + L_f}p_1,$$

$$= E - K'\frac{p_1}{L_a + L_f}\frac{p_2}{I_1} - \frac{R_a + R_f}{L_a + L_f}p_1,$$

and (5.32) will be modified to

$$\dot{p}_2 = -k(q_2 - q_3) + K'\dot{q}_1^2 - \frac{R_1}{I_1}p_2,$$

$$= -k(q_2 - q_3) + K'\frac{p_1^2}{(L_a + L_f)^2} - \frac{R_1}{I_1}p_2.$$

Shunt motor: In case of a shunt motor, the dynamics of the shunt field will have to be considered and hence an additional generalized coordinate (q_4), the flow of charge in the shunt field, will be included. In this case the back emf equation will be

$$E_b = K\phi\dot{q}_2 = K'\dot{q}_4\dot{q}_2,$$

and the torque equation will be

$$F = K\phi\dot{q}_1 = K'\dot{q}_4\dot{q}_1.$$

The complete set of state variable equations can be derived in a similar manner, which we leave for the reader. ◀

It is necessary to remember that the Hamiltonian function is not necessarily a constant, and the total energy function is not necessarily equal to the Hamiltonian function. That is why caution has to be exercised in applying the Hamiltonian procedure. Fortunately, however, in most engineering systems the two functions are equal, allowing a simple procedure of obtaining the first-order equations.

5.3 Chapter Summary

First-order equations can be obtained either by reduction of the Lagrangian equations, or by the Hamiltonian methodology. In the first method, one defines the *generalized momenta* as

$$p_i = \frac{\partial \mathcal{L}}{\partial \dot{q}_i},$$

which give the first-order equations in $\dot{q}_i$. Then, the first-order equations in $\dot{p}_i$ are obtained from the equations

$$\dot{p}_i - \frac{\partial \mathcal{L}}{\partial q_i} + \frac{\partial \mathfrak{R}}{\partial \dot{q}_i} = 0.$$

In the Hamiltonian formalism, the steps are:

- Define the elementary variables as the generalized coordinates q_i and the conjugate momenta p_i.
- Express the kinetic energy, potential energy and the Rayleigh potential in terms of q_i and p_i.
- Check if the potential energy is independent of the generalized velocities $\dot{q}_i$, and if the kinetic energy is a homogeneous function of degree 2. If these are true, express the Hamiltonian function as $H = T + V$.
- The first-order equations are then obtained from

$$\dot{q}_i = \frac{\partial H}{\partial p_i},$$

$$\dot{p}_i = -\frac{\partial H}{\partial q_i} - \frac{\partial \mathfrak{R}}{\partial \dot{q}_i}.$$

Problems

1. Obtain the first-order equations for the problems given in Chapter 4.
2. Obtain the differential equations of these two systems using the Hamiltonian method.

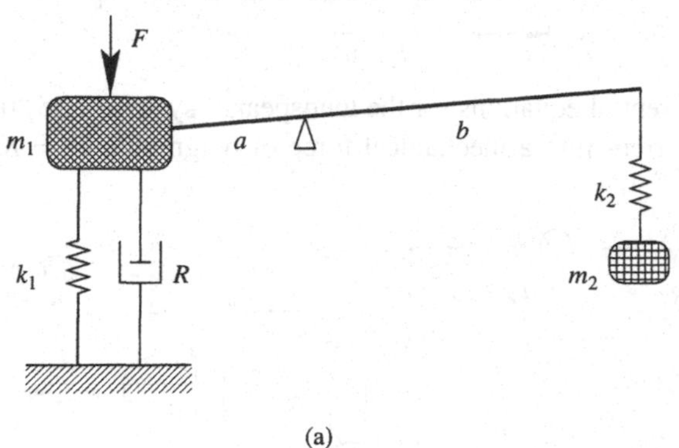

(a)

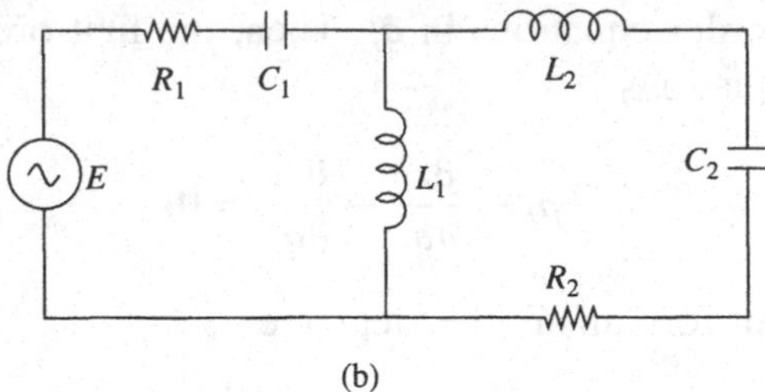

(b)

3. The mass M_2 moves on a platform of mass M_1, which slides on the ground under the action of the force F. The viscous friction coefficients between the surfaces are as shown. Obtain the first-order differential equations.

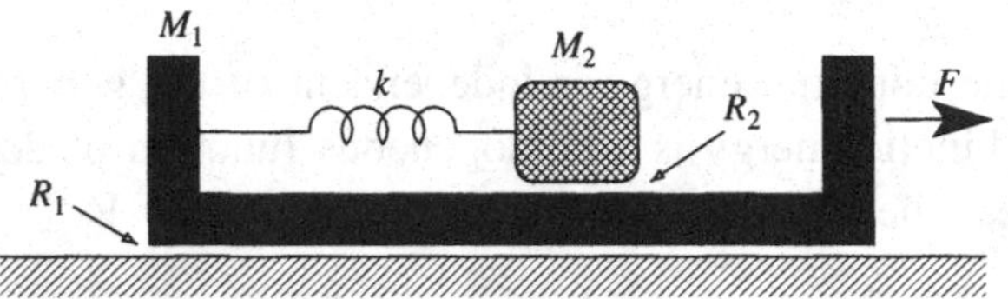

4. Obtain the first-order differential equations for the inverted pendulum system given below.

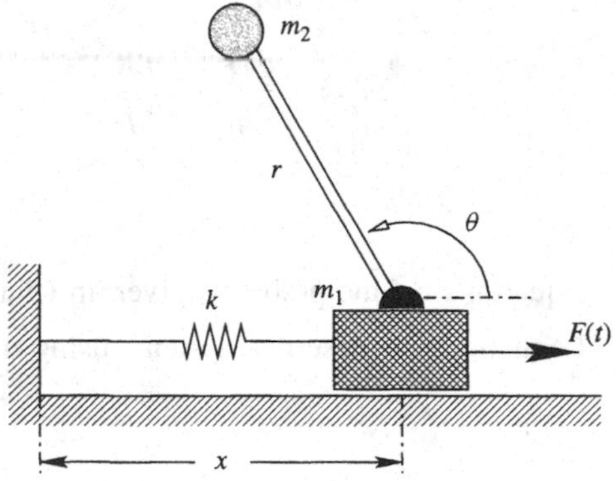

5. Obtain the differential equations for the loudspeaker system. The "transducer" transforms the electrical current into a mechanical force of magnitude $F = K_T i$.

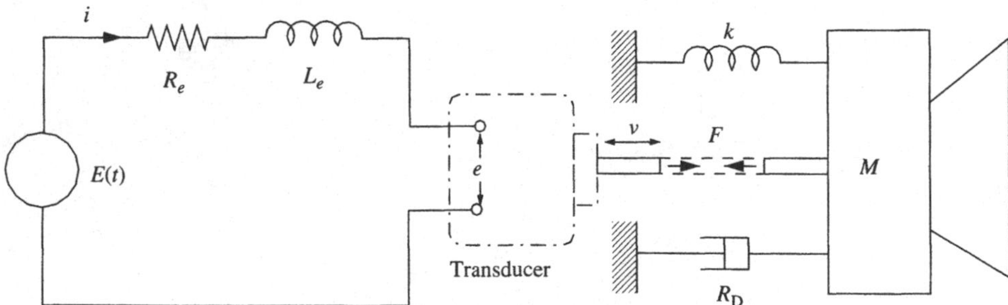

6. In the system shown, the ground is vibrating according to the relation $y_g = A \sin \omega t$. Find the equations of motion of the mass M in the first order.

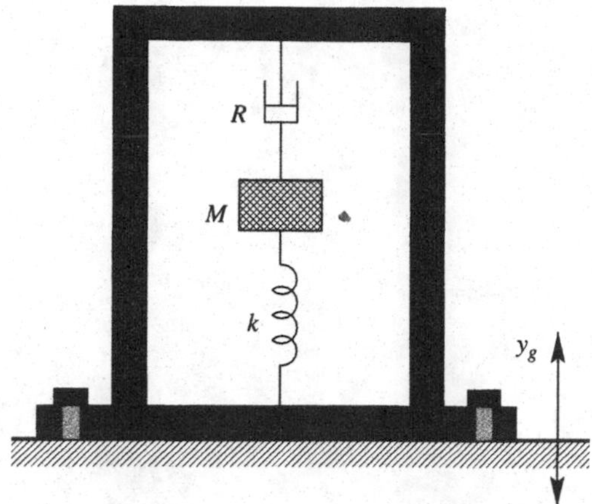

7. A locomotive is attached to two wagons through links that have spring-damper arrangement as shown. Obtain the first-order differential equations.

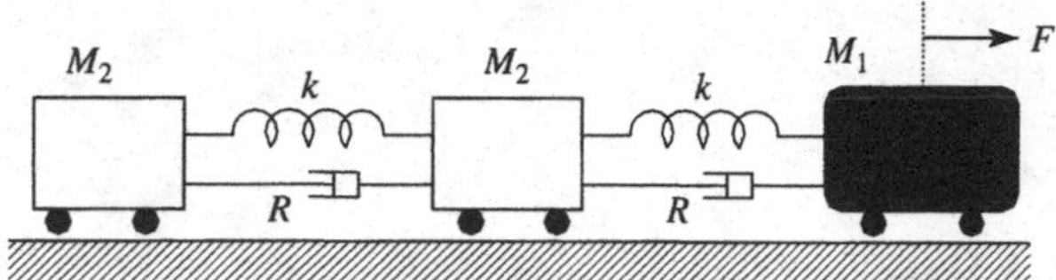

6

Unified Modelling of Systems Through the Language of Bond Graphs

6.1 Introduction

We have seen that the Lagrangian and Hamiltonian methods enable one to derive the differential equations of any given system. If this yields a set of linear equations, starting from any initial condition the time evolution of the system can be obtained in a closed form by solving the equations. But most often the differential equations are nonlinear, which can only be solved numerically with the help of computers. Thus, the job of deriving the differential equations has to be performed by a human being, while the job of solving them is most often performed by a computer.

Prof. H. M. Paynter of the Massachusetts Institute of Technology, US, invented in 1960 a language of system representation through exchange of power and information – by which the job of deriving differential equations can also be performed by a computer. For this, the description of the system has to be conveyed to the computer in clear and unambiguous terms, so that the machine can follow an algorithmic procedure to derive the differential equations. The equations can then be solved, the time evolution of the variables can be plotted, and many other operations can be performed by the computer to obtain a complete understanding of the system's behaviour. The language in which any system can be represented in a graphical form and can be conveyed to a computer is called *bond graph*.

Apart from offering an advantage in deriving system equations, the representation of a system in terms of bond graphs also brings out the hidden interactions

Dynamics for Engineers S. Banerjee

between elements, and offers the modeller a better understanding of the system dynamics.

6.2 The Basic Concept

Any system is composed of *elements*. They interact with each other by exchanging energy and information, and this exchange determines the dynamics of the system.

Energy, or its time derivative, power, is the fundamental quantity that is exchanged between elements of a system. And power is the product of two factors: the *effort* and the *flow*. In translational mechanics, power is the product of force and velocity; in electrical system power is the product of voltage and current; in hydraulic system power is the product of pressure and volume flow rate; in magnetic domain, power is the product of magneto-motive force (mmf) and flux rate; and in thermal domain the product of temperature and entropy exchange rate gives the power. Thus, in any system, be it mechanical, electrical, chemical or thermal, we can define a generalized effort variable and a generalized flow variable (also called the *across variable* and the *through variable*) – whose product gives the power exchanged between the elements. Effort and flow are the two *information* variables exchanged between elements.

Thus, if the exchange of energy and information between the elements can be properly represented in a graphical form, it would contain everything necessary to obtain the dynamical evolution of the system. The idea of bond graphs follows up this basic concept.

In bond graph representation, the flow of power between elements is represented by bonds. Through a bond, an element exchanges effort and flow with the rest of the system, and as we have seen earlier, their product determines the power exchange.

6.3 One-port Elements

Though an element like a capacitor or an inductor has two terminals, in bond graph representation we consider only the way it exchanges power with the rest of the system. This exchange of power is assumed to occur through abstract entities called *energy port*. Through this energy port, the effort and flow are exchanged with the system. For example, power flow into or out of an inductor is given by the voltage applied on it and the current flowing through it. These are the information variables associated with one bond. Therefore, the inductor has only one energy port and one bond associated with it carrying the information about the voltage and current.

By the same argument we conclude that the capacitor, resistor, voltage source and current source are *one-port elements*.

We can now obtain the general properties of these elements (see Section 1.1 to refresh the background). The inductor has the property:

$$v = L\frac{\mathrm{d}i}{\mathrm{d}t}$$

$$\text{or } i = \frac{1}{L}\int_{-\infty}^{\text{present}} v\,\mathrm{d}t.$$

Thus, the general property of an inductor can be said to be

$$\text{flow} = (\text{a constant or a function}) \times \int_{-\infty}^{\text{present}} \text{effort}\,\mathrm{d}t.$$

Note that the coefficient may be a constant – if the inductance does not vary with time. It may also be a function of time or of a system variable (like the current through it). Any element with the above relationship between the effort and the flow (e.g. the mass in translational mechanics) will be called an *inertial element* and will be denoted by the symbol

——— I

that is, an *I* element having one bond connected to it, through which it exchanges energy and information with the rest of the system.
Likewise, the capacitor has the property:

$$\text{effort} = (\text{a constant or a function}) \times \int_{-\infty}^{\text{present}} \text{flow}\,\mathrm{d}t.$$

Any element with this relationship between the effort and the flow (e.g. the spring in translational mechanics) will be called a *compliance element* and will be denoted by the symbol

——— C

The resistance has the relationship

$$\text{effort} = \text{a constant or a function} \times \text{flow},$$

$$\text{or} \quad \text{flow} = \text{a constant or a function} \times \text{effort},$$

and any element with this property will be called a *resistive element* and will be denoted by the symbol

——— R

A voltage source in electrical domain and an externally impressed force in mechanical domain are examples of *source of effort*, denoted by the symbol

——— SE

An *SE* element dictates the effort applied to the rest of the system. Therefore the effort in the bond connected to an *SE* element is an externally determined quantity that does not depend on the rest of the circuit. But the flow through that bond is not determined by the external agency; it is determined by the other circuit elements and their connection.

For a current source in an electrical circuit or a cam in a mechanical system, the flow variable is externally determined and the effort variable is decided by the rest of the system. Such an element is called a *source of flow* and is represented by the symbol

—— SF

6.4 The Junctions

To illustrate the formation of bond graphs with the above one-port elements, let us take the simple electrical circuit shown in Fig. 6.1(a), consisting of a resistor, a capacitor and an inductor in series with a voltage source. Each element interacts with the rest of the system in a particular way – the voltage source gives power, the resistor consumes power and dissipates it in thermal domain, the inductor and capacitor act as temporary reservoirs of energy. The power exchange of the individual elements with the rest of the system may be represented as in Fig. 6.1(b).

Is there any law followed in course of the interaction between elements? Yes, there is. Notice that in all these interactions, one aspect remains common to all the elements: they share the same current or *flow*. This is due to the constraint imposed by the series connection.

Such constrained interactions are represented by *junctions*. Where the constraint equalizes the flows in the elements, the resulting junction is called a *series* or **1** junction. This results in the bond graph shown in Fig. 6.1(c), which contains, in a graphical form, the essentials of the interaction between the elements of this system.

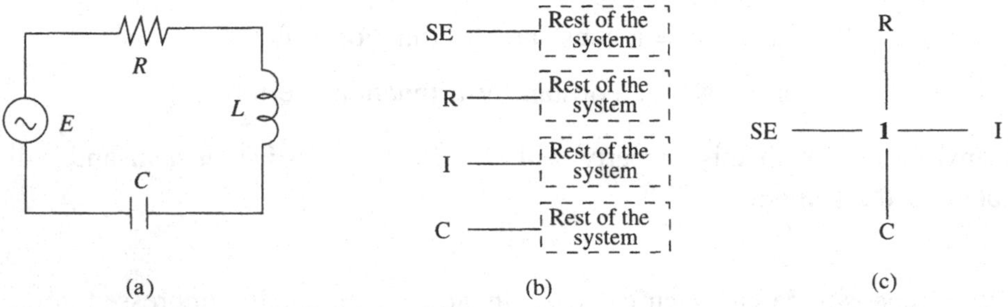

Figure 6.1 (a) The RLC series circuit, (b) the interaction of the elements with the rest of the system, and (c) the resulting bond graph.

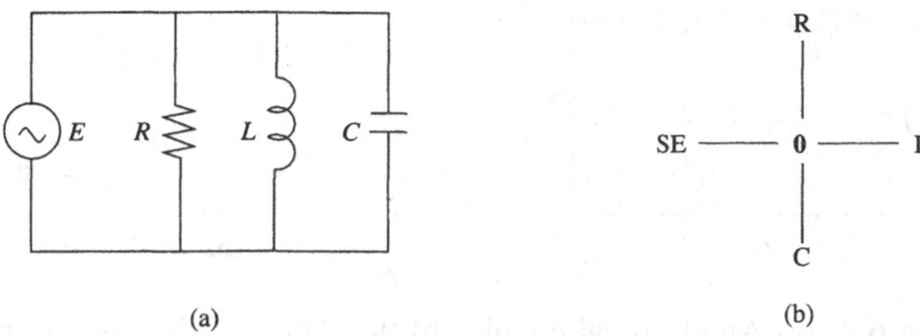

Figure 6.2 (a) The RLC parallel circuit, (b) the corresponding bond graph.

Similarly, we can take a parallel circuit as shown in Fig. 6.2. Here also the elements interact with the other elements through bonds, but now the law of interaction between elements is different. The constraint imposed by the parallel connection makes the elements share the same voltage or effort. This special type of interaction between the elements also has to be represented by a junction – this time it is the *parallel* or **0** junction. The resulting bond graph is shown in Fig. 6.2(b).

Thus, in general, the **0** junction is one that equalizes the efforts in all bonds connected to it, while the **1** junction equalizes the flows. It is interesting to note that there are only two ways in which elements of any system can interact and exchange power, and so there can be only two types of junctions: **0** and **1**. To form bond graphs for simple electrical circuits, one has to follow the junction structure in the interaction between elements. We illustrate this with the help of two examples below.

In the bond graphs, only the generic symbols for the elements are used. However, if there are more than one resistor in a circuit (as will be most often the case), it becomes necessary to state which resistor is represented by a particular R element in the bond graph. This is shown by writing the symbol of the circuit element (R_2, L_1, etc.) beside the generic symbol, separated by a colon. This convention is followed in the examples below.

► **Example 6.1** Obtain the bond graph of the electrical circuit shown in Fig. 6.3(a).

Solution: We start making the bond graph from one end, say, at the *SE* element or the voltage source. The *SE* element shares the same flow with the inductor and the rest of the circuit. Therefore, it has to be connected, along with the *I* element, to a **1** junction. Another bond is connected to this junction, which represents the power flow into the "rest of the circuit". This "rest of the circuit" consists of a parallel connection of a capacitor and a voltage source that share the same effort. So the *C* element and the *SF* element are connected to a **0** junction. This completes the bond graph as shown in Fig. 6.3(b). ◄

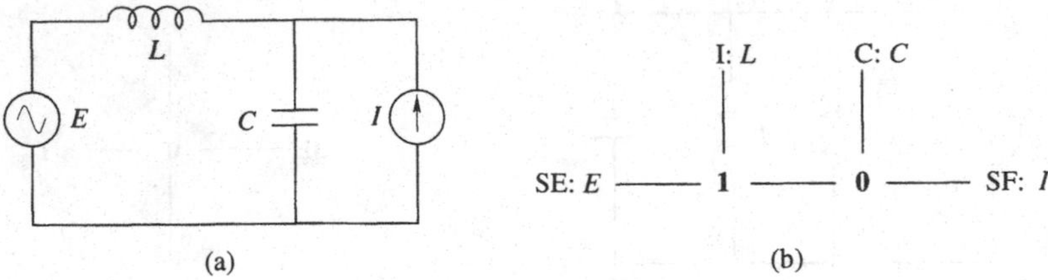

Figure 6.3 (a) An electrical circuit, (b) the corresponding bond graph.

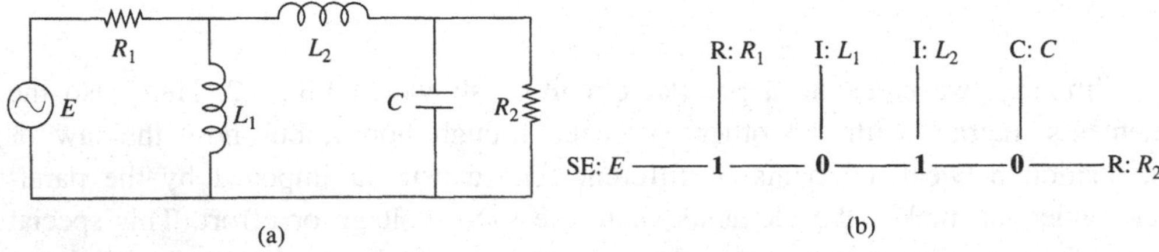

Figure 6.4 (a) An electrical circuit, (b) the corresponding bond graph.

► **Example 6.2** Obtain the bond graph of the simple electrical circuit shown in Fig. 6.4(a).

Solution: We start from the left end and argue as follows. The voltage source shares the same flow with R_1. This combination shares the same effort with the inductor L_1. This combination, in turn, shares the same flow with L_2, and the resulting combination shares the same effort with C and R_2. This gives the bond graph shown in Fig. 6.4(b). ◄

6.5 Junctions in Mechanical Systems

In mechanical systems, if some elements share the same velocity (flow), then these elements are attached to a **1** junction and if some elements share the same force (effort), then they are connected to a **0** junction. For example, if a force is applied on a mass, the agent applying the force has to move along with the mass with the same velocity. So the *SE* element and the *I* element would be connected through a **1** junction, as shown in Fig. 6.5.

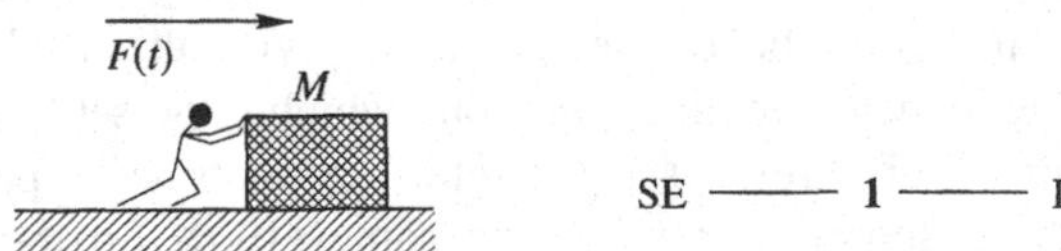

Figure 6.5 Elements sharing the same flow in a mechanical system.

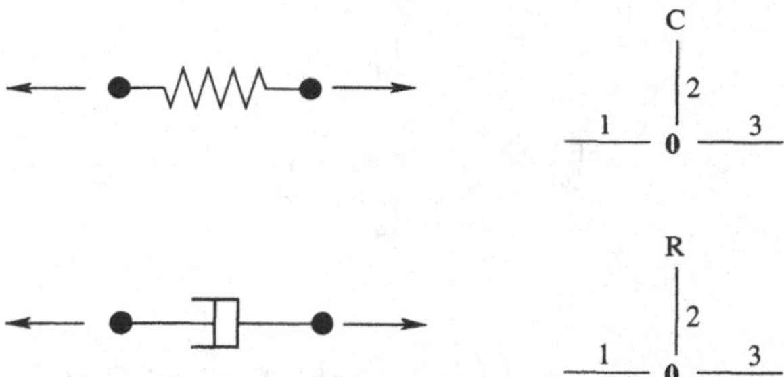

Figure 6.6 Elements sharing the same effort in a mechanical system.

Now take the spring as shown in Fig. 6.6. The elements connected to the two ends of the spring feel the same force, and therefore share the same effort (otherwise there will be a resultant force, causing an infinite acceleration in the ideally massless spring). The spring itself feels the same effort, as do the elements at the two ends of it. Moreover, the velocity (flow) felt by the spring is the relative velocity of the two ends. These are the typical characteristics of the **0** junction. Therefore, in the bond graph the spring will be represented by a C element connected through a **0** junction as shown in Fig. 6.6, satisfying the physical relationships $e_1 = e_2 = e_3$ and $f_1 + f_2 + f_3 = 0$. The sense (positive or negative) of the flow variables will come from the power directions which we shall discuss in Section 6.7.

It is easy to see that the mechanical R element or the damper will have the same junction structure as the C element.

▶ **Example 6.3** Take the mechanical system shown in Fig. 6.7(a). The mass shares the same flow with the agent applying the force, and with the top end of the spring-damper arrangement. Therefore, these are connected through a **1** junction. The junction structure of the rest of the system can be approached in two possible ways.

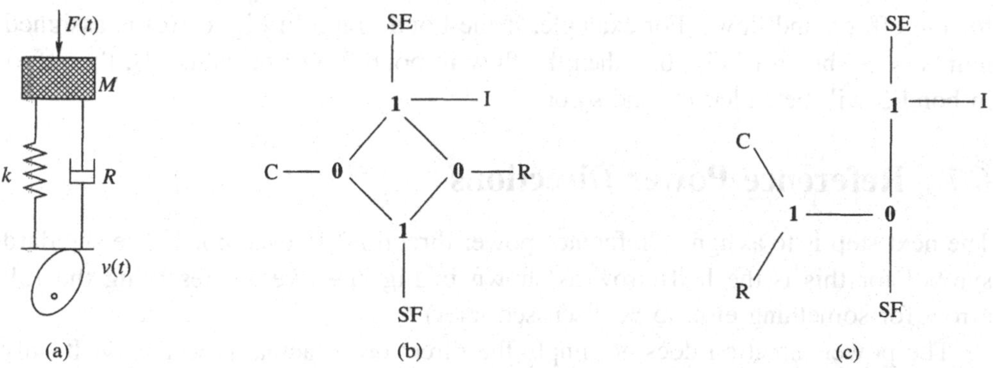

Figure 6.7 A mechanical system, and two possible ways of drawing the bond graph.

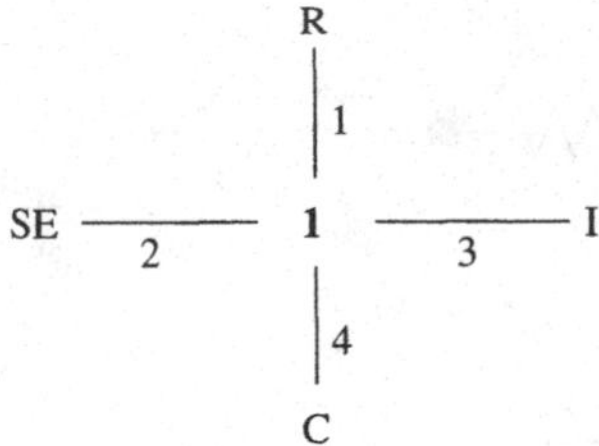

Figure 6.8 The numbered bond graph of the RLC series circuit.

The first line of argument is as follows. The top and the bottom ends of the spring share the same effort. Therefore, the C element is connected through a **0** junction as shown in Fig. 6.7(b). The same argument applies to the R element. The cam shares the same flow with the bottom end of both the spring and the damper. Therefore, these are connected through a **1** junction. This completes the bond graph.

The bond graph of the same system can also be conceived as follows. The spring-damper combine applies the same effort on the mass at the top and the cam at the bottom. Therefore, the mass and the cam are connected through a **0** junction. Now the spring and the damper share the same flow as they have to move together, and the effort felt at the top and the bottom are the sum of the efforts at the spring and the damper. Therefore, the C element and the R element are connected through a **1** junction. This bond graph is shown in Fig. 6.7(c). Both these bond graphs are valid representations of the system. ◀

6.6 Numbering of Bonds

After the basic bond graph is created, the next step is to assign a number to every bond. The numbering is arbitrary. There is a convention (though not any rule) to start from any bond and assign the number 1, and then go on successively assigning the numbers 2, 3, ... to the other bonds.

The assignment of a number to a bond allows the use of generic nomenclature for the efforts and flows. For example, if the bond graph in Fig. 6.1(c) is assigned numbers as shown in Fig. 6.8, then the flow in bond 2 will be called f_2, the effort in bond 3 will be called e_3, and so on.

6.7 Reference Power Directions

The next step is to assign a "reference power direction" to each bond. The standard symbol for this is the half-arrow as shown in Fig. 6.9 (we are reserving the full arrow for something else, to be discussed later).

The power direction does not imply the direction of actual power flow. It only specifies the direction in which the power flow is assumed to be positive. For example, in Fig. 6.9(a), bond 4 has been assigned a power direction from the

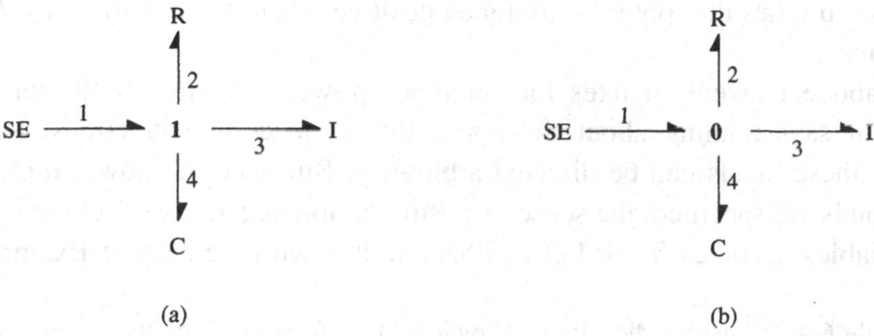

Figure 6.9 The bond graphs of the RLC series circuit and parallel circuit with reference power directions assigned.

1 junction towards the C element. This means that the variables e_4 and f_4 are chosen such that when both assume positive (or negative) values, the power goes from the system towards the capacitor. When these two variables have different signs, power flows in the opposite direction, that is, from the capacitor to the rest of the system.

With these power directions, we can state the power balance at the **1** junction of Fig. 6.9(a) (since a junction neither creates nor absorbs power) as

$$e_1 f_1 - e_2 f_2 - e_3 f_3 - e_4 f_4 = 0,$$

assuming the power going into the junction as positive and that coming out of the junction as negative. Since at a **1** junction the flows are equalized,

$$f_1 = f_2 = f_3 = f_4.$$

This gives

$$e_1 - e_2 - e_3 - e_4 = 0.$$

Note that the sense of these variables follow from the assigned power directions.

Similarly, for the **0** junction in Fig. 6.9(b) we have

$$\text{Power balance: } e_1 f_1 - e_2 f_2 - e_3 f_3 - e_4 f_4 = 0,$$

$$\text{Effort equalization: } e_1 = e_2 = e_3 = e_4,$$

$$\text{Resulting flow equation: } f_1 - f_2 - f_3 - f_4 = 0.$$

The standard convention is to assume positive direction of power when it flows out of the sources (i.e. SE and SF elements), and into the other elements. The power directions assigned in Fig. 6.9 follow this convention, where the positive power is directed out of the voltage source and into the R, L and C elements. This convention implies that power is assumed positive when it is *absorbed* by I, C and R elements.

The above convention fixes the reference power direction in the terminating bonds, but says nothing about the bonds that connect two junctions. Reference power in these bonds can be directed arbitrarily. But once the power directions in all the bonds are specified, the sense of positivity and negativity of all the effort and flow variables becomes fixed. Let us illustrate this with the help of Example 6.4.

► **Example 6.4** Consider the circuit shown in Fig. 6.10(a). The bond graph with two alternative power directions is shown in Fig. 6.10(b) and (c). The power directions in bonds 1, 2, 4 and 5 can be assigned as per convention, but bond number 3 can have two possible power directions. Let us now understand the implication of the given power directions on the assumed positive direction of the power variables in the actual circuit.

To start with, let us assign the positive direction in *one* of the variables. Suppose, the voltage across the *SE* element is positive when the upper terminal is positive and the lower terminal negative as shown in Fig. 6.11.

The power in the *SE* element is positive when it delivers power to the circuit. This gives the positive direction of current in the first loop, as shown in Fig. 6.11(a). The power in

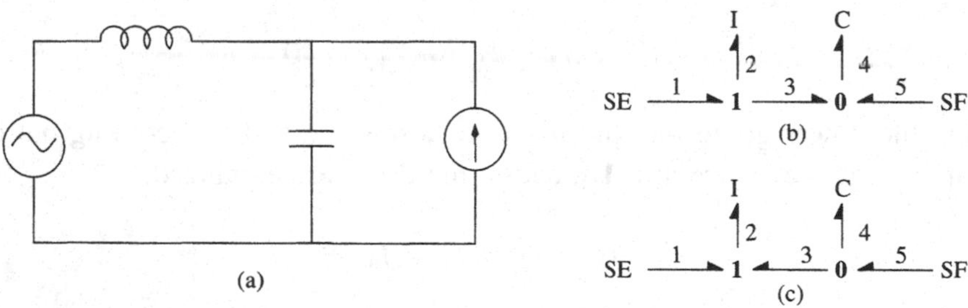

Figure 6.10 (a) The circuit pertaining to Example 6.4; (b) and (c) the bond graph with two alternative power directions.

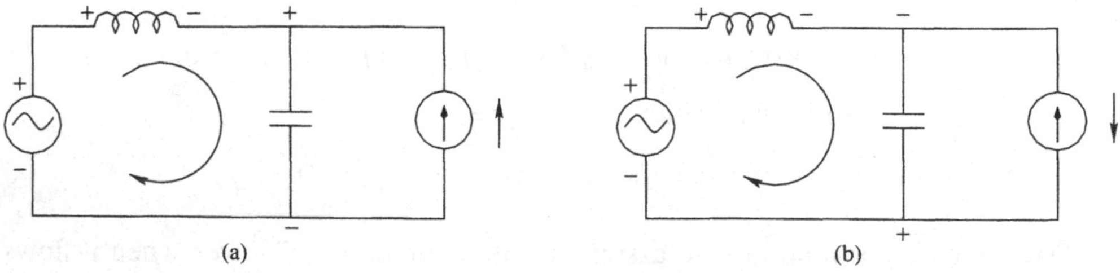

Figure 6.11 (a) The positive directions of the variables when the power directions are as in Fig. 6.10(b); (b) The positive directions corresponding to Fig. 6.10(c).

the inductor is positive when it goes into the inductor. This implies the positive direction of the voltage across the inductor.

So far there is no problem. But now we come to the bond number 3, which can have two possible power directions. If we take the power direction as shown in Fig. 6.10(b), then power is positive when it flows into the capacitor-current source combine. This gives the direction of positive voltage across this combine. This is shown in Fig. 6.11(a). This direction of positive voltage implies that in the current source, positive current is directed upward (since positive power flows out of the source).

On the other hand, if we take the power direction as shown in Fig. 6.10(c), then the voltage across the capacitor-current source combine takes positive value when the lower terminal is positive and the upper terminal is negative. This implies that the current through the *SF* element is positive when it flows downward. This is shown in Fig. 6.11(b). ◀

Thus, we see that the power directions assigned to the bonds decide the sense of the effort and flow variables. In mechanical systems, this implies the positive directions of motion of individual elements, forces, and so on.

To illustrate, let us take the mechanical system in Fig. 6.12. Notice that this system yields the same bond graph as shown in Fig. 6.10(b) and (c), and therefore can be power directed in two possible ways. We leave it for the reader to check, following the procedure outlined in Example 6.4, that the positive senses of the effort and flow variables (especially the direction of positive velocity of the cam) will be as shown in Fig. 6.12(a) and (b) for the two possible power directions in bond 3.

It may be noted, however, that it is not necessary to assign the positive and negative sense of all the variables before starting to model a system. One may assign arbitrary power directions to junction-to-junction bonds and proceed to obtain the system equations on that basis. The assignment of power directions, therefore, can be automated and handled by computers. Keeping track of the resulting direction of the variables becomes necessary only at the stage of interpreting the system dynamics.

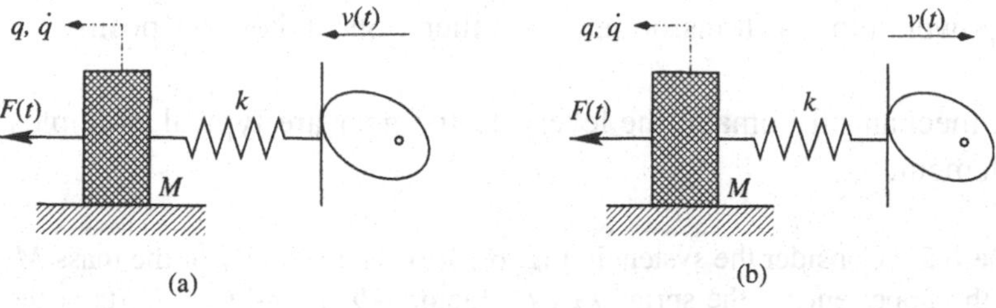

Figure 6.12 A mechanical system that gives the same bond graph as in Fig. 6.10. (a) and (b) show the positive direction of variables corresponding to the power directions in Fig. 6.10(b) and (c) respectively.

6.8 Two-port Elements

There are some elements that require two energy ports, and therefore two connected bonds, to represent the exchange of power. These are classified under two categories: the *transformer* and the *gyrator*.

The transformer element: An ideal electrical transformer has the property that it does not generate, store or consume power, but only transfers it from the primary side to the secondary side, which is at a different voltage. Such an element is represented by

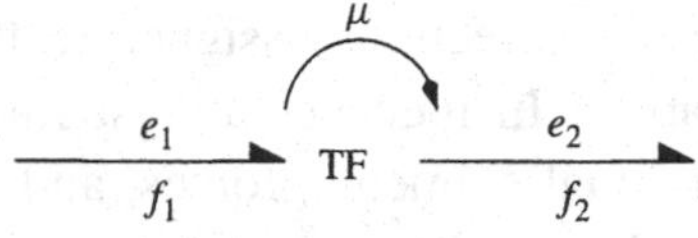

Since the power in the two sides must be equal,

$$e_1 f_1 = e_2 f_2.$$

It is a convention to write the transformation ratio of the *flows*, showing the direction of transformation. The notation shown above implies that the flows at the two sides are related by

$$f_2 = \mu f_1,$$

and the efforts by

$$e_2 = \frac{1}{\mu} e_1.$$

It is also a convention to set power direction in such a way that if one bond brings positive power into the transformer, the other bond takes the positive power out of it.

In the mechanical domain, the lever and the gear are typical examples of transformer element.

► **Example 6.5** Consider the system in Fig. 6.13(a). The velocity of the mass M_1 is related to that of the upper end of the spring k_2 by a factor $-b/a$, and the efforts at the two ends of the lever are related by a reciprocal factor. Therefore, the spring-mass duo at the right-hand side is connected to the rest of the system through a transformer element as shown in Fig. 6.13(b). The right end of the lever, the spring k_2, and the mass M_2 share the same effort; and so they are connected with a **0** junction. ◄

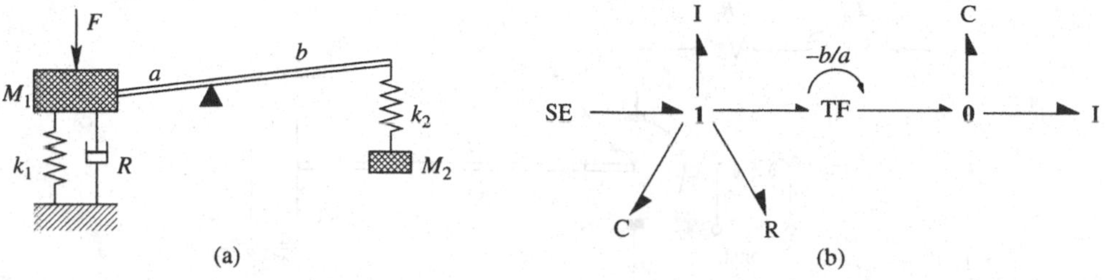

Figure 6.13 The system pertaining to Example 6.5.

The gyrator element: A two-port element that relates the effort at one side to the flow at another and vice versa, is called a gyrator, represented by the bond structure

$$\xrightarrow[f_1]{e_1} \overset{\mu}{\underset{}{\text{GY}}} \xrightarrow[f_2]{e_2}$$

Here μ is the gyrator modulus, and no sense of direction is necessary because the variables at the two sides are related by

$$e_2 = \mu f_1 \qquad \text{and} \qquad e_1 = \mu f_2.$$

These equations satisfy the power balance at the two sides $e_1 f_1 = e_2 f_2$.

The ideal dc machine is a typical example of a gyrator. The armature current and the back emf in the electrical domain are related to the speed and torque in the mechanical domain by

$$\tau = K\phi i_a \qquad \text{and} \qquad e_b = K\phi\omega_r,$$

where τ is the torque,
ϕ is the field flux,
ω_r is the angular speed of the rotor,
i_a is the armature current,
e_b is the back emf
and K is the motor constant.

This is typically the power relationship of a gyrator with modulus $K\phi$. In mechanical domain, a typical gyrator element is the gyroscope.

► **Example 6.6** Take the dc motor–load system shown in Fig. 6.14. In the electrical subsystem, the voltage source and the armature resistance and inductance share the same flow with the back emf. Therefore, these are connected through a **1** junction. The conversion from electrical domain to mechanical domain (and vice versa) is represented by the gyrator element. The rotor inertia and friction share the same flow (rotor speed) with that coming from the gyrator. So these are also connected through a **1** junction. The flexible shaft,

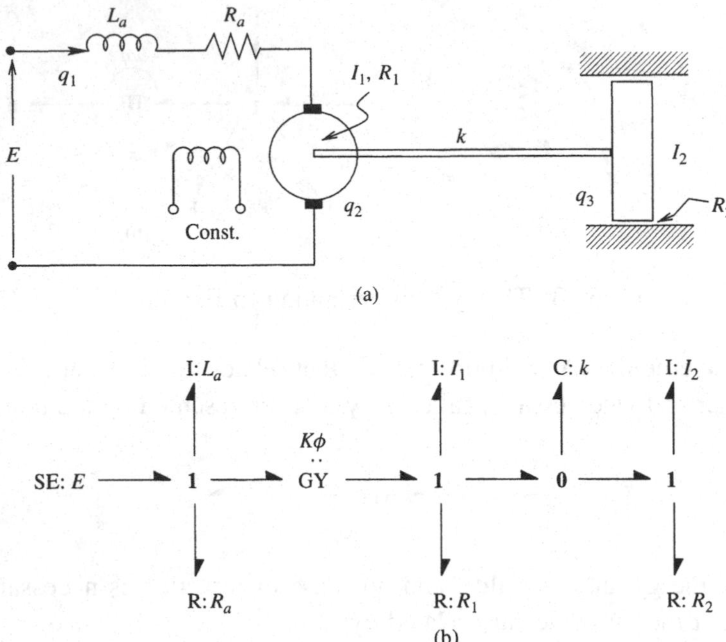

Figure 6.14 (a) A dc motor connected to a load, (b) its bond graph.

represented by the compliance element, equates the effort at both its ends, and (as per our discussion on the spring) is connected through a **0** junction. Finally, the load inertia and friction share the same flow with the right-hand end of the shaft. So these are connected through a **1** junction. This completes the bond graph as shown in Fig. 6.14(b). ◀

6.9 The Concept of Causality

We have said earlier that bond graph is a schematic representation of the exchange of power and information between elements of a system. So far we have dealt with the power aspect. Now we have to understand how information flows between elements, and we shall see in the next section that these two aspects taken together suffice in determining the differential equations of a system.

We start from the important observation that no element can determine both the power variables – effort and flow – at the bond attached to itself. If it determines the value of one of the variables, the magnitude of the other variable must be determined by the rest of the system. For every bond, if the *information* about effort is dictated by the element at one end, the *information* about flow must come from the opposite end, and vice versa.

To understand this point, imagine a voltage source connected to a resistance. The voltage source dictates the effort variable unilaterally, but cannot determine

Figure 6.15 The convention of the causal stroke.

the flow variable. The current flowing through the circuit is determined by the resistor. Therefore, the *SE* element gives out effort information and receives flow information, while the *R* element receives effort information and gives out flow information. Likewise, for a current source connected to a resistor, the *SF* element dictates the flow while the *R* element dictates the effort. In a slightly different language, we may say that the *SE* element *causes* the effort information and the *SF* element *causes* the flow information. The *R* element causes effort information or flow information depending on the information it receives. It is easy to see that the situation is the same for mechanical *SE*, *SF* and *R* elements.

Since in every bond the effort and flow information go in opposite directions, it suffices to show the direction of one of the variables. The general convention is to put a stroke at the end which gives the flow information. It is easy to remember this convention by picturing an injection syringe as shown in Fig. 6.15.

With this convention, the flow of information for the *SE*, *SF* and the *R* element can be represented by the *causal strokes* as shown in Fig. 6.16. Note that the *R* element can have both the directions of causality.

For storage elements, the concept of causality or the flow of information can be understood in a slightly different way. Take the governing equation of an *I* element:

$$\text{flow} = (\text{a constant or a function}) \times \int_{-\infty}^{\text{present}} \text{effort } dt.$$

This means that it needs the whole history of the effort information to determine the flow information at any moment. Since by definition the cause must precede the effect, it follows that for the *I* element effort is the cause and flow is the effect.

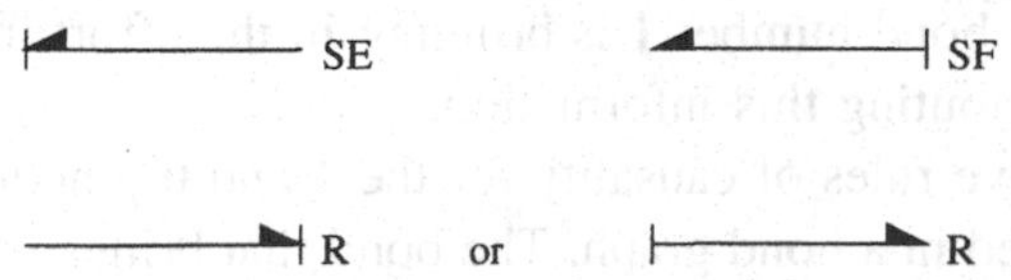

Figure 6.16 The natural causalities for the one-port elements.

Accordingly, the I element should receive effort information and give out flow information, giving the causal stroke:

$$\longrightarrow\!| \ \mathrm{I}$$

Likewise, the compliance element has the property:

$$\text{effort} = (\text{a constant or a function}) \times \int_{-\infty}^{\text{present}} \text{flow } \mathrm{d}t.$$

Thus, it needs the whole history of the flow information to determine the effort information at any given moment. This in turn implies that for the C element, flow is the cause and effort is the effect, and this element normally should receive flow information and should give out effort information. This will be represented by the causal stroke:

$$\longrightarrow\!| \ \mathrm{C}$$

Since the above causal relationship for the I and C elements follows from the integral relationship, these are called *integral causality*. We shall see later that under certain circumstances the integral causality may be violated.

For the transformer element, if the flow information comes in through one bond, the flow information goes out through the other. This implies the possible causal strokes:

$$|\!\longrightarrow \mathrm{TF}\ |\!\longrightarrow \quad \text{or} \quad \longrightarrow\!| \ \mathrm{TF} \longrightarrow\!|$$

For the gyrator, if one bond brings in the flow information, the other takes out the effort information. Therefore, the possible causalities are:

$$|\!\longrightarrow \mathrm{GY}\ |\!\longrightarrow \quad \text{or} \quad \longrightarrow\!| \ \mathrm{GY}\ |\!\longrightarrow$$

Now we come to the causal structures at the junctions. The **1** junction equalizes the flows in the bonds connected to it. Therefore, the information of flow must come from only one bond, and all other bonds must take out the flow information. Fig. 6.17(a) shows a typical causal structure at a **1** junction where bond number 2 is bringing in flow information, and the other bonds are taking this information out.

The **0** junction equalizes the efforts in the bonds connected to it. This implies that only one bond must bring in the effort information and all the other bonds must take this information out. Figure 6.17(b) shows a typical causal structure at a **0** junction where the bond number **1** is bringing in the effort information while the other bonds are distributing this information.

Note that the above rules of causality for the **1** and **0** junctions are "hard rules", and cannot be violated in a bond graph. The bond that brings some information into a junction is called *strong bond* while the other bonds taking out that information are called *weak bonds*.

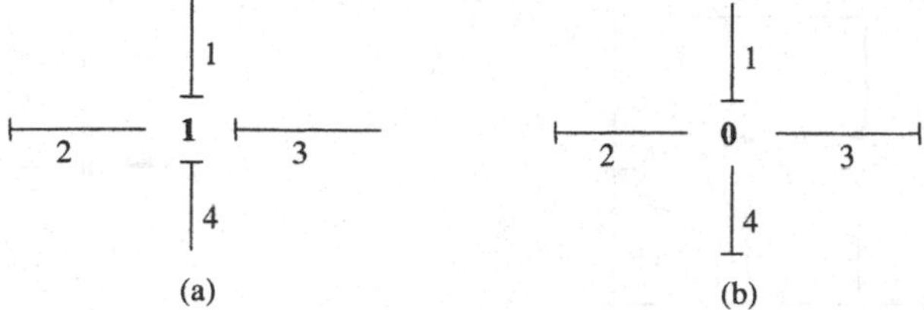

Figure 6.17 The "hard" rules of causality at the junctions.

6.10 Differential Causality

We have argued earlier that for storage elements, the *proper* causality should be given by the integral relationship. However, in some circumstances this may not be possible and we may be forced to put the causal stroke at the opposite end. Such a reversed causality is called differential causality.

Let us now look for the circumstances under which differential causality may occur. Consider a current source in series with an inductor, as shown in Fig. 6.18.

Notice that in the bond graph we have honoured the hard rule of the flow of information at the junction, and as a result have ended up with a differential causality at the *I* element, shown with a circle. The question is, what is wrong here?

We know that the inductor has the property that it does not allow its current to change instantaneously, and the voltage induced across it is given by

$$v = L\frac{\mathrm{d}i}{\mathrm{d}t}.$$

Thus, if the current dictated by the source changes instantaneously, the voltage induced would be infinite. This tells us that one should not connect a current source in series with an inductor.

Similar is the situation of a voltage source connected in parallel with a capacitor as shown in the circuit shown in Fig. 6.19. Here, again, we have honoured the causality of the source and the junction. Out of the weak bonds, *I* shows integral causality; *R* can take both the causalities, and so it is also properly causaled; but the *C* element now has differential causality. The reason is the same; the capacitor

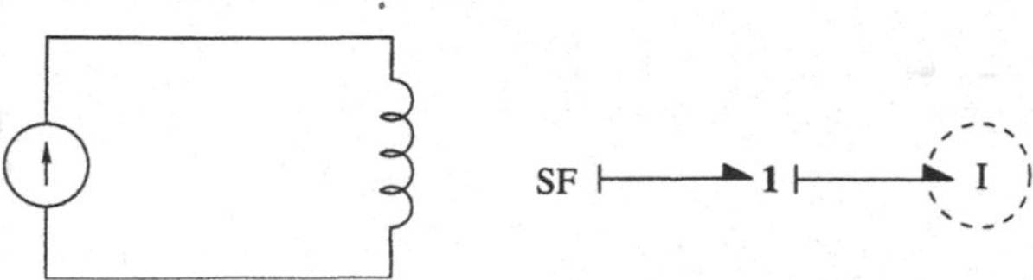

Figure 6.18 Example of differential causality.

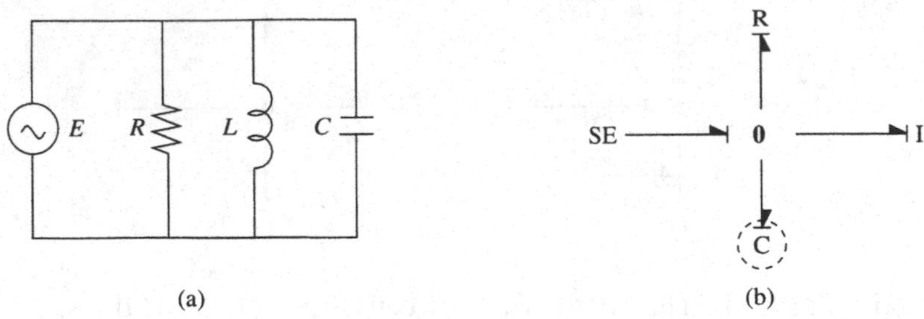

Figure 6.19 The parallel RLC circuit showing differential causality.

does not allow the voltage given by the source to vary independently. If, say, the voltage source gives a square wave, the current through the capacitor will be infinite at the transitions. The conclusion is that a capacitor should never be connected in parallel with an independent voltage source.

We have seen in Chapter 3 that in electrical circuits such situations arise, in general, if the branches of a cutset contain only inductors and current sources, or if the branches in a loop contain only capacitors and voltage sources.

Notice also that if a resistance is connected in parallel to the current source in Fig. 6.18 or a resistance is connected in series with the voltage source in Fig. 6.19, the resulting bond graphs no longer have any differential causality. This is shown in Fig. 6.20. The elements in question are now integrally causaled because for Fig. 6.20(a) the inductor current is not constrained to be the same as that dictated

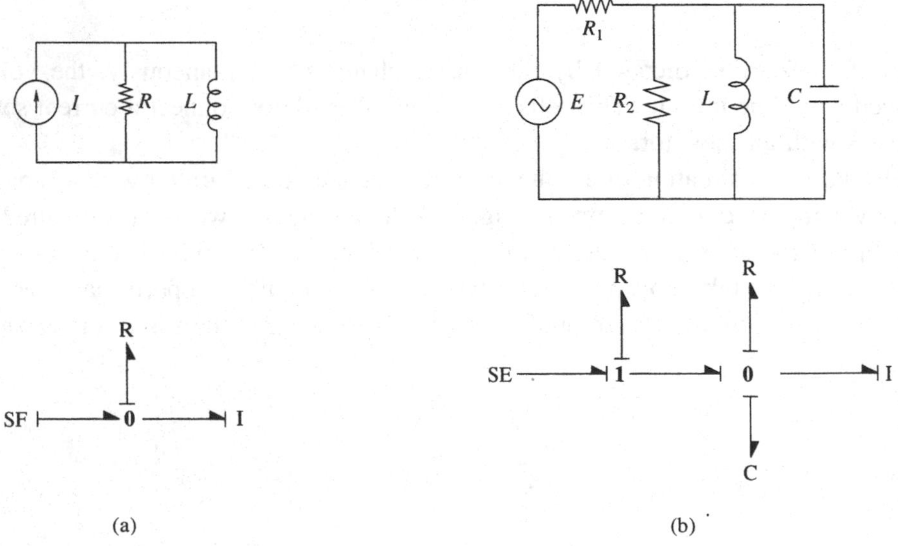

Figure 6.20 The removal of differential causality.

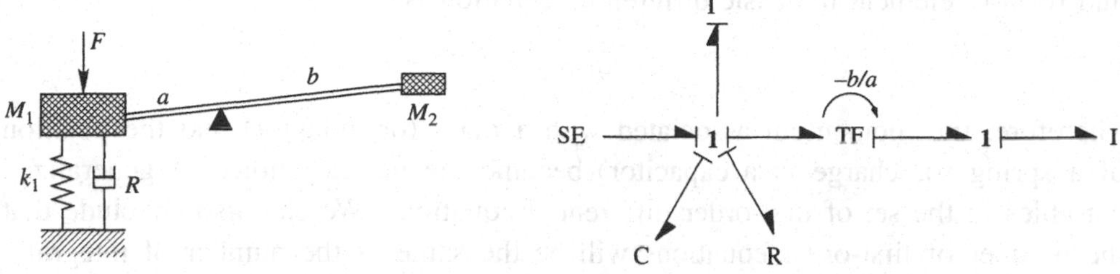

Figure 6.21 The system pertaining to Example 6.7.

by the current source, and for Fig. 6.20(b) the capacitor voltage is allowed to vary depending on the current through it.

The above tells us that for the sources of effort and flow, if we take into account their internal resistance, the problem of differential causality is often removed. However, there are also other situations that can give rise to differential causality.

► **Example 6.7** Take the system in Fig. 6.21. Here the mass M_2 is attached directly to the end of the lever, and share the same flow with that end. Hence, it is connected to the transformer element through a **1** junction. If we integrally causal the mass M_1, the flow information comes to the transformer from the left side as shown, and as a result the I element representing mass M_2 becomes differentially causaled. The reader may check that if we assign integral causality to the mass M_2, then the other mass gets differential causality.

In this case, the two masses do not have independent freedom of movement – freedom to decide their flows depending on the efforts they receive. This causes the differential causality. There are two ways of getting rid of this problem. First, we can assume some flexibility in the lever rod – a situation same as having a spring attached as in Fig. 6.13. Second, we can take the two masses together and consider the combined inertia in place of mass M_1 (and neglect the lever). The actual motion of the mass M_2 can be obtained separately by a linear relationship with the motion of mass M_1. ◄

Differential causality also arises if the shaft in Fig. 6.14 is assumed to be rigid (the reader may check this). The motor rotor and the load are then forced to move together, so that they cannot decide their flows independently. One way of overcoming the problem is to assume some flexibility in the shaft; another is to take the inertia of the rotor and the shaft together – as if they were a single body.

6.11 Obtaining Differential Equations from Bond Graphs

Once a bond graph has been numbered, power directed and causaled, one has the basic inputs necessary to obtain the differential equations. We first note that *each* (integrally causaled) storage element gives rise to a first-order differential equation. We know that for an I element the basic differential equation is

$$\dot{p} = e,$$

and for a C element the basic differential equation is

$$\dot{q} = f.$$

Therefore, the momentum associated with a mass (or inductor) and the position of a spring (or charge in a capacitor) become the natural choice of generalized variables in the set of first-order differential equations. We can also conclude that the number of first-order equations will be the same as the number of integrally causaled storage elements (we shall see later what happens in case of differentially causaled storage elements). Thus, the integrally causaled storage elements provide the state variables of the system.

The steps to obtain the differential equations follow from the answers to the two questions:

1. *What do the elements give to the system?*
2. *What do the integrally causaled storage elements receive from the system?*

We illustrate the procedure with the following examples.

► **Example 6.8** Let us take the system in Example 6.3, whose bond graph is reproduced in Fig. 6.22, now properly numbered, power directed and causaled. The variables of the system are the momentum of the mass and the position of the spring. Since the mass is the element connected to bond 2, and the spring is connected to bond 7, we shall call the variables p_2 and q_7.

We now ask the first question to the elements of the bond graph.

1. What does I_2 give to the system? Answer: the flow f_2. We now express it in terms of the state variable

$$f_2 = \frac{p_2}{M}.$$

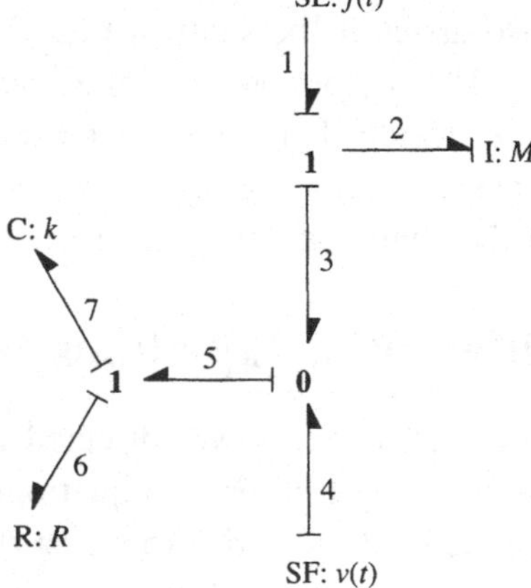

Figure 6.22 The bond graph pertaining to Example 6.3.

2. What does C_7 give to the system? Answer: the effort e_7. Expressed in terms of the state variables, we have

$$e_7 = kq_7.$$

3. What does R_6 give to the system? Answer: the effort e_6. First, we express it in terms of the property of the element by writing $e_6 = Rf_6$. But f_6 is not a state variable. In order to express the right-hand side in terms of the state variables, we have to see from where the information about f_6 comes. This information comes from the **1** junction where the strong bond 5 brings in the flow information. Therefore we write:

$$e_6 = Rf_6 = Rf_5.$$

This information, in turn, comes from the **0** junction. At this junction, power balance gives (notice the power directions)

$$e_3 f_3 + e_4 f_4 - e_5 f_5 = 0.$$

Since at the **0** junction, $e_3 = e_4 = e_5$, we have $f_5 = f_3 + f_4$. This gives

$$e_6 = Rf_5 = R(f_3 + f_4).$$

The information for f_3 comes from f_2 through the **1** junction. Now we are in a position to express the RHS in terms of the state variables:

$$\begin{aligned} e_6 &= R(f_2 + f_4), \\ &= R\left(\frac{p_2}{M} + v(t)\right). \end{aligned}$$

4. For the sources the question has a trivial answer. The *SE* element gives the effort $e_1 = F(t)$ and the *SF* element gives the flow $f_3 = v(t)$.

Notice that in obtaining answer to this question for any particular element, we first have to relate the variable to the property of that element. When asking the question to, say, the resistance at bond 6, we get the answer e_6. We do not immediately go about looking for the source of this information in the bond graph. Instead, we first relate it to the property of the resistance, by writing $e_6 = Rf_6$, and then scout the bond graph to find the source of the information f_6.

Now let us ask the second question to the storage elements.

1. What does I_2 receive from the system? Answer: the effort e_2. By the property of the I element, $e_2 = \dot{p}_2$. We now reverse the sides and try to express e_2 in terms of the state variables:

$$\begin{aligned} \dot{p}_2 &= e_2, \\ &= e_1 - e_3 \quad \text{[from power balance at } \mathbf{1} \text{ junction]}, \\ &= e_1 - e_5 \quad \text{[information coming from the strong bond at } \mathbf{0} \text{ junction]}, \\ &= e_1 - e_6 - e_7 \quad \text{[from power balance at } \mathbf{1} \text{ junction]}. \end{aligned}$$

Notice that the values of all the quantities in the RHS are available from the answers to the first question as already obtained. Substituting, we get

$$\dot{p}_2 = F(t) - R\left(\frac{p_2}{M} + v(t)\right) - kq_7.$$

2. What does C_7 receive from the system? Answer: the flow f_7. We know $f_7 = \dot{q}_7$. We now reverse the sides and express f_7 in terms of the state variables.

$$\begin{aligned}\dot{q}_7 &= f_7,\\ &= f_5 \quad \text{[from power balance at } \mathbf{1} \text{ junction]},\\ &= f_3 + f_4 \quad \text{[information coming from the strong bond at } \mathbf{0} \text{ junction]},\\ &= v(t) + f_2 \quad \text{[information coming from the strong bond at } \mathbf{1} \text{ junction]},\\ &= v(t) + \frac{p_2}{M} \quad \text{[Already obtained from the first question]}.\end{aligned}$$

These are the two first-order differential equations of the system. The system of equations being linear, it can be expressed as matrix equations as

$$\begin{pmatrix}\dot{p}_2\\ \dot{q}_7\end{pmatrix} = \begin{pmatrix}-R/M & -k\\ 1/M & 0\end{pmatrix}\begin{pmatrix}p_2\\ q_7\end{pmatrix} + \begin{pmatrix}1 & -R\\ 0 & 1\end{pmatrix}\begin{pmatrix}F(t)\\ v(t)\end{pmatrix}.$$ ◄

► **Example 6.9** Consider the circuit whose bond graph was derived earlier in Example 6.2. In Fig. 6.23 we present the bond graph with power directions and causalities added. In this circuit there are three storage elements, all integrally causaled, and hence the number of state variables will be three: p_4, p_6, and q_8.

Now we ask: *what do the elements give to the system?*

SE: $e_1 = E$.

I_4: $f_4 = p_4/L_1$.

I_6: $f_6 = p_6/L_2$.

C_8: $e_8 = q_8/C$.

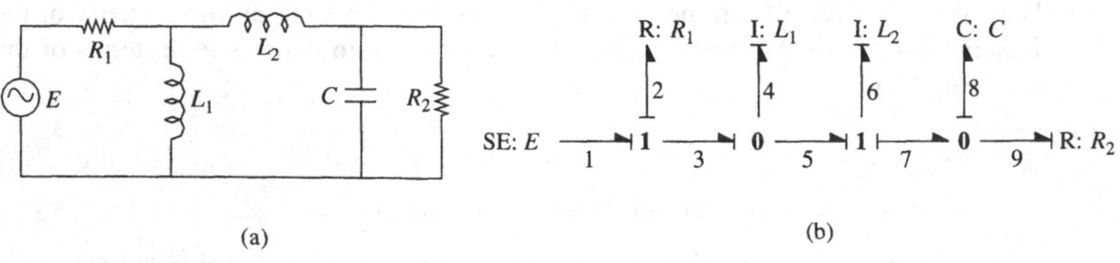

Figure 6.23 The bond graph pertaining to Example 6.2.

R_2: For this element we derive as follows:

$$\begin{aligned} e_2 &= R_1 f_2, \\ &= R_1 f_3 \quad \text{[Where does the information } f_2 \text{ come from? Ans: Strong bond 3]}, \\ &= R_1(f_4 + f_5) \quad \text{[The information } f_3 \text{ comes from } f_4 \text{ and } f_5 \text{ at the } \mathbf{0} \text{ junction]}, \\ &= R_1(f_4 + f_6) \quad [f_5 \text{ comes from the strong bond 6 at the } \mathbf{1} \text{ junction]}, \\ &= R_1\left(\frac{p_4}{L_1} + \frac{p_6}{L_2}\right). \end{aligned}$$

R_9: $f_9 = e_9/R_2 = e_8/R_2 = \frac{q_8}{R_2 C}$ [since e_8 brings in the information e_9].

Now that the answers to the first question have been obtained in terms of the state variables, we ask the second question: *What do the three integrally causaled storage elements receive from the system?*

I_4: $e_4 = \dot{p}_4$. Now we interchange the sides, and find where the e_4 information comes from.

$$\dot{p}_4 = e_4 = e_3 = e_1 - e_2 = E - R_1\left(\frac{p_4}{L_1} + \frac{p_6}{L_2}\right).$$

I_6: $e_6 = \dot{p}_6$. This gives

$$\dot{p}_6 = e_6 = e_5 - e_7 = e_3 - e_8 = e_2 - e_8 = R_1\left(\frac{p_4}{L_1} + \frac{p_6}{L_2}\right) - \frac{q_8}{C}.$$

C_8: $f_8 = \dot{q}_8$. This gives

$$\dot{q}_8 = f_8 = f_7 - f_9 = f_6 - f_9 = \frac{p_6}{L_2} - \frac{q_8}{R_2 C}.$$

These are the differential equations of the system. ◀

► **Example 6.10** Now we consider the dc-motor system of Fig. 6.14. The bond graph is reproduced in Fig. 6.24 with numbers and causalities assigned. In this system, the state variables are p_2, p_6, q_9 and p_{11}.

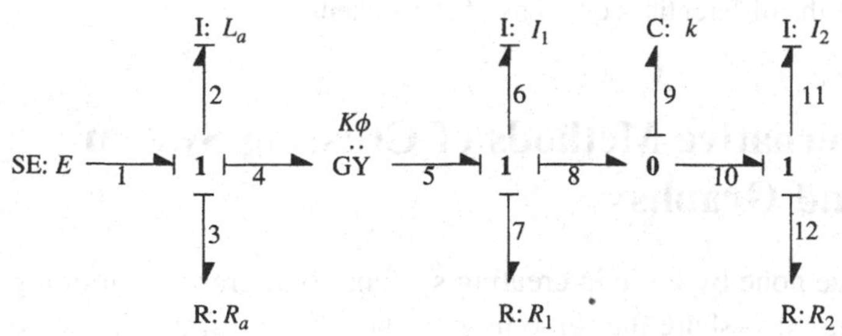

Figure 6.24 The bond graph of the motor-load system of Fig. 6.14.

We ask the first question: *what do the elements give to the system?* For the source and the storage elements the answers to the question are readily obtained:

$$e_1 = E, \quad f_2 = \frac{p_2}{L_a}, \quad f_6 = \frac{p_6}{I_1}, \quad e_9 = kq_9, \quad f_{11} = \frac{p_{11}}{I_2}.$$

For the resistances, we have to locate the source of the relevant information at the strong bonds.

$$e_3 = R_a f_3 = R_a f_2 = \frac{R_a}{L_a} p_2,$$

$$e_7 = R_1 f_7 = R_1 f_6 = \frac{R_1}{I_1} p_6,$$

$$e_{12} = R_2 f_{12} = R_2 f_{11} = \frac{R_1}{I_2} p_{11}.$$

Now, the gyrator element gives two different information to the two bonds connected to it. These are

At bond 4: $e_4 = K\phi f_5 = K\phi f_6 = K\phi p_6/I_1$.

At bond 5: $e_5 = K\phi f_4 = K\phi f_2 = K\phi p_2/L_a$.

Thus, the first question is answered for all the elements. Now we ask the second question to all the storage elements: *What do you receive from the system*?

The armature inductance receives effort information: $e_2 = \dot{p}_2$. This gives
$\dot{p}_2 = e_2 = e_1 - e_3 - e_4 = E - R_a p_2/L_a - K\phi p_6/I_1$.

The rotor inertia receives effort information: $e_6 = \dot{p}_6$. This gives
$\dot{p}_6 = e_6 = e_5 - e_7 - e_8 = e_5 - e_7 - e_9 = K\phi p_2/L_a - R_1 p_6/I_1 - kq_9$.

The shaft compliance receives flow information: $f_9 = \dot{q}_9$. This gives
$\dot{q}_9 = f_9 = f_8 - f_{10} = f_6 - f_{11} = p_6/I_1 - p_{11}/I_2$.

The load inertia receives effort information: $e_{11} = \dot{p}_{11}$. This gives
$\dot{p}_{11} = e_{11} = e_{10} - e_{12} = e_9 - e_{12} = kq_9 - R_2 p_{11}/I_2$.

These four are the differential equations of the system. ◄

6.12 Alternative Methods of Creating System Bond Graphs

So far we have gone by logic in creating system bond graphs – looking for elements (or subsystems) that share the same flow or the same effort. This method works well in simple systems where the junction structures are clearly evident in the system description. However, in some systems the relationships between elements may not be so salient, and it may become difficult to create the bond graph of such a system. There are alternative methods suited to electrical and mechanical systems that enable one to bring the hidden relationships to the fore.

6.12.1 Electrical systems

For electrical systems a straightforward method is to identify the *nodes* or points that are at different voltages. The terminals of all elements connected to a node must share the same voltage; and so a node may be assigned a **0** junction. All elements connected between two nodes share the same flow. Therefore, the elements can be connected to the nodes through **1** junctions.

The bond graph thus obtained will contain many unnecessary bonds. The next step is therefore to *reduce* it to the minimum number necessary. This is done by assuming one of the nodes to be at zero potential. This is represented by adding an *SE* element to that **0** junction, with the effort given by the *SE* element set to zero. As a result, the effort variable in all the bonds connected to that **0** junction would also be zero. Since power is the product of effort and flow, these bonds carry no power. Therefore, these bonds may be eliminated.

At this stage, there will be some junctions that have only two bonds connected. If the power directions in the two bonds are the same, the two bonds connected to a junction carry the same effort and flow information, and hence the junction serves no purpose. These "through" junctions may also be eliminated.

Finally, the resulting bond graph is suitably laid out in a handsome way.

▶ **Example 6.11** Consider the system in Fig. 6.25(a). In the first step, we identify the nodes. These are marked in the figure as points A, B, C and D. In the bond graph we put **0** junctions in the positions of the nodes. Then we connect all the elements between nodes by **1** junctions as shown in Fig. 6.25(b).

The next step is to reduce the bond graph. In this step, we first ground any one of the nodes (say node A), that is, assume that it is at zero potential. All the other node voltages are now measured in relation to this node. Grounding a node is equivalent to connecting a source of effort of zero magnitude, as shown in the bond graph of Fig. 6.25(c).

At bond number 2, the effort information is zero, and the **0** junction equalizes the efforts $e_2 = e_1 = e_3 = e_{12}$. Hence the bonds 1, 3 and 12 also carry zero effort, and hence zero power. Therefore, we can eliminate these bonds and reduce the bond graph as shown in Fig. 6.25(d).

Now we find that there are some junctions attached to only two bonds, and so the two bonds carry the same information (bonds 13 and 14, bonds 4 and 5, bonds 18 and 19, etc.). We can eliminate these "through" junctions also. Finally we rearrange the bond graph layout. This results in the bond graph in Fig. 6.26. ◀

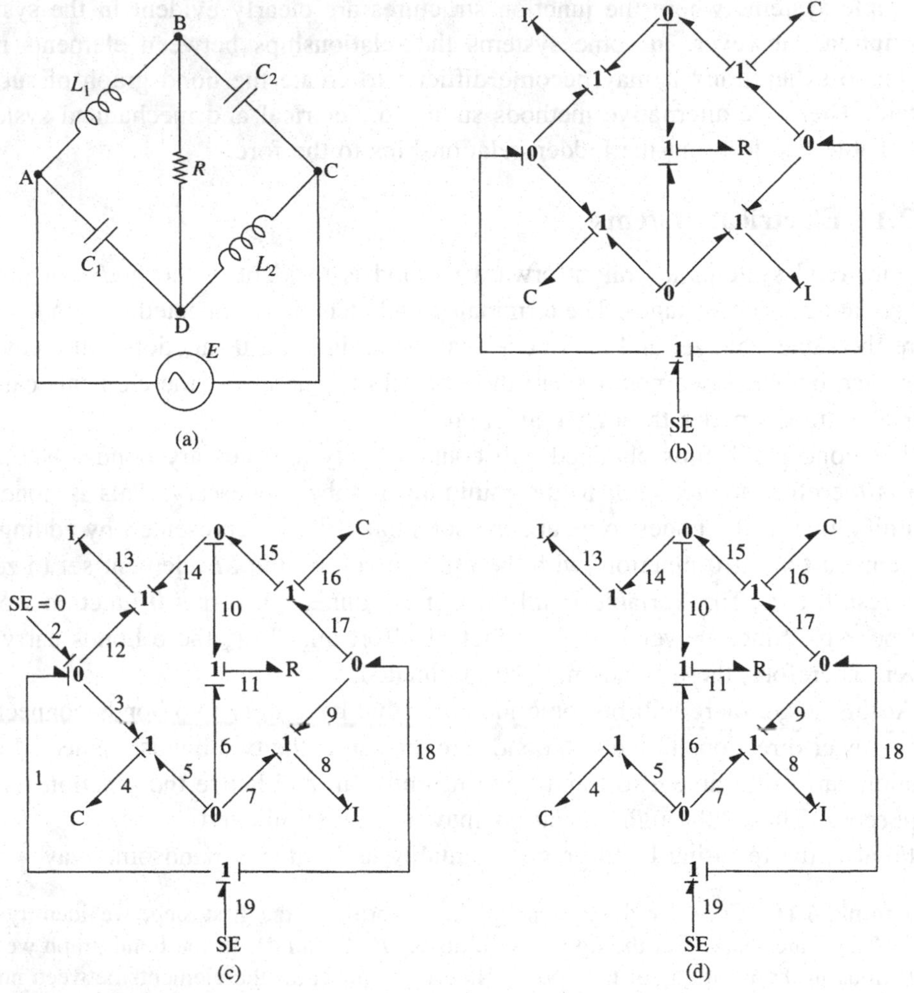

Figure 6.25 (a) The electrical circuit, (b) the first stage in creating the bond graph, (c) identification of the grounded node and (d) reduction of the bond graph.

6.12.2 Mechanical systems

For mechanical systems a convenient method is to identify the points that are at different velocities. All elements attached to these "velocity points" share the same flow. Therefore, the velocity points may be represented by **1** junctions. The masses should generally be at different velocities (otherwise there will be differential causality), and these may be connected directly to the appropriate **1** junctions. The

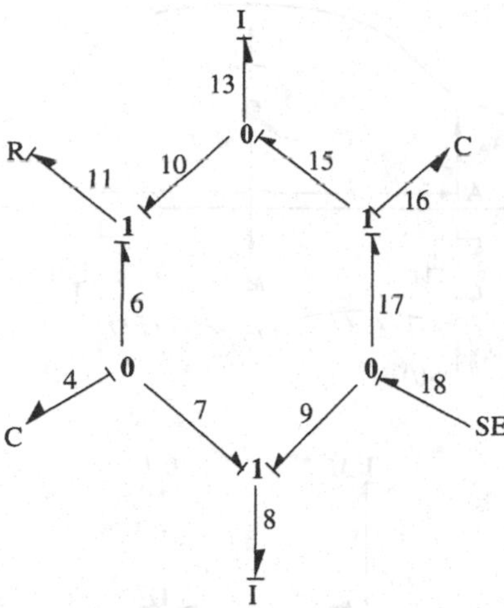

Figure 6.26 The final form of the bond graph of the system in Fig. 6.25.

extremities of the *C* and *R* elements share the same effort. Hence, the *C* and *R* elements may be connected through **0** junctions. The rest of the bond graph may be connected logically.

If there is a point in the system at zero velocity – as in an anchored point – it offers scope for reduction of the bond graph. In some cases, a velocity point can be identified as the "reference" with respect to which the other velocities are measured. These situations are represented by adding a *SF* element to that **1** junction, and setting its value equal to zero. The further reduction of the bond graph follows the same procedure as outlined above.

► **Example 6.12** Consider the mechanical system in Fig. 6.27. This is actually a simplified representation of a car manoeuvering a bumpy road.

We identify the following velocity points in this system. The two points where the sources of flow are applied and the two points of contact of the suspensions with the mass are easily identifiable as distinct velocity points. Since mass has two types of motions – translational and rotational – these are also two velocity points. So we first locate the six **1** junctions.

The springs and dampers are connected through **0** junctions between the appropriate **1** junctions. Forces are applied on the mass from below, through the points of contact with the suspensions. These forces, available as the effort variables at the **1** junctions, cause both translational and rotational motion. The distribution of the efforts is accomplished by two **0** junctions. We now have to decide how these efforts are applied on the vertical motion and the rocking motion, represented by the two **1** junctions. The efforts coming from the two suspensions are simply added up to yield the total force causing the up and down motion.

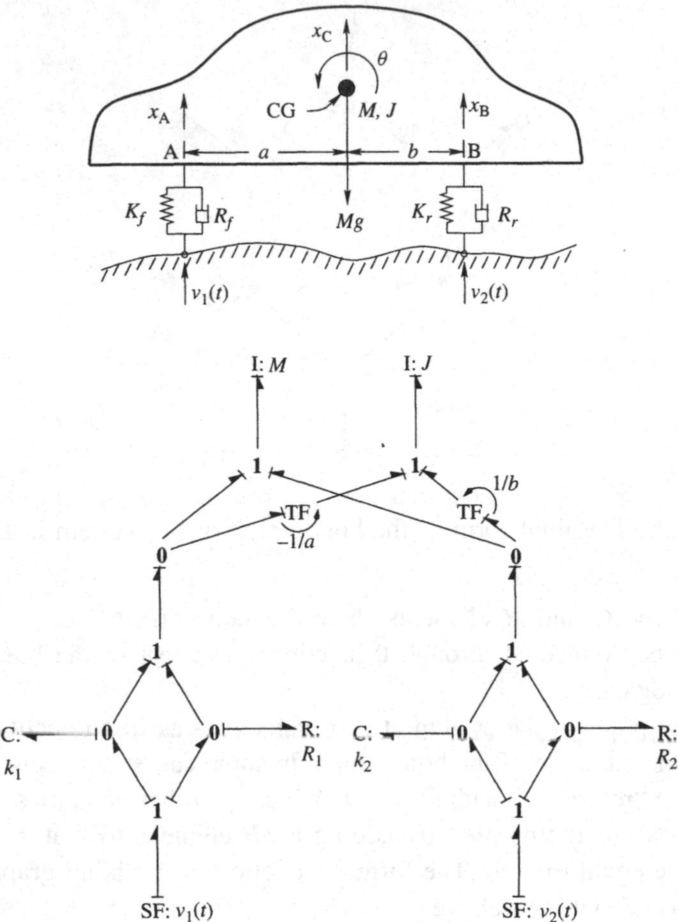

Figure 6.27 The mechanical system of Example 6.12 and its bond graph.

So two bonds from the **0** junctions go straight to the **1** junction to which the inertial element M is connected. The forces translate into the moments by factors $-\frac{1}{a}$ and $\frac{1}{b}$ respectively, causing the rotational motion. So the **0** junctions are connected to the **1** junction representing rotational motion through transformers with the above transformation ratios. The inertial element J is connected to this junction. This completes the bond graph. ◀

6.13 Algebraic Loops

In some systems, while assigning the causalities, one may encounter a problem of ambiguity of causal structure. Take, for example, the electrical circuit shown in Fig. 6.28.

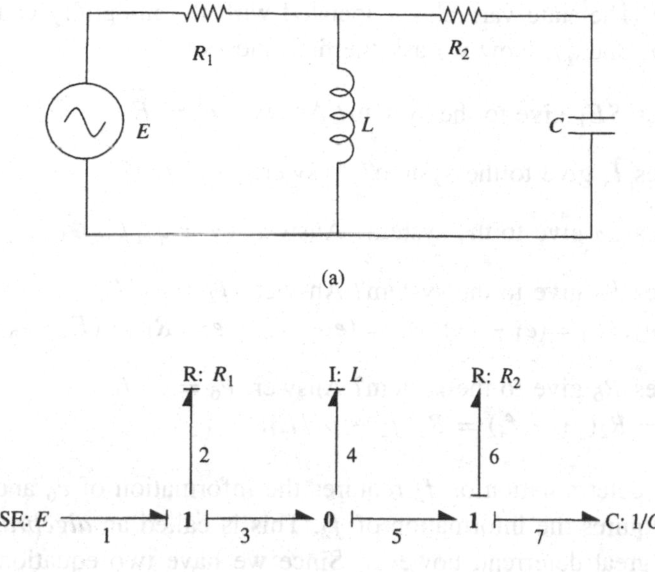

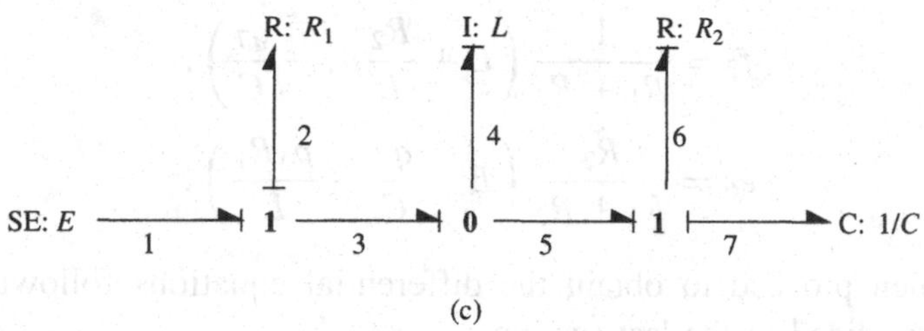

Figure 6.28 The electrical circuit and its bond graph, causaled in two possible ways.

Notice that after the sources and storage elements are integrally causaled, none of the junctions has a strong bond. Therefore, to complete the causal structure, we have to arbitrarily assign a causality to one of the resistances. This provides the necessary strong bond and the rest of the bond graph can be causaled. But if we had assigned the opposite causality to the resistance, the causalities of the rest of the bonds would also be reversed – leading to a completely different causal structure. The two possible ways of causaling the bond graph are shown in Fig. 6.28(b) and (c).

Though both these causal structures are valid representation of the system, the ambiguity gives rise to a typical problem in deriving the system equations. Let us illustrate this by deriving the equations of this system from the bond graph of Fig. 6.28(b). The state variables associated with the integrally causaled storage elements are p_4 and q_7. Now we ask the first question.

- What does SE_1 give to the system? Answer: $e_1 = E$.
- What does I_4 give to the system? Answer: $f_4 = p_4/L$.
- What does C_7 give to the system? Answer: $e_7 = q_7/C$.
- What does R_2 give to the system? Answer: $f_2 = e_2/R_1$ $= (e_1 - e_3)/R_1 = (e_1 - e_5)/R_1 = (e_1 - e_6 - e_7)/R_1 = (E - e_6 - q_7/C)/R_1$.
- What does R_6 give to the system? Answer: $e_6 = R_2 f_6$ $= R_2 f_5 = R_2(f_3 - f_4) = R_2(f_2 - p_4/L)$.

Notice that determination of f_2 requires the information of e_6 and the determination of e_6 requires the information of f_2. This is called an *algebraic loop*.

This is no great deterrent, however. Since we have two equations for the two unknowns, we can easily solve the equations to obtain these two quantities. In this problem, we obtain

$$f_2 = \frac{1}{R_1 + R_2}\left(E + \frac{R_2}{L}p_4 - \frac{q_7}{C}\right),$$

$$e_6 = \frac{R_2}{R_1 + R_2}\left(E - \frac{q_7}{C} - \frac{p_4 R_1}{L}\right).$$

We can then proceed to obtain the differential equations following the same procedure as outlined in the last section.

If three or more resistances are coupled in such an algebraic loop, it becomes more convenient to solve the resulting coupled algebraic equations by the matrix inversion method.

6.14 Fields

So far we have assumed that a variable (effort or flow) of an I, C or R element is related to the other variables of the same element. That is why these were said to be one-port elements.

There are situations where one variable of an element is determined by a linear combination of information from more than one bond. In such cases the element is called a *field* or a multi-port I, C or R element.

As an example, consider the electrical circuit in Fig. 6.29 containing a mutual inductance. The voltage induced in the inductor L_1 is dependent on the current in both the inductances, and so the inductor cannot be represented by a one-port

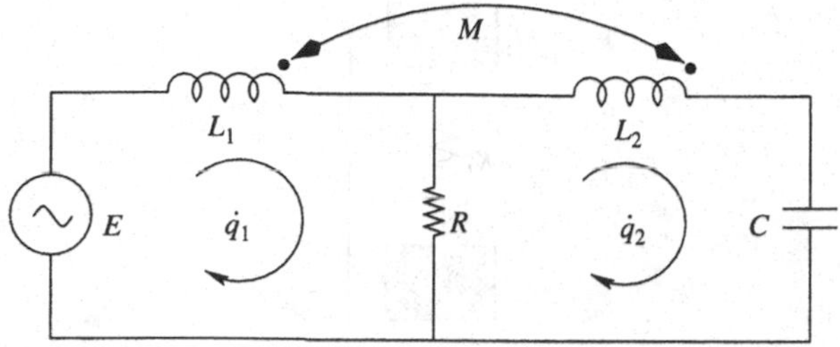

Figure 6.29 The electrical circuit with a mutual inductance.

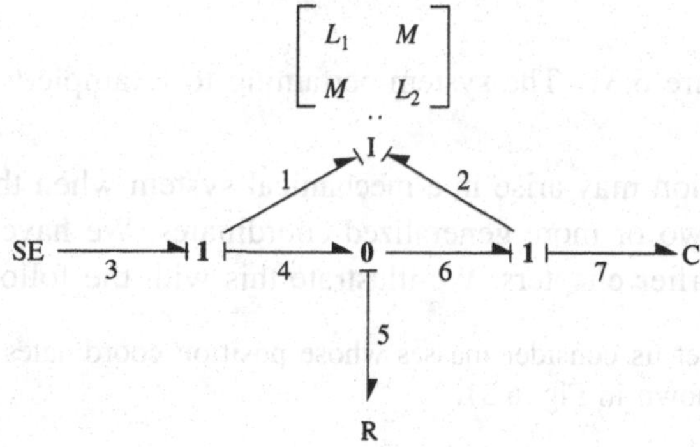

Figure 6.30 The bond graph of the system in Fig. 6.29.

element. To obtain the relationships between the flows and the momenta, we proceed via the kinetic energy:

$$T = \frac{1}{2}L_1\dot{q}_1^2 + \frac{1}{2}L_2\dot{q}_2^2 + M\dot{q}_1\dot{q}_2.$$

Therefore,

$$p_1 = \left(\frac{\partial \mathcal{L}}{\partial \dot{q}_1}\right) = \left(\frac{\partial T}{\partial \dot{q}_1}\right) = L_1\dot{q}_1 + M\dot{q}_2, \tag{6.1}$$

$$p_2 = \left(\frac{\partial \mathcal{L}}{\partial \dot{q}_2}\right) = \left(\frac{\partial T}{\partial \dot{q}_2}\right) = L_2\dot{q}_2 + M\dot{q}_1 \tag{6.2}$$

or, in matrix form,

$$\begin{pmatrix} p_1 \\ p_2 \end{pmatrix} = \begin{pmatrix} L_1 & M \\ M & L_2 \end{pmatrix} \begin{pmatrix} \dot{q}_1 \\ \dot{q}_2 \end{pmatrix}.$$

This relationship is represented by a two-port I-field as shown in the bond graph of Fig. 6.30.

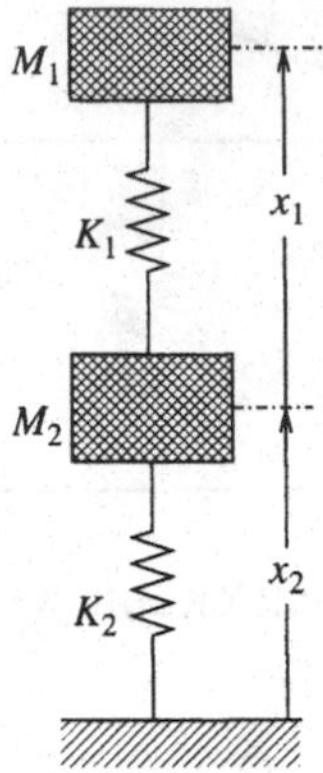

Figure 6.31 The system pertaining to Example 6.13.

A similar situation may arise in a mechanical system when the momentum of a mass is given by two or more generalized coordinates. We have seen examples of such systems in earlier chapters. We illustrate this with the following example.

► **Example 6.13** Let us consider masses whose position coordinates are defined relative to one another, as shown in Fig. 6.31.

In this case

$$T = \frac{1}{2}m_1(\dot{x}_1 + \dot{x}_2)^2 + \frac{1}{2}m_2\dot{x}_2^2.$$

The generalized momenta are:

$$p_{x1} = \left(\frac{\partial T}{\partial \dot{x}_1}\right) = m_1(\dot{x}_1 + \dot{x}_2),$$

$$p_{x2} = \left(\frac{\partial T}{\partial \dot{x}_2}\right) = m_1(\dot{x}_1 + \dot{x}_2) + m_2\dot{x}_2.$$

The momenta are therefore obtained as a linear combination of both the variables. ◄

To generalize, an element is represented by an I-field if the momenta in a set of bonds are related to the flows in the same set of bonds by the relation

$$p_i = \sum_{j=1}^{n} I_{ij} f_j.$$

Likewise, the C-field is one where the effort variables in a set of bonds are related to the position coordinates in the same set of bonds by

$$e_i = \sum_{j=1}^{n} k_{ij} q_j,$$

and the R-field is one where the effort and flow variables in a set of bonds are related by

$$e_i = \sum_{j=1}^{n} R_{ij} f_j.$$

C-fields occur in beam vibration problems and other areas involving cross-coupled springs. R-fields are common in representing analog electronic devices. In representing a transistor, we need a R-field because there are two effort variables and two flow variables associated with it sharing a resistance-type relationship (in the common-emitter configuration, the effort variables are the base-emitter voltage and the collector-emitter voltage, while the flow variables are the base current and the collector current).

In obtaining the differential equations in a system involving fields, we proceed in the same way as outlined in the last section.

► **Example 6.14** Let us obtain the equations for the system given in Fig. 6.30. Here the state variables are p_1, p_2 and q_7. The answer to the first question give the algebraic relationships:

$$e_3 = E,$$

$$f_1 = \frac{L_2 p_1 - M p_2}{L_1 L_2 - M^2}, \qquad \text{[from (6.1) and (6.2)]}$$

$$f_2 = \frac{M p_1 - L_1 p_2}{M^2 - L_1 L_2}, \qquad \text{[do]}$$

$$e_7 = \frac{q_7}{C},$$

$$e_5 = R f_5 = R(f_4 - f_6) = R(f_1 - f_2) = R\left(\frac{L_2 p_1 - M p_2}{L_1 L_2 - M^2} - \frac{M p_1 - L_1 p_2}{M^2 - L_1 L_2}\right),$$

$$= R\left(\frac{L_2 p_1 - L_1 p_2 + M p_1 - M p_2}{L_1 L_2 - M^2}\right).$$

The answers to the second question give the differential equations:

$$\dot{q}_7 = f_7 = f_2 = \frac{M p_1 - L_1 p_2}{M^2 - L_1 L_2},$$

$$\dot{p}_1 = e_1 = e_3 - e_4 = e_3 - e_5 = E - R\left(\frac{L_2 p_1 - L_1 p_2 + M p_1 - M p_2}{L_1 L_2 - M^2}\right),$$

$$\dot{p}_2 = e_2 = e_6 - e_7 = e_5 - e_7 = R\left(\frac{L_2 p_1 - L_1 p_2 + M p_1 - M p_2}{L_1 L_2 - M^2}\right) - \frac{q_7}{C}.$$ ◄

6.15 Activation

So far we have presented bond graphs where all the bonds carry power. However, in practical systems there often arises the necessity to represent bonds that carry only a certain information and no power.

Instrumentation used in a system only *observe* certain information variables – for example, an ammeter observes the flow information in a certain branch of an electrical circuit. By definition an ideal observation instrument should not affect the dynamics of a system, and so the bond representing such an element should not impart any power to the rest of the system. And if some part of the system is to be controlled depending on such an information variable, as is often the case in feedback controlled systems, one needs to use bonds that carry only information and no power. Such bonds are called *activated bonds* and are specified by a full arrow.

Bonds can be either flow activated or effort activated. A flow activated bond carries only flow information, without taking cognizance of the effort variable. An effort activated bond disregards the flow information and carries only the effort information to the other end of the bond. One needs to specify which variable is activated in a bond, and this is done by writing either e or f next to the full arrow. In Fig. 6.32(a) the flow information goes from the junction to the element and in Fig. 6.32(b) the effort information goes from the junction to the element. Note that the causalities are given accordingly.[1]

It is to be noted that the natural state variable associated with an I element is the momentum, and that associated with a C element is its charge (or position). Therefore, in a bond graph, the information of the *position* of a mass or the *current* through a capacitance is not available. However, often we do need such information and place instruments or observers of such information. This can be accomplished by adding a suitable element with an activated bond.

For example, suppose we want to observe the *position* of the mass in Fig. 6.7. This information is not normally available from the system equations obtained from the bond graph. So, in order to observe this variable, we can add an activated bond to the **1** junction and connect a dummy C element to it, as shown in Fig. 6.33(b).

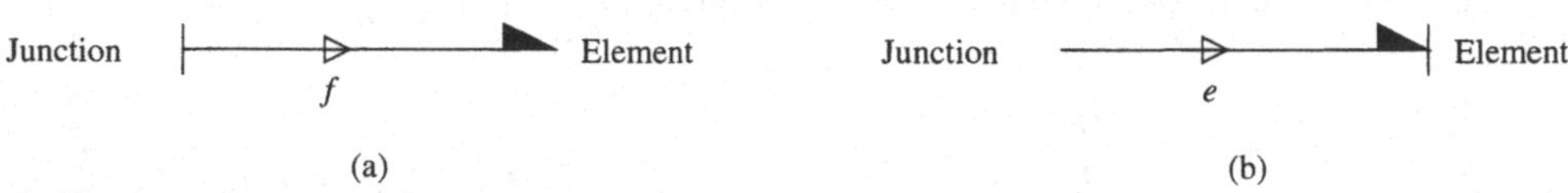

Figure 6.32 Representations of activated bonds.

[1] Some authors follow the opposite convention, that is, if the letter e is placed next to the full arrow, it would mean that the effort information is masked and the flow information is carried by the bond. In this book, we will not follow that convention.

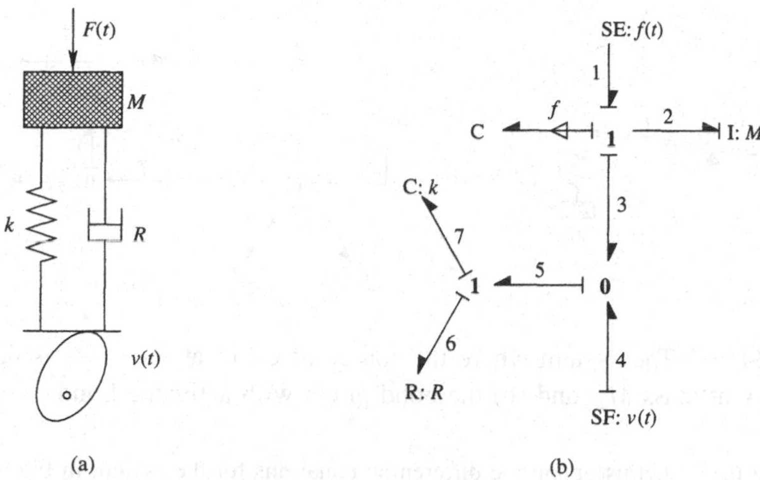

Figure 6.33 (a) The system requiring a observer for the position of the mass, and (b) the bond graph with an activated bond with a C element to do the job.

Since it is an activated bond, its effort information is zero, and so this bond does not carry any power and does not affect the dynamics of the rest of the system. It only makes the position variable available in the system of equations.

Similarly, if it is required to observe the current through a capacitor in an electrical circuit or the flow at a spring in a mechanical system, one can add an activated bond with a dummy I element connected to it.

Often it is necessary to control a system depending on the value of such an observed variable. This can also be done with an activated bond.

► **Example 6.15** Referring to the system in Fig. 6.13, suppose we want to control the force F depending on the velocity of the mass M_2. The feedback loop is schematically shown with a broken line in Fig. 6.34(a). In this case, the flow information of the mass M_2 is available at the **1** junction, and so an activated bond connected to this junction will take that information. Since the force on mass M_1 is determined by the flow at mass M_2, we need to convert the flow information into effort information. This is done by the gyrator element, and the output of this gyrator is connected in place of the SE element (Fig. 6.34(b)). ◄

This case illustrates that the activated bonds are not power conserving. Bond 9 does not carry any power as the effort information is zero. But at bond 10, the effort information is given by αf_9, and the flow information at this bond is determined by the rest of the system.

Obtaining system equations: One follows the same method as outlined earlier when obtaining bond graphs of systems with activated bonds. Only, one has to remember which information is masked in an activated bond.

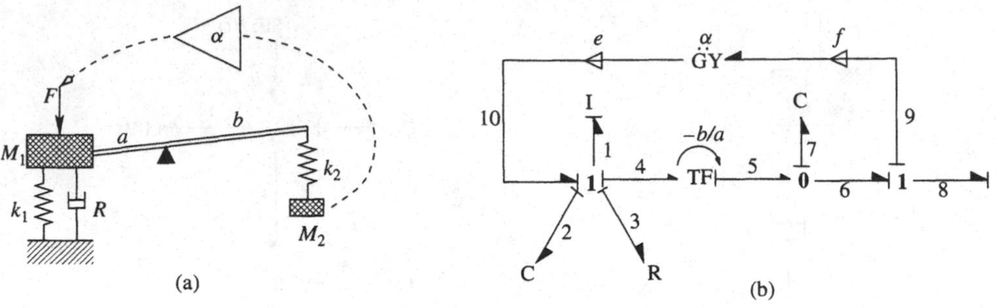

Figure 6.34 (a) The system where the source of effort at mass M_1 is determined by the flow at mass M_2, and (b) the bond graph with activated bonds.

► **Example 6.16** Let us obtain the differential equations for the system in Fig. 6.34. Here the state variables are p_1, q_2, q_7, and p_8. We ask the first question and get the answers:

$$
\begin{aligned}
f_1 &= \frac{p_1}{M_1},\\
e_2 &= k_1 q_2,\\
e_7 &= k_2 q_7,\\
f_8 &= \frac{p_8}{M_2},\\
e_3 &= R f_3 = R f_1 = R\frac{p_1}{M_1},\\
e_4 &= -\frac{b}{a}e_5 = -\frac{b}{a}e_7 = -\frac{b}{a}k_2 q_7,\\
f_5 &= -\frac{b}{a}f_4 = -\frac{b}{a}f_1 = -\frac{b}{a}\frac{p_1}{M_1},\\
e_{10} &= \alpha f_9 = \alpha f_8 = \alpha\frac{p_8}{M_2}.
\end{aligned}
$$

The answers to the second question give

$$
\begin{aligned}
\dot{p}_1 &= e_1 = e_{10} - e_2 - e_3 - e_4 = \alpha\frac{p_8}{M_2} - k_1 q_2 - R\frac{p_1}{M_1} + \frac{b}{a}k_2 q_7,\\
\dot{q}_2 &= f_2 = f_1 = \frac{p_1}{M_1},\\
\dot{q}_7 &= f_7 = f_5 - f_6 = f_5 - f_8 = -\frac{b}{a}\frac{p_1}{M_1} - \frac{p_8}{M_2},\\
\dot{p}_8 &= e_8 = e_6 \qquad \text{(since } e_9 = 0\text{)},\\
&= e_7 = k_2 q_7.
\end{aligned}
$$

These are the differential equations of the system. ◄

6.16 Equations for Systems with Differential Causality

We have said earlier that differential causality can often be avoided by suitably specifying the system so that all the elements have freedom in deciding one of the information variables in the bond connected to it. However, in some rare cases this may not be possible. For example, in the system of Fig. 6.21 if some flexibility of the shaft is assumed, the stiffness will have to be very high. Therefore, there will be a wide disparity between the two compliance elements of the system – one will have a fast swing and the other will have a slow swing. The resulting differential equations in such systems, called *stiff systems*, are rather difficult to solve by numerical methods.

In such situations, we may be forced to obtain the differential equations from the differentially causaled bond graph. We illustrate the method in the following example.

► **Example 6.17** Let us obtain the differential equations for the system in Fig. 6.21. The bond graph of the system is reproduced in Fig. 6.35.

The state variables of this system are p_1 and q_3. Note that I_7 is differentially causaled and so does not generate a state variable.

Answers to the first question are obtained as

$$e_2 = F(t),$$

$$f_1 = p_1/M_1,$$

$$e_3 = kq_3,$$

$$e_4 = Rf_3 = Rf_1 = Rp_1/M_1,$$

$$e_5 = -\frac{b}{a}e_6 = -\frac{b}{a}e_7,$$

$$f_6 = -\frac{b}{a}f_5 = -\frac{b}{a}f_1 = -\frac{b}{a}p_1/M_1.$$

Now we come to a stage where we have to ask that question to the differentially causaled I element. It gives e_7, and expressing this in terms of the state variables needs a

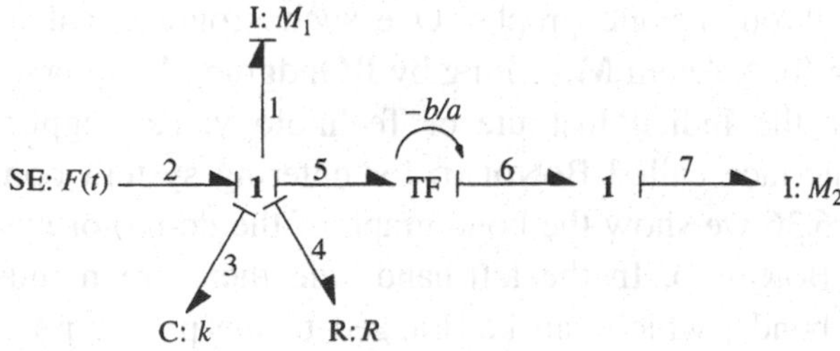

Figure 6.35 The bond graph of the system in Fig. 6.21.

different treatment:

$$\begin{aligned}
e_7 &= M_2 \dot{f}_7, \\
&= -M_2 \frac{b}{a} \dot{f}_5, \\
&= -M_2 \frac{b}{a} \dot{f}_1, \\
&= -M_2 \frac{b}{a} \dot{p}_1 / M_1, \\
&= -\frac{bM_2}{aM_1} e_1, \\
&= -\frac{bM_2}{aM_1}(e_2 - e_3 - e_4 - e_5), \\
&= -\frac{bM_2}{aM_1}\left(F(t) - kq_3 - \frac{Rp_1}{M_1} + \frac{b}{a} e_7\right).
\end{aligned}$$

Notice that e_7 appears both in the LHS and RHS. Solving, we get

$$e_7 = -\frac{abM_2}{a^2 M_1 + b^2 M_2}\left(F(t) - kq_3 - \frac{Rp_1}{M_1}\right).$$

Now we ask the second question to the storage elements with integral causality:

$$\begin{aligned}
\dot{q}_3 &= f_3 = f_1 = p_1/M_1, \\
\dot{p}_1 &= e_1 = e_2 - e_3 - e_4 - e_5 = F(t) - kq_3 - \frac{Rp_1}{M_1} + \frac{b}{a} e_7, \\
&= F(t) - kq_3 - \frac{Rp_1}{M_1} - \frac{b^2 M_2}{a^2 M_1 + b^2 M_2}\left(F(t) - kq_3 - \frac{Rp_1}{M_1}\right).
\end{aligned}$$

◀

6.17 Bond Graph Software

There are many types of software for obtaining the differential equations of complicated systems through bond graphs. One such program, called SYMBOLS (The acronym standing for **SY**stem **M**odelling by **BO**ndgraph **L**anguage and **S**imulation) was developed at the Indian Institute of Technology, Kharagpur, India. It has a graphical user interface called BONDPAD for entering system bond graphs. As an example, in Fig. 6.36 we show the bond graph of the dc-motor system of Fig. 6.14 entered through BONDPAD. In the left-hand side there are menus showing icons of elements and bonds, which can be dragged to the proper position to draw the bond graph. There are icons at the top of the screen with which the numbers, power directions and causalities can be automatically assigned. The user can change these,

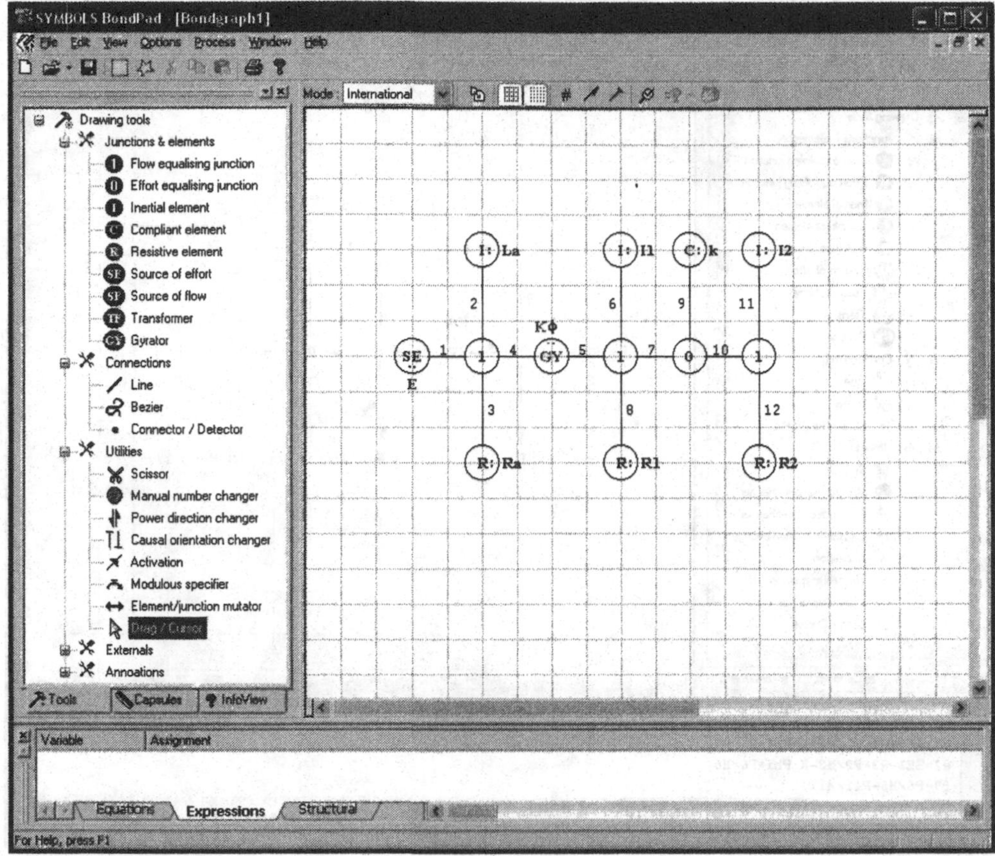

Figure 6.36 The graphical user interface called BONDPAD, with which a bond graph of a system can be communicated to a computer.

if necessary. There is another icon that generates the differential equations at the click of a mouse button. The equations are displayed in the area at the bottom of the screen (Fig. 6.37).

Then the user has to give the parameter values and the initial conditions. The program then solves the differential equations, and plots the graphs. Many other analyses and control functions can also be performed.

The program is free and can be downloaded from the website http://www.symbols2000.com/. The program is sufficient for the complexity of systems presented in this book. A professional version that can handle industrial-size systems, is however priced.

The reader should get himself/herself acquainted with at least one of the programs with which modelling and simulation of dynamical systems can be done through the language of bond graphs.

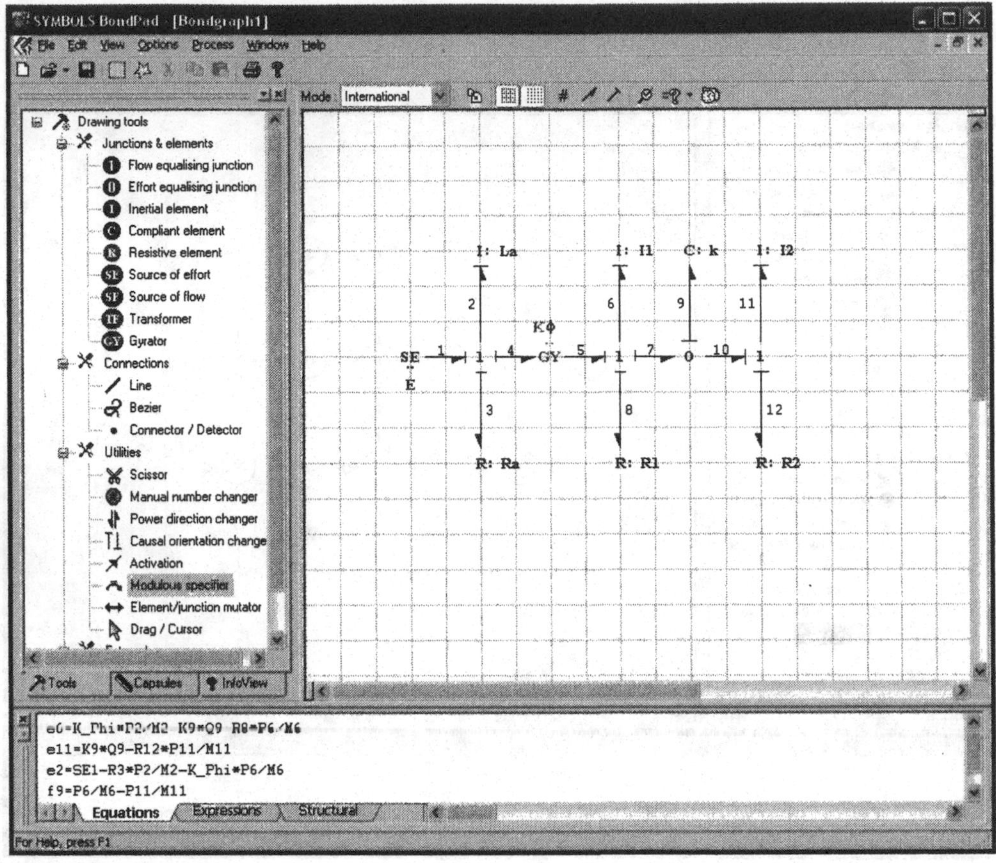

Figure 6.37 The final form of the bond graph and the generated equations in SYMBOLS.

6.18 Chapter Summary

Bond graph is a systematic graphical representation of any dynamical system, capturing the interaction of the various elements through exchange of energy and information. This representation allows one to obtain the first-order differential equations following an algorithmic procedure. The procedure is so algorithmic that even computers can obtain the system equations. Many computer programs have been written to do this job. This becomes particularly convenient in obtaining models of relatively complicated systems, in which deriving the equation by hand becomes a cumbersome exercise.

Further Reading

D. C. Karnopp, D. L. Margolis, and R. C. Rosenberg, *System Dynamics: A Unified Approach*, 2nd Ed., Wiley, New York, 1990.

A. Mukherjee and R. Karmakar, *Modeling and Simulation of Engineering Systems Through Bondgraphs*, CRC Press, Boca Raton, Florida, 1999.

Problems

1. Obtain the bond graph for the following systems.

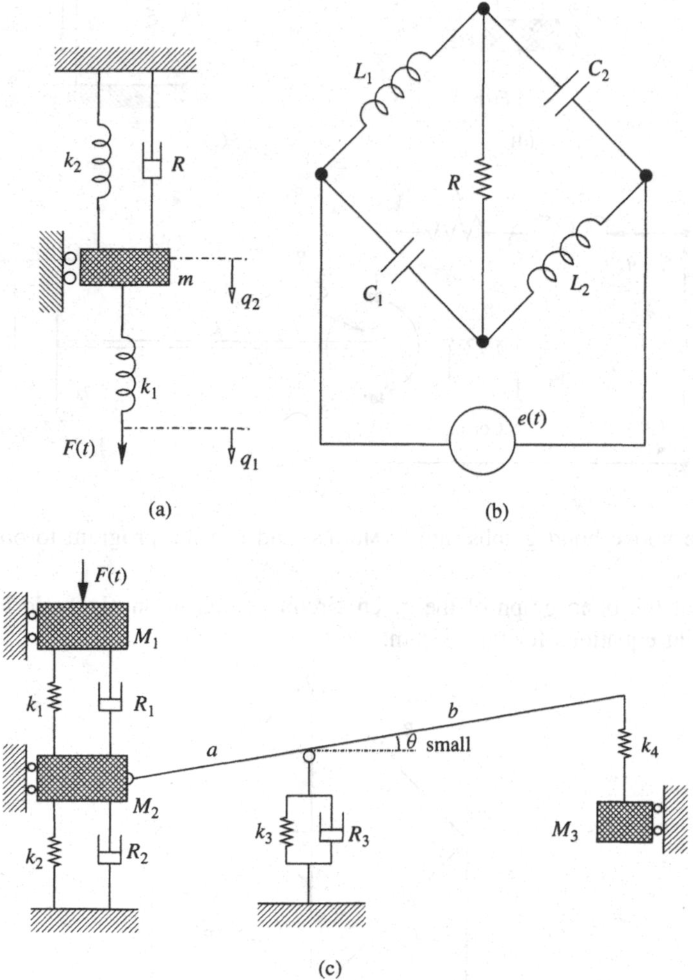

(a) (b)

(c)

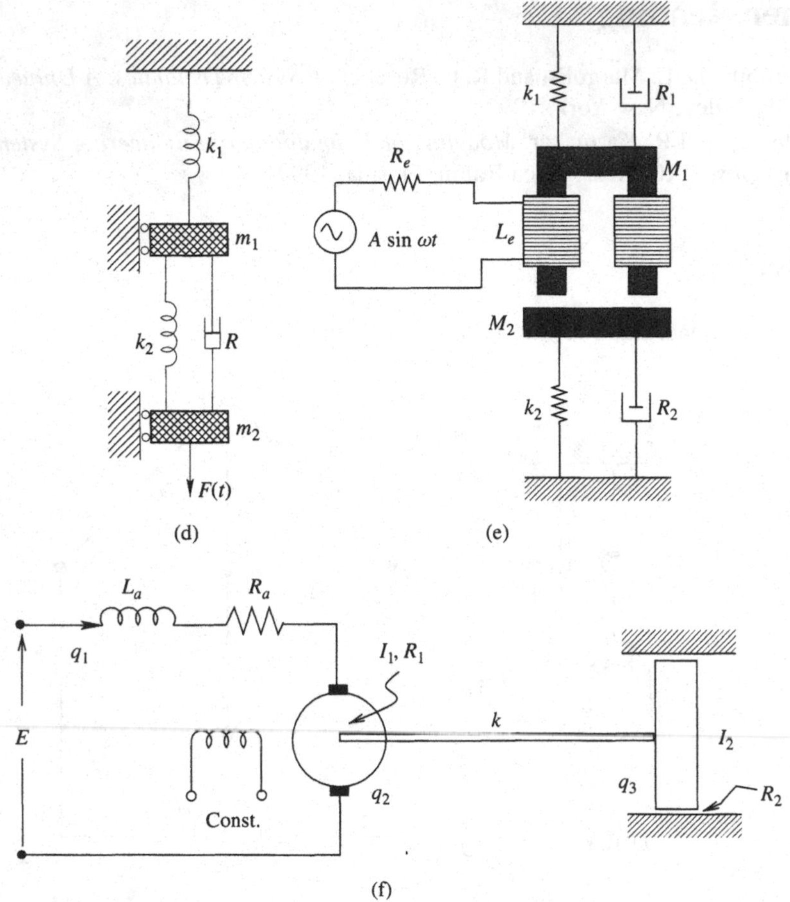

2. Enter the above bond graphs into SYMBOLS, and use the program to obtain the system equations.
3. Show that the bond graph of the given circuit results in an algebraic loop. Obtain the differential equations for this system.

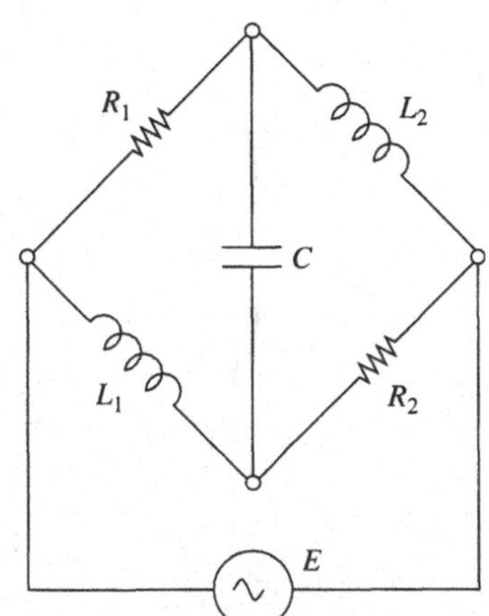

4. Show that the following circuits have at least one differentially causaled element. What causes the differential causality in these systems?

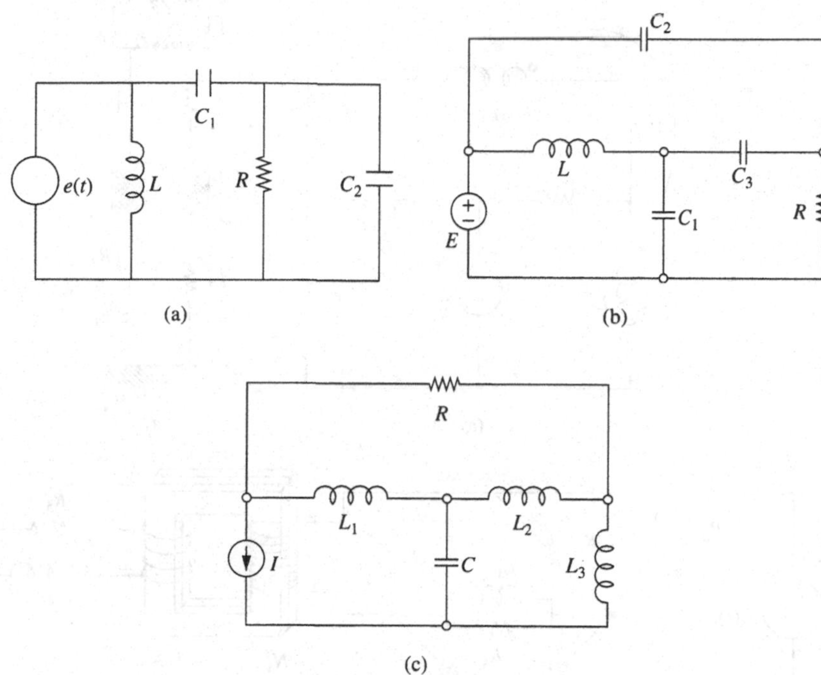

5. Obtain the bond graph and show the activated bonds in the following system. From the bond graph, obtain the differential equations.

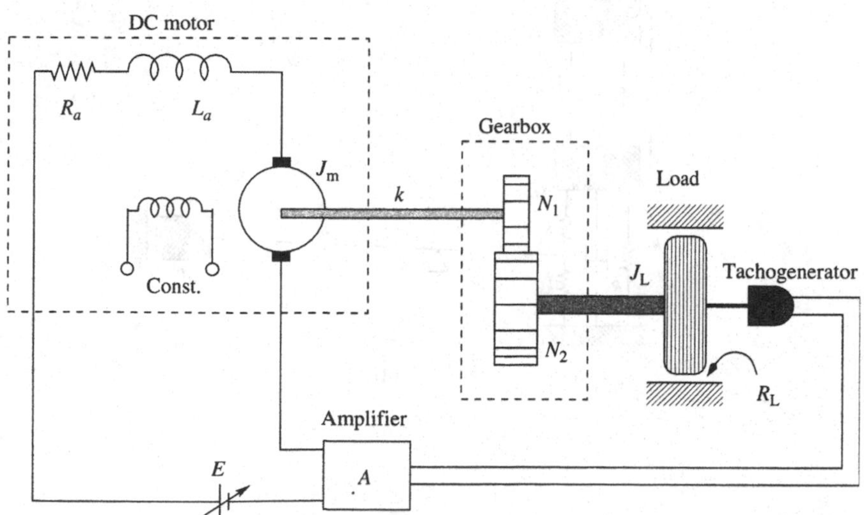

6. Obtain the bond graphs, and the system of differential equations for the following systems.

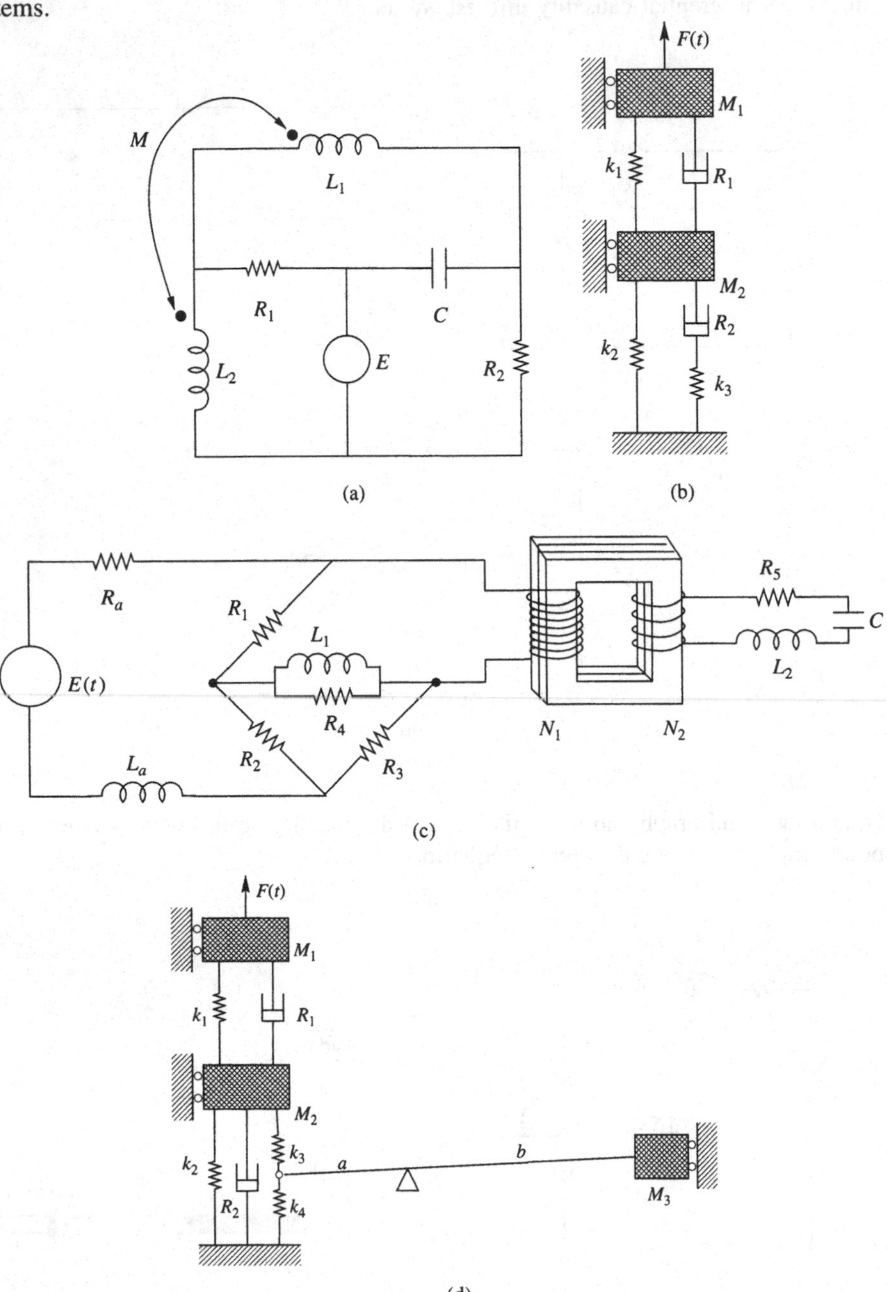

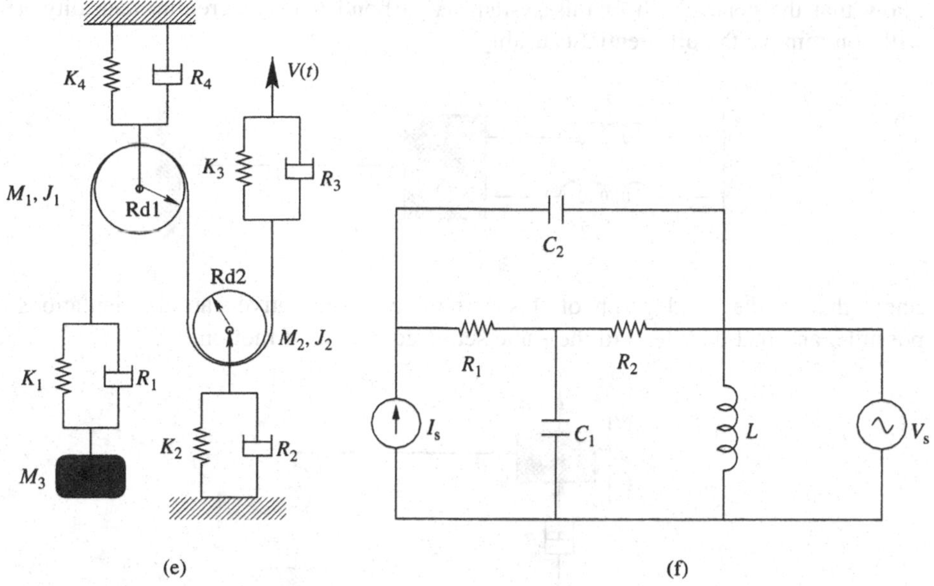

7. In this problem, the cylinders contain incompressible fluid that can pass from one cylinder to the other. Other than the two masses shown, all components are massless. Draw the complete bond graph.

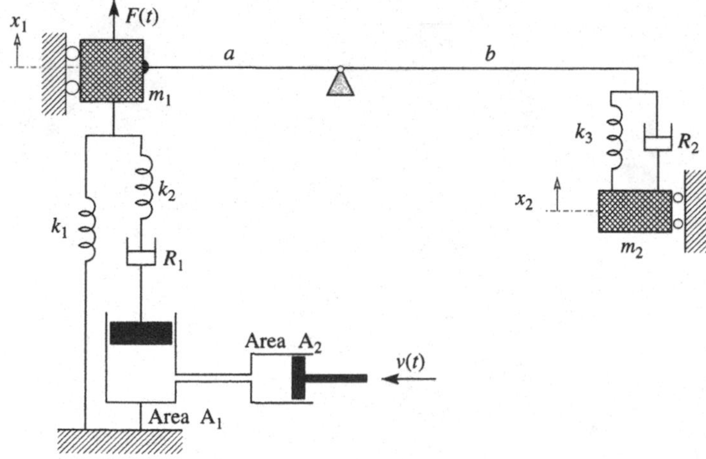

8. Show that the bond graph of this system has a bond with differential causality. How will you remove the differential causality?

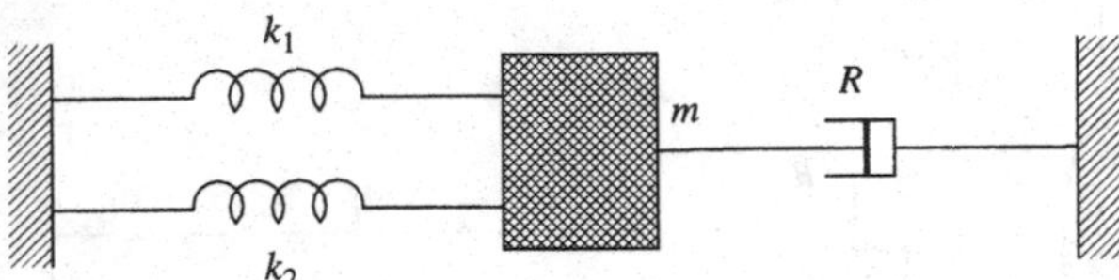

9. Show that in the bond graph of this system, two alternative causal orientations are possible, and that they lead to the same set of equations of motion.

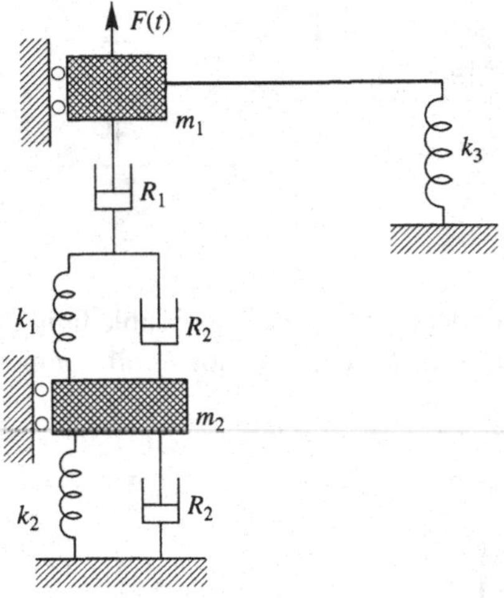

Part II

Solving differential equations and understanding dynamics

We have seen that in studying a dynamical system, one identifies a minimum number of variables that uniquely specify the state of the system. These are the *state variables*. The study of dynamics is essentially an investigation of how these state variables change with time. Mathematically, this is done by relating the rate of change of these state variables to their current values, via a system of first-order differential equations $\dot{\mathbf{x}} = f(\mathbf{x})$. In Part I, we have introduced the basic methodologies of obtaining the differential equations of physical systems.

Once the state-variable equations for a system are obtained, the next step is to solve the set of equations. In Part II, we will learn how to do it. In that process, we will develop an understanding of dynamics in linear as well as nonlinear systems through a geometric view of the state space.

Dynamics for Engineers S. Banerjee

7

Numerical Solution of Differential Equations

In Part I, we have learnt how to obtain differential equations of physical and engineering systems. If $\{x_i, i = 1 \text{ to } n\}$ are the state variables, the system of n first-order equations is obtained in the form

$$\dot{x}_i = f_i(x_1, x_2, \ldots, x_n) \tag{7.1}$$

or, in vector form,

$$\dot{\mathbf{x}} = f(\mathbf{x}). \tag{7.2}$$

In the next step, given the initial set of values of the state variables, we would like to find out how the state variables change with time. That is, we would like to solve the set of differential equations with a given initial condition to obtain the values of the state variables at future times.

In some special cases, particularly if the right-hand side of (7.2) is linear, it may be possible to solve the differential equations analytically – as we shall see in Chapter 9. However, in most of the cases of realistic systems, it is not possible to obtain a closed form solution giving the states of the system as functions of time. In such cases, the solution has to be obtained numerically.

7.1 The Basic Method and the Techniques of Approximation

Most of the books on numerical analysis give extensive treatment of the various methods of obtaining solution of ordinary differential equations. In this section, we

Dynamics for Engineers S. Banerjee

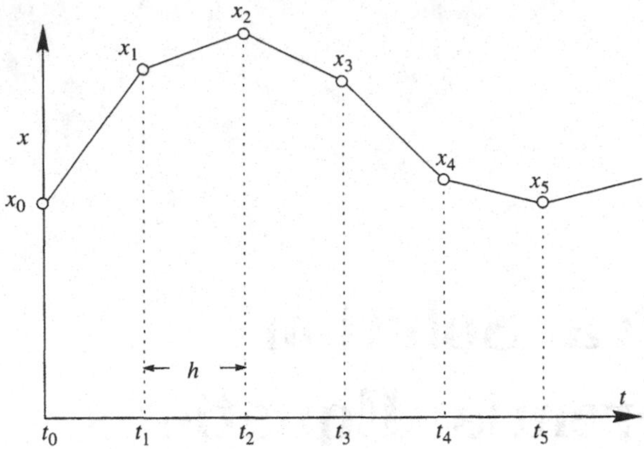

Figure 7.1 Illustration of numerical solution of the differential equation in steps of $t = h$.

shall briefly discuss the methods without going into the technical detail. The reader may consult any book devoted to numerical analysis for the proofs of the statements and other details.

We consider a general first-order differential equation given in the form

$$\dot{x} = f(x, t), \tag{7.3}$$

with the initial condition

$$x(t_0) = x_0.$$

In the numerical method, we seek to obtain the solution in steps at points $x_r = x_0 + rh$ where $r = 1, 2, 3 \ldots$ (see Fig. 7.1 for illustration).

How do we obtain the value of x after the first step? Ideally, the value should be obtained by integration of (7.3):

$$x_1 = x_0 + \int_{t_0}^{t_1} f(x, t)\,\mathrm{d}t. \tag{7.4}$$

But in this case, the numerical value of the second term may not be exactly known, and we have to make some approximation to obtain x_1. The various methods of numerical solution of differential equations actually differ in the way this approximation is done.

7.1.1 The Euler method

As a start, assume that the function remains constant through the interval $[t_0, t_1]$, that is, $f(x, t) = f(x_0, t_0)$ for $t_0 \leq t \leq t_1$. This gives $x_1 \approx x_0 + hf(x_0, t_0)$. Similarly, in the range $t_1 \leq t \leq t_2$ we have $x_2 \approx x_1 + hf(x_1, t_1)$. Proceeding this way, we obtain, in general,

$$x_{n+1} \approx x_n + hf(x_n, t_n).$$

This is known as the *Euler method*.

► **Example 7.1** Let us solve the equation

$$\dot{x} = \sin x + x^2,$$

with the initial condition $x_0 = 1.0$ at $t = t_0$, and step length $h = 0.1$. The first three steps will be:

$$x_1 = 1 + 0.1\left(\sin 1 + 1^2\right) = 1.184,$$

$$x_2 = 1.184 + 0.1\left(\sin 1.184 + 1.184^2\right) = 1.4169,$$

$$x_3 = 1.4169 + 0.1\left(\sin 1.4169 + 1.4169^2\right) = 1.7166.$$ ◄

One drawback of this method is that every step incurs some error that tends to build up. Therefore, in order to achieve a reasonable accuracy, one has to take a very small value of h. This makes the computation very slow.

7.1.2 The trapezoidal rule

In order to reduce the error, instead of approximating $f(x, t)$ by $f(x_0, t_0)$, one can approximate $f(x, t)$ by $\frac{1}{2}[f(x_0, t_0) + f(x_1, t_1)]$. This is called the trapezoidal rule, which obtains

$$x_1 = x_0 + \frac{h}{2}\left[f(x_0, t_0) + f(x_1, t_1)\right].$$

But then, we need to assume some approximate value of x_1. A possible way of doing that is to take clue from the Euler method, that is, by assuming $x_1 = x_0 + hf(x_0, t_0)$. Substituting this approximate value of x_1 and $t_1 = t_0 + h$, we get

$$x_1 = x_0 + \frac{h}{2}[f_0 + f(x_0 + hf_0, t_0 + h)], \tag{7.5}$$

where $f_0 = f(x_0, t_0)$.

If we now set

$$k_1 = hf_0,$$

$$k_2 = hf(x_0 + k_1, t_0 + h),$$

then (7.5) becomes

$$x_1 = x_0 + \frac{1}{2}(k_1 + k_2).$$

This is also called the second-order Runge-Kutta formula. It can be shown that the error per step in this method is of the order of h^3, which is considerably smaller than that in the Euler's method.

For further iterates, one has to obtain k_1 and k_2 for each step, and x_{n+1} is calculated as

$$x_{n+1} = x_n + \frac{1}{2}(k_1 + k_2).$$

► **Example 7.2** Let us solve the same problem as Example 7.1 using the trapezoidal rule. In the first step, the values of k_1 and k_2 are calculated as

$$k_1 = 0.1\left(\sin 1 + 1^2\right) = 0.184,$$
$$k_2 = 0.1\left(\sin 1.184 + 1.184^2\right) = 0.2328.$$

This gives

$$x_1 = 1 + \frac{1}{2}(0.184 + 0.2328) = 1.2084.$$

In the next step, k_1 and k_2 are again calculated as

$$k_1 = 0.1\left(\sin 1.2084 + 1.2084^2\right) = 0.23952,$$
$$k_2 = 0.1\left(\sin 1.44792 + 1.44792^2\right) = 0.30889,$$

since $x_1 + k_1 = 1.44792$. With these values, we calculate

$$x_2 = 1.2084 + \frac{1}{2}(0.23952 + 0.30889) = 1.482605.$$

In the succeeding steps, the same procedure is repeated. ◄

7.1.3 The fourth-order Runge-Kutta formula

To obtain a greater degree of accuracy, one can approximate the integral in (7.4) by more accurate expressions of $f(x, t)$. This gives the Runge-Kutta formulae of higher orders. Of particular interest is the fourth-order Runge-Kutta formula

$$x_1 = x_0 + \frac{1}{6}(k_1 + 2k_2 + 2k_3 + k_4), \tag{7.6}$$

where

$$k_1 = hf(x_0, t_0),$$
$$k_2 = hf\left(x_0 + \frac{1}{2}k_1, t_0 + \frac{1}{2}h\right),$$
$$k_3 = hf\left(x_0 + \frac{1}{2}k_2, t_0 + \frac{1}{2}h\right),$$
$$k_4 = hf\left(x_0 + k_3, t_0 + h\right).$$

It can be shown that in this formula the error is of the order of h^5. The derivation of this formula is somewhat involved. The reader may consult some book dedicated to numerical analysis for the derivation.

► **Example 7.3** We show the calculation of x_1 by the fourth-order Runge-Kutta method for the same problem as in Example 7.1. The k's are calculated as

$$k_1 = 0.1\left(\sin 1 + 1^2\right) = 0.18141,$$

$$k_2 = 0.1\left(\sin 1.09071 + 1.09071^2\right) = 0.20766,$$

$$k_3 = 0.1\left(\sin 1.10383 + 1.10383^2\right) = 0.21114,$$

$$k_4 = 0.1\left(\sin 1.10557 + 1.10557^2\right) = 0.21160.$$

Then x_1 is obtained:

$$x_1 = 1 + \frac{1}{6}(0.18141 + 2 \times 0.20766 + 2 \times 0.21114 + 0.21160) = 1.20510.$$ ◄

The above formula can easily be extended to higher dimensions. For example, if we have a two-dimensional system given by

$$\dot{x} = f(x, y, t),$$

$$\dot{y} = g(x, y, t),$$

and the initial condition is (x_0, y_0) at $t = t_0$, then the fourth-order Runge-Kutta method gives

$$x_1 = x_0 + \frac{1}{6}(k_1 + 2k_2 + 2k_3 + k_4),$$

$$y_1 = y_0 + \frac{1}{6}(l_1 + 2l_2 + 2l_3 + l_4),$$

where

$$k_1 = hf(x_0, y_0, t_0),$$

$$l_1 = hg(x_0, y_0, t_0),$$

$$k_2 = hf\left(x_0 + k_1/2, y_0 + l_1/2, t_0 + h/2\right),$$

$$l_2 = hg\left(x_0 + k_1/2, y_0 + l_1/2, t_0 + h/2\right),$$

$$k_3 = hf\left(x_0 + k_2/2, y_0 + l_2/2, t_0 + h/2\right),$$

$$l_3 = hg\left(x_0 + k_2/2, y_0 + l_2/2, t_0 + h/2\right),$$

$$k_4 = hf\left(x_0 + k_3, y_0 + l_3, t_0 + h\right),$$

$$l_4 = hg\left(x_0 + k_3, y_0 + l_3, t_0 + h\right).$$

The method can be extended to even higher dimensions in a similar manner.

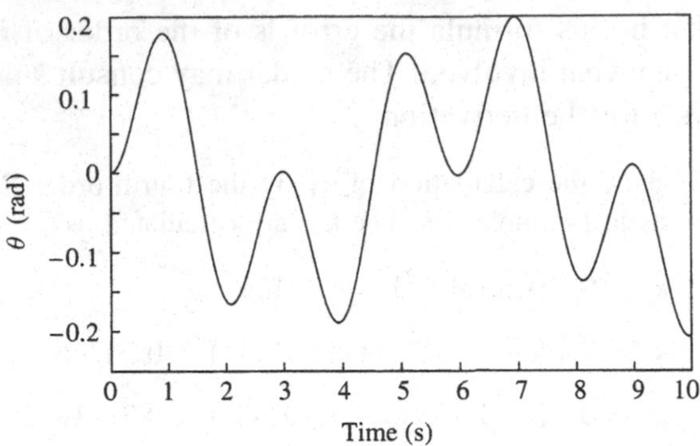

Figure 7.2 The time evolution of the angle θ for the pendulum with an oscillating support, computed with the fourth-order Runge-Kutta method.

► **Example 7.4** Let us consider the pendulum with oscillating support, whose differential equation was derived in Example 4.5 as

$$\ddot{\theta} + \frac{g}{l}\sin\theta - \frac{A}{l}\omega^2\cos\theta\,\cos\omega t = 0.$$

Let us assume the parameter values:
Length of chord (l) = 1 m,
Acceleration due to gravity (g) = 9.81 m/s^2,
Amplitude of oscillation of the support (A) = 1 m,
Angular frequency of oscillation of the support (ω) = 1 rad/s.
Let the bob be initially at rest, that is, $\theta = 0$ and $\dot{\theta} = 0$ at $t = 0$. We compute the evolution in the angle θ for 10 s, with the fourth-order Runge-Kutta method for a step length of $h = 0.01$ s. The result is shown in Fig. 7.2. ◄

7.2 Methods to Balance Accuracy and Computation Time

Even if the fourth-order Runge-Kutta method ensures accuracy of the order of h^5, some error still remains. In order to reduce the error, one has to use a sufficiently small value of the step length h. This increases the computation time. On the other hand, if a larger value of h is chosen to economize the computation, accuracy is sacrificed.

One way to achieve a balance between the demands of accuracy and computation time is to use a variable step length. When the solution progresses in almost a linear fashion, a higher step length suffices to ensure a desired accuracy, and when the solution takes quick turns a smaller step length must be used. But how to decide the appropriate value of h at a given step in the computation?

For this, one generally uses two different methods of estimating the value of the state variable in the next step. If the two estimates differ significantly, a small step size must be used. On the other hand, if they differ by a small amount – smaller than a specified tolerance limit – the step size can be increased.

The Runge-Kutta methods of the third order and the fourth order can be used to obtain the two estimates of the state variable in the next step – which are to be compared. For a greater accuracy, the Runge-Kutta methods of the fourth and fifth orders can also be compared. And the step length h in each step can be obtained as a function of the difference of the two estimates.

Various functional forms have been proposed to obtain the optimal step size. For example, in the Runge-Kutta–Fehlberg method, the optimal step size is obtained by multiplying the current step size h by a factor s given by

$$s = \left(\frac{\text{tol}\, h}{2|z_{k+1} - y_{k+1}|} \right)^{1/4},$$

where "tol" is the specified error control tolerance, and z_{k+1} and y_{k+1} are the estimates of the state variables at the $(k+1)$th step obtained by two different methods.

7.3 Chapter Summary

Differential equations can be solved numerically by proceeding in small time-steps, starting from a given initial condition. Since, the exact position of the next step is not known beforehand, one tries to get an approximate value. The various techniques of numerically solving differential equations involve different methods of approximation.

Computational tools like MATLAB, MAPLE, SCILAB, and MATHEMATICA have built-in subroutines for solving differential equations that incorporate the Runge-Kutta method of various orders. The reader should get acquainted with at least one of them, and should be able to solve any given set of differential equations using these software.

Further Reading

K. E. Atkinson, *An Introduction to Numerical Analysis*, John Wiley & Sons, New York, 1989.

R. W. Hamming, *Numerical Methods for Scientists and Engineers*, Dover Publications, New York, 1973.

J. H. Mathews, *Numerical Methods for Mathematics, Science and Engineering*, Prentice Hall, Englewood Cliffs, New Jersey, 1992.

Problems

1. Write a program to solve a set of n first-order differential equations (the program should be able to take any value of n). Take any five systems from the exercise problems of Chapters 5 and 6, and compute the values of the state variables for 10 s starting from a feasible initial condition.
2. A second-order system is given by

$$\begin{bmatrix} \dot{x}_1 \\ \dot{x}_2 \end{bmatrix} = \begin{bmatrix} -4 & -3 \\ 2 & 3 \end{bmatrix} \begin{bmatrix} x_1 \\ x_2 \end{bmatrix}.$$

3. With the initial condition (2, 2), calculate the state vector at 0.2 s, with the second-order Runge-Kutta method, and the fourth-order Runge-Kutta method. After you study Chapter 9, obtain the exact solution and compare the results from the above numerical methods with those obtained from the exact solution.

8

Dynamics in the State Space

We have seen that the mathematical model of any continuous-time dynamical system is obtained in the form

$$\dot{x}_i = f_i(x_1, x_2, \ldots, x_n) \tag{8.1}$$

with $i = 1, 2, 3, \ldots, n$, or in vector form

$$\dot{\mathbf{x}} = f(\mathbf{x}). \tag{8.2}$$

If the system equations do not have any externally applied time-varying input or time-varying parameters, the system is said to be *autonomous*. In such systems, the right-hand side of (8.2) does not contain any time-dependent term. A typical example is the simplified model of atmospheric convection, known as the *Lorenz system*:

$$\begin{aligned}\dot{x} &= -3(x - y),\\ \dot{y} &= -xz + rx - y,\\ \dot{z} &= xy - z,\end{aligned} \tag{8.3}$$

where r is a parameter.

Systems with external inputs or forcing functions or time variations in their definition, are called *non-autonomous* systems. In such systems, the right-hand side of (8.2) contains time dependent terms. As a typical example one may consider a periodically forced pendulum, the equations of which are

$$\begin{aligned}\dot{x} &= y,\\ \dot{y} &= -g \sin x + F \cos \omega t.\end{aligned} \tag{8.4}$$

8.1 The State Space

Geometrically, the dynamics can be visualized by constructing a space with the state variables as the coordinates. In engineering sciences, this space is called the *state space*, and in physical sciences it is called the *phase space*. Here we will follow the former nomenclature. The state of the system at any instant is represented by a point in this space (Fig. 8.1).

Starting from any given initial position, the state-point moves in the state space, and this movement is completely determined by the state equations. The path of the state-point is called the *orbit* or the *trajectory* of the system that starts from the given initial condition. These trajectories are obtained as the solution of the set of

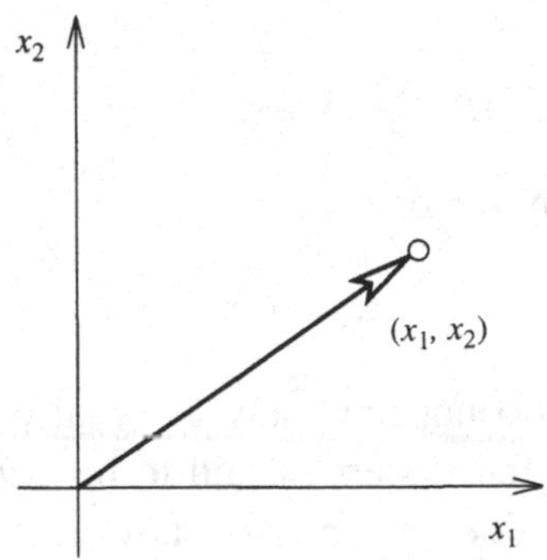

Figure 8.1 The state vector in a two-dimensional state space.

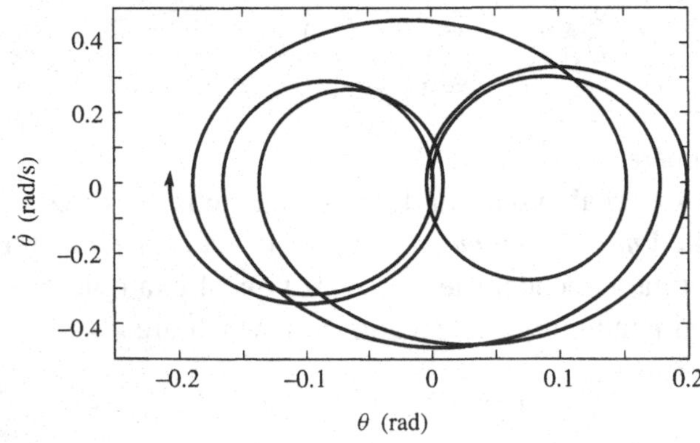

Figure 8.2 The trajectory of the pendulum with oscillating support for 10 s starting from a standstill position. The parameters are $l = 1$ m, $g = 9.81$ m/s^2, $A = 1$ m and $\omega = 1$ rad/s.

differential equations (8.2). As an example, a typical trajectory of the pendulum with oscillating support (whose equation was derived in Example 4.5, and the waveform was plotted in Fig. 7.2) is shown in Fig. 8.2.

8.2 Vector Field

In studying the dynamical behaviour of a given system, one has to compute the trajectory starting from a given initial condition. We have seen that this can be done numerically. However, it is generally not necessary to compute all possible trajectories (which may be a cumbersome exercise) in order to study a given system. It may be noted that the left-hand side of (8.2) gives the rate of change of the state variables. For example, if a two-dimensional system is given by

$$\dot{x} = f_1(x, y), \tag{8.5}$$

$$\dot{y} = f_2(x, y), \tag{8.6}$$

then, for any given state-point (x, y), (8.5) defines how x will change with time at the next instant and (8.6) tells how y will change with time. Thus, $[\dot{x}, \dot{y}]^T$ is a vector[1], which is expressed as a function of the state variables. The equation $\dot{\mathbf{x}} = f(\mathbf{x})$ thus defines a vector at every point of the state space. This is called the *vector field.* To give an example, the vector field for the system

$$\dot{x} = y,$$
$$\dot{y} = (1 - x^2)\dot{x} - x$$

is shown in Fig. 8.3. A solution starting from any initial condition follows the direction of the vectors, that is, the vectors are tangent to the solutions. The properties of a system can be studied by studying this vector field.

8.3 Local Linearization Around Equilibrium Points

The points where the $\dot{\mathbf{x}}$ vector has zero magnitude, that is, where $\dot{\mathbf{x}} = f(\mathbf{x}) = 0$, are called the *equilibrium points*. Since the velocity vector at the equilibrium point has magnitude zero, if an initial condition is placed there, the state-point will forever remain there. However, this does not guarantee that the equilibrium state will be stable, that is, any deviation from it will die down.

It is therefore important to study the *local* behaviour of the system in the neighbourhood of an equilibrium point. Since it is straightforward to obtain the solutions of a set of *linear* differential equations, the local properties of the state space in

[1] $\mathbf{A}^T$ means the transpose of the matrix $\mathbf{A}$.

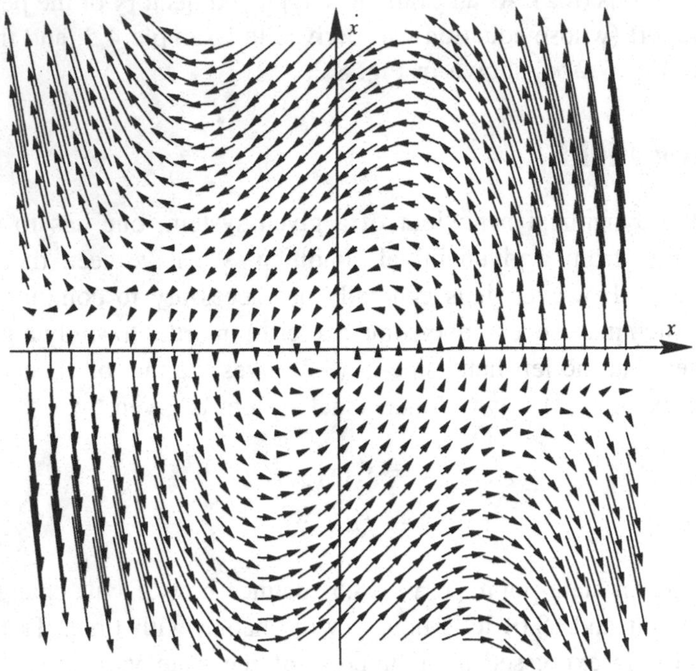

Figure 8.3 The vector field for the equation $\ddot{x} - (1 - x^2)\dot{x} + x = 0$.

the neighbourhood of an equilibrium point can be studied by locally linearizing the differential equations at that point. Indeed, most of the tools for the design and analysis of engineering systems concentrate only on the local behaviour – because in general, the nominal operating point of any system is located at an equilibrium point, and if perturbations are small then the linear approximation gives a simple workable model of the dynamical system.

To understand the process of local linearization, let us first consider a one-dimensional system $\dot{x} = f(x)$. The function $y = f(x)$ is represented by a curve, and the tangent at a given point x^* represents the linear approximation of the function in the neighbourhood of that point (Fig. 8.4). Therefore, the local linearization of the equation $\dot{x} = f(x)$ is given by $\delta\dot{x} = \frac{\mathrm{d}y}{\mathrm{d}x}\delta x$. Note that the fixed point of the original system becomes the origin in the local linear representation, and so the variable changes to $\delta x = x - x^*$.

In a two-dimensional system, the same procedure is followed. The function at the right-hand side of (8.2) represents a surface, and so the local linear representation should geometrically be a tangential plane. This is obtained by using the Jacobian matrix of the functional form $f(\mathbf{x})$ at an equilibrium point. If the state space is two-dimensional, given by equations of the form (8.5) and (8.6), then the local

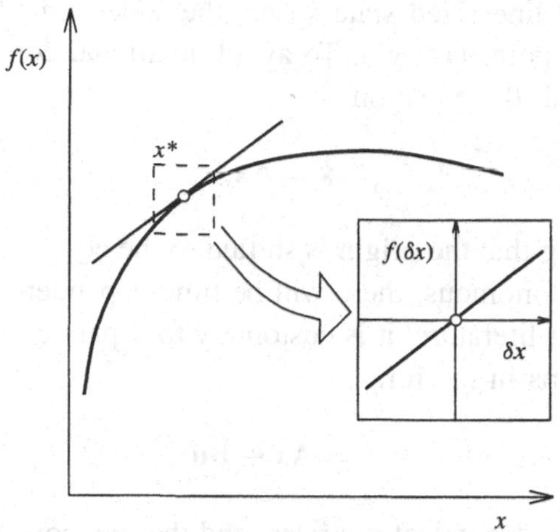

Figure 8.4 The local linear approximation of a function $f(x)$ at the point x^*.

linearization at an equilibrium point (x^*, y^*) is given by

$$\begin{pmatrix} \delta\dot{x} \\ \delta\dot{y} \end{pmatrix} = \begin{pmatrix} \frac{\partial f_1}{\partial x} & \frac{\partial f_1}{\partial y} \\ \frac{\partial f_2}{\partial x} & \frac{\partial f_2}{\partial y} \end{pmatrix} \begin{pmatrix} \delta x \\ \delta y \end{pmatrix}, \tag{8.7}$$

where $\delta x = x - x^*$, $\delta y = y - y^*$. The matrix containing the partial derivatives is called the Jacobian matrix and the numerical values of the partial derivatives are calculated at the equilibrium point. This is really just a (multivariate) Taylor series expanded to first order.

► **Example 8.1** For the second-order equation $\ddot{x} - (1 - x^2)\dot{x} + x = 0$, let the state variables be x and $\dot{x}$. Thus, the state variable equations are:

$$\dot{x} = y,$$
$$\dot{y} = (1 - x^2)y - x.$$

The equilibrium point is $(0, 0)$ (obtained by putting $\dot{x} = 0$ and $\dot{y} = 0$). The Jacobian matrix of (8.7) is obtained as

$$\begin{pmatrix} 0 & 1 \\ -2xy - 1 & 1 - x^2 \end{pmatrix}.$$

Substituting the values of x and y at the equilibrium point, we get the linearized system as

$$\begin{pmatrix} \delta\dot{x} \\ \delta\dot{y} \end{pmatrix} = \begin{pmatrix} 0 & 1 \\ -1 & 1 \end{pmatrix} \begin{pmatrix} \delta x \\ \delta y \end{pmatrix}.$$ ◄

Notice that in the linearized state space, the state variables are the *deviations* from the equilibrium point (x^*, y^*). To avoid notational complexity, we will drop the δ and proceed with the equation

$$\dot{\mathbf{x}} = \mathbf{A}\mathbf{x}, \tag{8.8}$$

with the understanding that the origin is shifted to the equilibrium point. If the original system is non-autonomous, there will be time-dependent terms in the Jacobian matrix. In engineering literature, it is customary to separate the time-dependent and time-independent terms in the form

$$\dot{\mathbf{x}} = \mathbf{A}\mathbf{x} + \mathbf{B}\mathbf{u}, \tag{8.9}$$

where $\mathbf{A}, \mathbf{B}$ are time-independent matrices, and the components of the vector $\mathbf{u}$ are the externally imposed inputs of the system.

8.4 Chapter Summary

For a given dynamical system, the minimum number of variables needed to specify the dynamical status of the system are called the state variables, and the space constructed with the state variables as coordinates is called the state space. The set of differential equations $\dot{\mathbf{x}} = f(\mathbf{x})$ defines a vector field in the state space. The solution of the system starting from any initial condition follows the flow in this vector field.

The points in the state space satisfying the condition $\dot{\mathbf{x}} = f(\mathbf{x}) = 0$ are the equilibrium points. The local linear approximation in the neighbourhood of the equilibrium points yields a set of linear differential equations. In the next chapter, we shall see how the solutions of linear differential equations can be obtained in closed form.

Problems

1. For the following systems, locate the equilibrium points and obtain the linear differential equations that describe the local behaviour around the equilibrium points.

   (a) $\dot{x} = y,$
   $\dot{y} = x^3 - x,$

   (b) $\dot{x} = y(x^2 + 1),$
   $\dot{y} = x^2 - 1,$

   (c) $\dot{x} = y(1 - x^2),$
   $\dot{y} = -x(1 - y^2).$

2. Write a program to draw the vector field over a given region of the state space of a second-order system. Use it to draw the vector field for the equations in Problem 1.

3. Find the equilibrium points of the following systems, and obtain the linear differential equations that represent the neighbourhood of these equilibrium points.

   (a) $\dot{x} = y^2 - 4x + 7, \quad \dot{y} = x - y,$

   (b) $\ddot{x} + \dot{x} - x^3 - x^2 + 2x = 0,$

   (c) $\dot{x} = x, \quad \dot{y} = -x + 2y,$

   (d) $\dot{x} = y(x^2 + 1), \quad \dot{y} = 2xy^2,$

   (e) $\dot{x} = -x, \quad \dot{y} = 2x^2y^2.$

4. The mass in this system slides on an uneven surface whose frictional coefficient is given by $R = x^2 - 1$. Obtain the differential equations, and the location of the equilibrium points. Obtain the linear descriptions of the system around the equilibrium points.

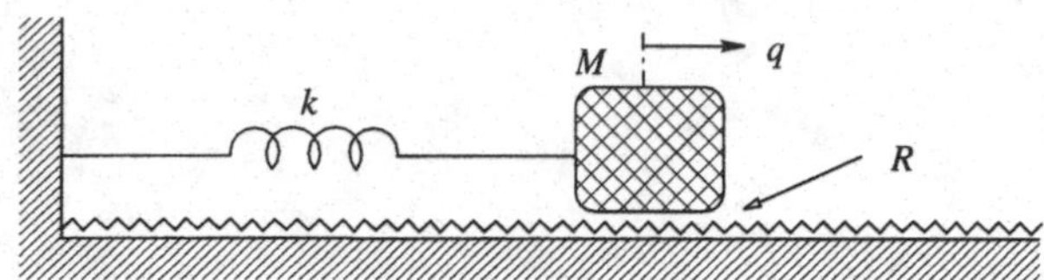

5. The following equations represent the dynamics of the population of two competing species, known as the Lotka-Volterra equations.

$$\dot{x} = ax - xy,$$
$$\dot{x} = xy - by.$$

Plot the vector field and comment on its salient features.

9

Solutions for a System of First-order Linear Differential Equations

In the last chapter, we have seen that in the neighbourhood of an equilibrium point, the local behaviour of a vector field is represented by linear differential equations of the form

$$\dot{\mathbf{x}} = \mathbf{A}\mathbf{x} + \mathbf{B}\mathbf{u}, \tag{9.1}$$

where $\mathbf{A}, \mathbf{B}$ are time-independent matrices, and the components of the vector $\mathbf{u}$ are the externally imposed inputs of the system.

Note that in this equation if $\mathbf{B}\mathbf{u} = 0$, the location of the equilibrium point is at the origin, and if $\mathbf{B}\mathbf{u} \neq 0$ the location of the equilibrium point depends on the term $\mathbf{B}\mathbf{u}$. If $\mathbf{u}$ is a time varying vector (as happens, e.g. when there is a sinusoidal voltage source in a circuit), the location of the equilibrium point shifts continuously. However, if you ask the question "Is the equilibrium point stable?", you will have to look at how the deviations from the equilibrium point evolve with time. For this, it is necessary to put the movement of the equilibrium point out of consideration. Therefore, while studying the stability of the equilibrium point, one considers the unforced system

$$\dot{\mathbf{x}} = \mathbf{A}\mathbf{x}. \tag{9.2}$$

If this equation is to tell about the character of orbits starting in the neighbourhood of the equilibrium point, that information must be contained in the matrix $\mathbf{A}$.

Dynamics for Engineers S. Banerjee

9.1 Solution of a First-order Linear Differential Equation

Let us first take a single first-order differential equation in the form

$$\dot{x} = ax. \tag{9.3}$$

In this simple system, the variables are easily separated to give

$$\frac{1}{x}\,\mathrm{d}x = a\,\mathrm{d}t.$$

We now integrate both the sides, assuming x_0 to be the initial condition at $t = 0$, and x_t to be the value at time t.

$$\int_{x_0}^{x_t} \frac{1}{x}\,\mathrm{d}x = \int_0^t a\,\mathrm{d}t,$$

$$\ln x \,\Big|_{x_0}^{x_t} = at,$$

$$\ln x_t - \ln x_0 = at,$$

$$\ln \frac{x_t}{x_0} = at,$$

$$x_t = x_0\, e^{at}. \tag{9.4}$$

Thus, (9.4) is the solution of the first-order equation (9.3). It is clear that if a is a positive number, the solution will be an exponentially increasing function, and if a is a negative number, the solution will decrease exponentially to the final value zero (Fig. 9.1).

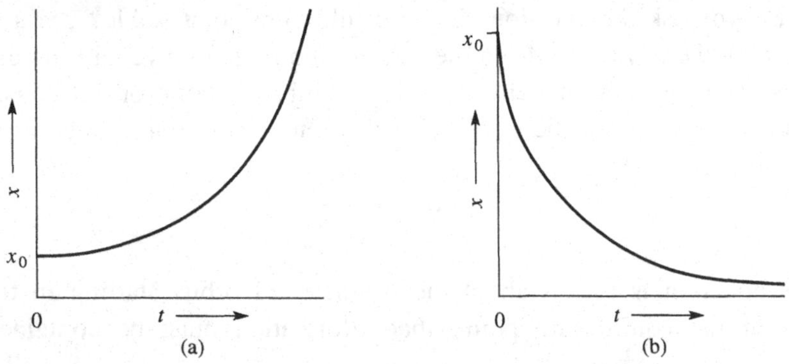

Figure 9.1 The solution of a first-order linear differential equation (9.3) (a) when a is positive, and (b) when a is negative.

9.2 Solution of a System of Two First-order Linear Differential Equations

We now proceed to obtain the solution of the differential equation system of the form (9.2) with an initial condition $\mathbf{x}_0$ at time t_0. We first develop the understanding of the dynamics of a two-dimensional linear system, which can easily be extended to higher dimensions. Thus, we consider

$$\mathbf{x} = \begin{bmatrix} x \\ y \end{bmatrix},$$

and we seek solutions of the form

$$x = x(t),$$
$$y = y(t).$$

Notice first of all, that the solution of the equation with a single initial condition is unique – because at every point the $\dot{\mathbf{x}}$ vector is unique. In the next step, we make use of a theorem that states that if we can find any two linearly independent[1] solutions

$$\begin{aligned} x &= x_1(t) \qquad & x &= x_2(t) \\ & & \text{and} & \\ y &= y_1(t) \qquad & y &= y_2(t) \end{aligned} \tag{9.5}$$

then the general solution of the system of equations (9.2) starting from any given initial condition is the linear combination

$$\begin{aligned} x &= c_1 x_1(t) + c_2 x_2(t), \\ y &= c_1 y_1(t) + c_2 y_2(t), \end{aligned} \tag{9.6}$$

where the constants c_1 and c_2 are given by the initial condition.

It is clear that (9.6) is a general solution if and only if the constants c_1 and c_2 can be uniquely determined for every initial condition (x_0, y_0), that is, if the equations

$$c_1 x_1(t_0) + c_2 x_2(t_0) = x_0,$$
$$c_1 y_1(t_0) + c_2 y_2(t_0) = y_0,$$

can be solved for every (x_0, y_0). This is possible if the determinant

$$\begin{vmatrix} x_1(t) & x_2(t) \\ y_1(t) & y_2(t) \end{vmatrix}$$

[1] Two solutions are said to be linearly dependent if one is a constant multiple of the other, that is, if $x_1(t) = kx_2(t)$ and $y_1(t) = ky_2(t)$. The solutions are said to be *linearly independent* if neither is a constant multiple of the other in the above sense.

is non-zero in the range of state space under consideration. It is easy to see that if the two solutions are linearly independent, then this determinant cannot be zero. Hence, (9.6) is a general solution of the system of equations (9.2).

The first step towards obtaining solutions of (9.2), therefore, is to seek a couple of convenient solutions that can be used for constructing the general solution.

9.3 Eigenvalues and Eigenvectors

Notice that in the equation

$$\dot{\mathbf{x}} = \mathbf{A}\mathbf{x},$$

the matrix **A** *operates* on the vector **x** to give the vector $\dot{\mathbf{x}}$. This is the basic function of any matrix – mapping one vector on to another vector. Generally, the derived vector is different from the source vector, both in magnitude and direction. But there may be some special directions in the state space such that if the vector **x** is in that direction, the vector $\dot{\mathbf{x}}$ (obtained by operating **x** with **A**) also lies along the same direction. It only gets stretched or squeezed. These special directions are called *eigendirections*. Any vector along an eigendirection is called *eigenvector* and the factor by which any eigenvector expands or contracts when it is operated on by the matrix **A**, is called the *eigenvalue*.

To find the eigenvectors, we need to find their eigenvalues first. When the matrix **A** operates on the vector **x**, and **x** happens to be an eigenvector, the result of the operation is obtained simply by multiplying **x** with a number λ. Thus, we can write

$$\mathbf{A}\mathbf{x} = \lambda\mathbf{x},$$

where λ is the eigenvalue. This yields

$$(\mathbf{A} - \lambda\mathbf{I})\mathbf{x} = 0,$$

where **I** is the identity matrix of the same dimension as **A**. This condition would be satisfied if the determinant $|\mathbf{A} - \lambda\mathbf{I}| = 0$. Thus,

$$\left|\begin{bmatrix} A_{11} & A_{12} \\ A_{21} & A_{22} \end{bmatrix} - \begin{bmatrix} \lambda & 0 \\ 0 & \lambda \end{bmatrix}\right| = 0,$$

$$(A_{11} - \lambda)(A_{22} - \lambda) - A_{12}A_{21} = 0,$$

$$\lambda^2 - (A_{11} + A_{22})\lambda + (A_{11}A_{22} - A_{12}A_{21}) = 0. \tag{9.7}$$

This is called the *characteristic equation*, whose roots are the eigenvalues. Thus, for a 2×2 matrix one gets a quadratic equation – which in general yields two eigenvalues.

Now take any of these eigenvalues (say λ_1) and proceed to find its corresponding eigenvector as follows. By the definition of eigenvector,

$$\mathbf{Ax} = \lambda_1 \mathbf{x}$$

or

$$(\mathbf{A} - \lambda_1 \mathbf{I})\mathbf{x} = \mathbf{0}$$

or

$$\begin{bmatrix} A_{11} - \lambda_1 & A_{12} \\ A_{21} & A_{22} - \lambda_1 \end{bmatrix} \begin{bmatrix} x \\ y \end{bmatrix} = 0.$$

This leads to the two equations

$$\begin{aligned} (A_{11} - \lambda_1)x + A_{12}y &= 0, \\ A_{21}x + (A_{22} - \lambda_1)y &= 0. \end{aligned} \tag{9.8}$$

These two equations always turn out to be identical. This is expected. For each eigenvalue there is one eigendirection, and *any vector* in that direction is an eigenvector. Hence, the two equations above should be identical so that the direction of the eigenvector is determinate but the magnitude is indeterminate.

To summarize, the eigenvalues are obtained by solving the equation $|\mathbf{A} - \lambda\mathbf{I}| = 0$, and the eigenvectors are obtained, for each real eigenvalue, from the equation $(\mathbf{A} - \lambda\mathbf{I})\mathbf{x} = 0$.

► **Example 9.1** For the linearized state variable equations

$$\begin{bmatrix} \dot{x} \\ \dot{y} \end{bmatrix} = \begin{bmatrix} -2 & 1 \\ 1 & -2 \end{bmatrix} \begin{bmatrix} x \\ y \end{bmatrix},$$

the solution of the quadratic equation (9.7) yields the eigenvalues -1 and -3. If we put $\lambda_1 = -1$, both the equations in (9.8) yield the same equation, $x = -y$. This is the eigenvector corresponding to the eigenvalue -1. Similarly, we get the eigenvector $x = y$ for $\lambda_2 = -3$. ◄

Since the eigenvalues are obtained from the matrix **A**, and this matrix, in turn, is obtained from the system of linear differential equations, one often says that λ_1 and λ_2 are the eigenvalues of the system of equations.

9.4 Using Eigenvalues and Eigenvectors for Solving Differential Equations

The definition of eigenvector tells us that if an initial condition is located on an eigenvector, then the $\dot{\mathbf{x}}$ vector remains along the same eigenvector and therefore the whole solution also remains along that eigenvector. This provides an easy way of writing down two solutions from which the general solution can be constructed.

Notice that the equation (9.7) may yield three different types of results:

1. eigenvalues real and distinct,
2. eigenvalues complex conjugate,
3. eigenvalue real and equal.

We now take up each case separately and demonstrate how the general solution can be formulated.

9.4.1 Eigenvalues real and distinct

Let λ_1 and λ_2 be the eigenvalues and $\mathbf{v}_1$ and $\mathbf{v}_2$ be the corresponding eigenvectors. If we place an initial condition on $\mathbf{v}_1$, by definition of eigenvector,

$$\dot{\mathbf{x}} = \mathbf{A}\mathbf{v}_1 = \lambda_1\mathbf{v}_1$$

since the evolution is constrained along the eigendirection (see Fig. 9.2). This is like a one-dimensional differential equation $\dot{x} = \lambda x$ whose solution is $x(t) = e^{\lambda t}x_0$.

Therefore, the solution of the differential equation $\dot{\mathbf{x}} = \lambda_1\mathbf{v}_1$ along the eigendirection is

$$\mathbf{x}_1(t) = e^{\lambda_1 t}\mathbf{v}_1.$$

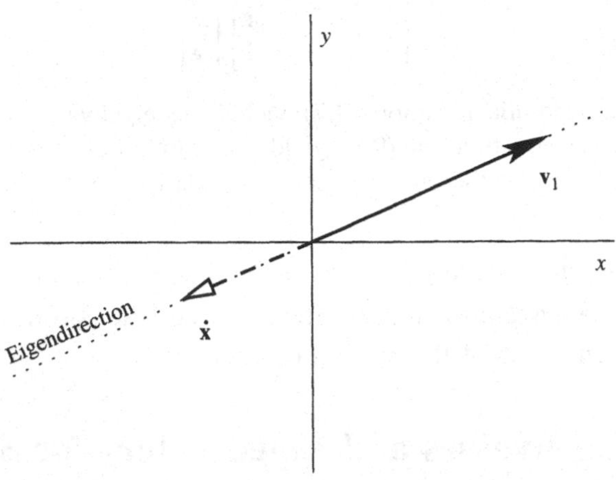

Figure 9.2 If the state vector is an eigenvector, $\dot{\mathbf{x}}$ also lies in the same eigendirection. Therefore, the dynamics becomes confined to that direction, making it effectively one-dimensional.

Similarly, for any initial condition placed along $\mathbf{v}_2$ we have another solution

$$\mathbf{x}_2(t) = e^{\lambda_2 t}\mathbf{v}_2.$$

Therefore, the general solution can be constructed as

$$\mathbf{x}(t) = c_1 e^{\lambda_1 t}\mathbf{v}_1 + c_2 e^{\lambda_2 t}\mathbf{v}_2. \tag{9.9}$$

► **Example 9.2** Let the system of equations be given by

$$\dot{\mathbf{x}} = \begin{bmatrix}\dot{x}\\ \dot{y}\end{bmatrix} = \begin{bmatrix}-4 & -3\\ 2 & 3\end{bmatrix}\begin{bmatrix}x\\ y\end{bmatrix}. \tag{9.10}$$

The matrix $\mathbf{A}$ has eigenvalues $\lambda_1 = 2$ and $\lambda_2 = -3$. For λ_1, the eigenvector is given by $2x = -y$. To choose any point on this eigenvector, set $x = 1$. This gives $y = -2$. Thus,

$$\mathbf{v}_1 = \begin{bmatrix}1\\ -2\end{bmatrix}.$$

For this initial condition, the solution is

$$\mathbf{x}_1(t) = e^{2t}\begin{bmatrix}1\\ -2\end{bmatrix}.$$

Similarly, for $\lambda_2 = -3$, the eigenvector is $x = -3y$. To choose a point on this eigenvector, take $x = 3$. This gives $y = -1$. Thus, the second eigenvector becomes

$$\mathbf{v}_2 = \begin{bmatrix}3\\ -1\end{bmatrix},$$

and the solution along this eigenvector is

$$\mathbf{x}_2(t) = e^{-3t}\begin{bmatrix}3\\ -1\end{bmatrix}.$$

Hence, the general solution of the system of differential equations is

$$\mathbf{x}(t) = c_1 e^{2t}\begin{bmatrix}1\\ -2\end{bmatrix} + c_2 e^{-3t}\begin{bmatrix}3\\ -1\end{bmatrix},$$

where the constants c_1 and c_2 are to be determined from the initial condition.

For example, if the initial condition is (1, 1) at $t = 0$, then this equation gives

$$c_1\begin{bmatrix}1\\ -2\end{bmatrix} + c_2\begin{bmatrix}3\\ -1\end{bmatrix} = \begin{bmatrix}1\\ 1\end{bmatrix}.$$

Solving, we get $c_1 = -4/5$ and $c_2 = 3/5$. Thus, the solution of the differential equation with this initial condition is

$$\mathbf{x}(t) = -\frac{4}{5}e^{2t}\begin{bmatrix}1\\ -2\end{bmatrix} + \frac{3}{5}e^{-3t}\begin{bmatrix}3\\ -1\end{bmatrix},$$

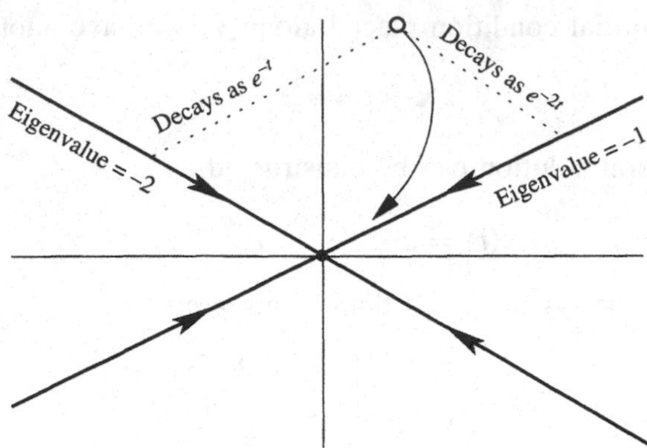

Figure 9.3 The dynamics in the system with eigenvalues −1 and −2.

or, in terms of the individual coordinates,

$$x(t) = -\frac{4}{5}e^{2t} + \frac{9}{5}e^{-3t},$$

$$y(t) = \frac{8}{5}e^{2t} - \frac{3}{5}e^{-3t}. \quad \blacktriangleleft$$

From the form of equation (9.9), it is evident that the solution $\mathbf{x}(t)$ is composed of two components along the two eigenvectors. These components evolve independently as $e^{\lambda_1 t}$ and $e^{\lambda_2 t}$ respectively.

This gives us a way of intuitively visualizing the dynamics in the state space. For example, consider a system with eigenvalues −1 and −2, and eigenvectors as shown in Fig. 9.3. The coordinate of the initial condition can be resolved into two components along the two eigenvectors, and these components will evolve as e^{-1t} and e^{-2t} respectively. Since the exponents are negative, both these components will progressively decay with time. The rates of decay will be given by the respective eigenvalues and hence, as time progresses, the state will move towards the eigenvector associated with the larger eigenvalue, and will finally converge onto the equilibrium point.

It follows that if the eigenvalues are real and negative, the system is stable in the sense that any perturbation from an equilibrium point decays exponentially and the system settles back to the equilibrium point. Such a stable equilibrium point is called a *node*. If the real parts of the eigenvalues are positive, any deviation from the equilibrium point grows exponentially, and the system is unstable.

If one eigenvalue is real and negative while the other is real and positive, the system is stable along the eigenvector associated with the negative eigenvalue, and is unstable away from this. The trajectory starting from any initial condition progressively converges on the eigenvector associated with the positive eigenvalue,

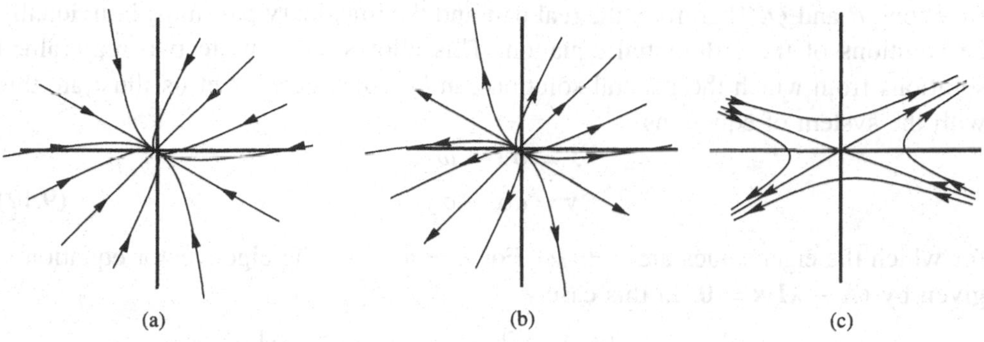

Figure 9.4 Vector fields of linear systems with real eigenvalues, (a) both eigenvalues negative, (b) both eigenvalues positive and (c) one eigenvalue negative and the other positive.

and moves to infinity along that line in the state space. Such an equilibrium point is called a *saddle*, and a system with a saddle equilibrium point is globally unstable. The vector fields of the three types of systems are shown in Fig. 9.4.

9.4.2 Eigenvalues complex conjugate

Complex eigenvalues always occur as complex conjugate pairs.[2] If $\lambda = \sigma + j\omega$ is an eigenvalue, $\bar{\lambda} = \sigma - j\omega$ is also an eigenvalue. Let $\mathbf{v}$ be an eigenvector corresponding to the eigenvalue $\lambda = \sigma + j\omega$. This is a complex-valued vector. It is easy to check that $\bar{\mathbf{v}}$, the conjugate of the vector $\mathbf{v}$, is associated with the eigenvalue $(\sigma - j\omega)$.

Though complex-valued eigenvectors cannot represent any specific direction in the real-valued state space, their physical significance derives from the fact that the eigenvector equation $\mathbf{Av} = \lambda\mathbf{v}$ holds, which allows us to obtain a solution as

$$\mathbf{x}_1(t) = e^{\lambda t}\mathbf{v}. \tag{9.11}$$

Here the left-hand side is a real-valued function and the right-hand side is complex-valued, expressible in the form $(P + jQ)$ – which is a linear combination of the

[2]The eigenvalues are obtained as roots of the characteristic equation of the form (9.7). For two-dimensional system, it is a quadratic equation, and from the expression of the roots of a quadratic it is easy to see that these can either be real or complex conjugate. For higher dimensional systems, the eigenvalues are in general obtained as the roots of a polynomial equation. D'Alembert showed in 1746 that for the polynomial equation

$$a_n\lambda^n + a_{n-1}\lambda^{n-1} + \cdots + a_1\lambda + a_0 = 0,$$

if the coefficients are all real numbers, then the roots can only be real numbers or complex conjugate pairs.

functions P and Q. Therefore, the real part and the imaginary part must individually be solutions of the differential equation. This allows us to write two real-valued solutions from which the general solution can be constructed. Let us illustrate this with the system of equations

$$\begin{aligned} \dot{x} &= \sigma x - \omega y, \\ \dot{y} &= \omega x + \sigma y, \end{aligned} \tag{9.12}$$

for which the eigenvalues are $\sigma \pm j\omega$. For $\lambda = \sigma + j\omega$ the eigenvector equation is given by $(\mathbf{A} - \lambda\mathbf{I})\mathbf{x} = \mathbf{0}$. In this case,

$$\mathbf{A} - \lambda\mathbf{I} = \begin{bmatrix} \sigma & -\omega \\ \omega & \sigma \end{bmatrix} - (\sigma + j\omega)\begin{bmatrix} 1 & 0 \\ 0 & 1 \end{bmatrix}.$$

Therefore, the eigenvector equation becomes

$$\begin{bmatrix} -j\omega & -\omega \\ \omega & -j\omega \end{bmatrix}\begin{bmatrix} v_1 \\ v_2 \end{bmatrix} = \begin{bmatrix} -j\omega v_1 - \omega v_2 \\ \omega v_1 - j\omega v_2 \end{bmatrix} = \begin{bmatrix} 0 \\ 0 \end{bmatrix}.$$

Thus, the eigendirection is given by $v_1 = jv_2$. To choose a specific eigenvector, let us take $v_2 = 1$, so that $\mathbf{v} = [j, 1]^T$ is an eigenvector. Thus, a complex-valued solution is

$$\mathbf{x}(t) = e^{(\sigma + j\omega)t}\begin{bmatrix} j \\ 1 \end{bmatrix}.$$

This can be written in terms of the sines and cosines by the Euler's formula[3] as

$$\begin{aligned} \mathbf{x}(t) &= e^{\sigma t}(\cos\omega t + j\sin\omega t)\begin{bmatrix} j \\ 1 \end{bmatrix}, \\ &= e^{\sigma t}\begin{bmatrix} j\cos\omega t - \sin\omega t \\ \cos\omega t + j\sin\omega t \end{bmatrix}, \\ &= e^{\sigma t}\left[\begin{pmatrix} -\sin\omega t \\ \cos\omega t \end{pmatrix} + j\begin{pmatrix} \cos\omega t \\ \sin\omega t \end{pmatrix}\right]. \end{aligned} \tag{9.13}$$

[3]The Euler's formula is

$$e^{jx} = \cos x + j\sin x.$$

If you have not come across this wonderful equation in the mathematics classes, here is how to prove it. First expand e^{jx} in a power series, and then manipulate as

$$\begin{aligned} e^{jx} &= \sum_{n=0}^{\infty}\frac{(jx)^n}{n!}, \\ &= 1 + jx - \frac{x^2}{2!} - j\frac{x^3}{3!} + \frac{x^4}{4!} + j\frac{x^5}{5!} - \cdots, \\ &= \left(1 - \frac{x^2}{2!} + \frac{x^4}{4!} - \cdots\right) + j\left(x - \frac{x^3}{3!} + \frac{x^5}{5!} - \cdots\right), \\ &= \cos x + j\sin x. \end{aligned}$$

We have thus separated it out into the real and imaginary parts. We now note that the above expression for $\mathbf{x}(t)$ is a linear combination of the real part and the imaginary part.

Hence, the two linearly independent real-valued solutions are

$$e^{\sigma t}\begin{pmatrix}-\sin\omega t\\ \cos\omega t\end{pmatrix} \quad \text{and} \quad e^{\sigma t}\begin{pmatrix}\cos\omega t\\ \sin\omega t\end{pmatrix}.$$

Therefore, a general solution is

$$\mathbf{x}(t) = c_1 e^{\sigma t}\begin{pmatrix}-\sin\omega t\\ \cos\omega t\end{pmatrix} + c_2 e^{\sigma t}\begin{pmatrix}\cos\omega t\\ \sin\omega t\end{pmatrix}. \tag{9.14}$$

The other eigenvalue supplies no new information, as it is the complex conjugate of the first one.[4] As a final check, one can differentiate $x(t)$ and $y(t)$ from (9.14) to obtain back (9.12) – which would imply that (9.14) is really a solution for (9.12).

By (9.14), the variable $y(t)$ can be expressed as

$$\begin{aligned} y(t) &= e^{\sigma t}\left(c_1 \cos\omega t + c_2 \sin\omega t\right), \\ &= \sqrt{c_1^2 + c_2^2}\; e^{\sigma t}\left(\frac{c_1}{\sqrt{c_1^2 + c_2^2}} \cos\omega t + \frac{c_2}{\sqrt{c_1^2 + c_2^2}} \sin\omega t\right), \\ &= \sqrt{c_1^2 + c_2^2}\; e^{\sigma t}\left(\sin\theta \cos\omega t + \cos\theta \sin\omega t\right), \\ &= A\, e^{\sigma t}\ \sin(\omega t + \theta), \end{aligned} \tag{9.15}$$

[4]If the reader is not convinced, there is another way to approach the problem. For $\lambda_1 = \sigma + j\omega$, let an eigenvector be $\mathbf{v}$, so that for $\lambda_2 = \sigma - j\omega$, the eigenvector will be its complex conjugate $\bar{\mathbf{v}}$. Therefore, the two basic solutions should be

$$\mathbf{v}e^{(\sigma + j\omega)t}, \quad \text{and} \quad \bar{\mathbf{v}}e^{(\sigma - j\omega)t}.$$

This will give the general (complex) solution as

$$\mathbf{x}(t) = c_1\mathbf{v}e^{(\sigma + j\omega)t} + c_2\bar{\mathbf{v}}e^{(\sigma - j\omega)t},$$

where c_1 and c_2 are constants that are in general complex. However, the actual physical solution is real – which is possible if and only if these constants are also complex conjugate

$$c_2 = \bar{c}_1.$$

This gives the real solution as

$$\mathbf{x}(t) = 2\mathrm{Re}\left\{c_1\mathbf{v}e^{(\sigma + j\omega)t}\right\}$$

or

$$\mathbf{x}(t) = \mathrm{Re}\left\{c\mathbf{v}e^{(\sigma + j\omega)t}\right\},$$

where c (=$2c_1$) is a complex constant that has to be determined from the initial condition. This gives the same result as (9.14).

where

$$A = \sqrt{c_1^2 + c_2^2},$$

$$\sin\theta = \frac{c_1}{\sqrt{c_1^2 + c_2^2}},$$

$$\cos\theta = \frac{c_2}{\sqrt{c_1^2 + c_2^2}},$$

$$\text{i.e. } \theta = \tan^{-1}\frac{c_1}{c_2},$$

which are obtained from the initial conditions. Similarly, it can be shown that

$$x(t) = Ae^{\sigma t}\cos(\omega t + \theta).$$

This implies that the system response for a 2D linear system with complex conjugate roots is a damped sinusoid (see Fig. 9.5). The rate of damping (given by the $e^{\sigma t}$ term) depends on the real part, and the frequency is given by the imaginary part of the complex eigenvalues. The amplitude and the phase are given by the initial condition.

9.4.3 Eigenvalues purely imaginary

The case of imaginary eigenvalues is a special case of this solution, where $\sigma = 0$. Thus, the solutions for imaginary eigenvalues $\lambda = \pm j\omega$ are

$$x(t) = -c_1 \sin\omega t + c_2 \cos\omega t,$$

$$y(t) = c_1 \cos\omega t + c_2 \sin\omega t. \tag{9.16}$$

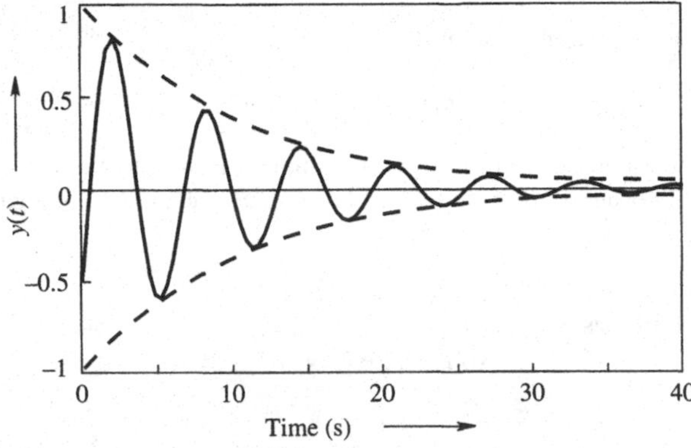

Figure 9.5 The solution of a system with complex conjugate eigenvalues, with $\sigma = -0.1$, $\omega = 1$ Hz, $A = 1$ and $\theta = \pi/6$. The envelope given by $e^{\sigma t}$ is also shown.

Here also, the values of c_1 and c_2 have to be obtained from the initial condition. To illustrate, let an initial condition be $(1, 0)$. Substituting this value of $\mathbf{x}$ at $t = 0$ in (9.16), we get $c_1 = 0$ and $c_2 = 1$. The trajectory is given by

$$x(t) = \cos \omega t \quad \text{and} \quad y(t) = \sin \omega t.$$

Thus, each state variable follows a sinusoidal variation, while $y(t)$ lags behind $x(t)$ by $\pi/2$. The equation for the trajectory in the state space is

$$x^2 = \cos^2 \omega t = 1 - \sin^2 \omega t = 1 - y^2,$$

which is the equation of a circle with radius 1. For initial conditions at various distances from the origin, the trajectories are circles of various radii, and the imaginary part ω of the eigenvalue gives the period of rotation. Thus, the vector field in the state space has the structure shown in Fig. 9.6(a). An equilibrium point with imaginary eigenvalues is called a *centre*.

In general, if the eigenvalues are purely imaginary, the orbits are elliptical. For initial conditions at different distances from the equilibrium point, the orbits form a

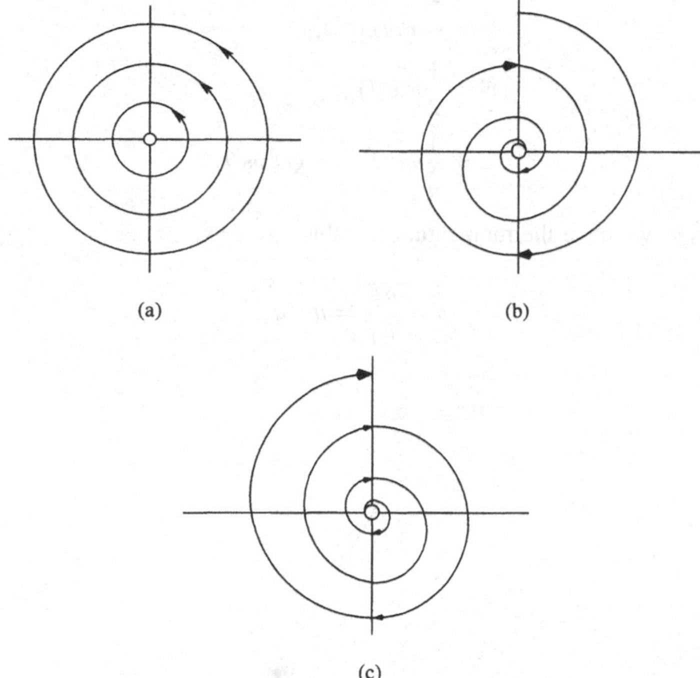

Figure 9.6 The structure of the vector field in the state space for (a) imaginary eigenvalues, (b) complex eigenvalues with negative real part and (c) complex eigenvalues with positive real part.

family of geometrically similar ellipses that are inclined at a constant angle to the axes, but having the same cyclic frequency.

When the eigenvalues are complex, with σ non-zero, the sinusoidal variation of the state variables will be multiplied by an exponential term $e^{\sigma t}$. If σ is negative, this term will decay as time progresses. Therefore, the waveform in time domain will be a damped sinusoid, and in the state space the state will spiral in towards the equilibrium point (Fig. 9.6b). If σ is positive, the $e^{\sigma t}$ term will increase with time, and so in the state space the behaviour will be an outgoing spiral (Fig. 9.6c).

► **Example 9.3** Galileo observed that the period of oscillation of a simple pendulum (Fig. 9.7) is independent of the mass of the bob and the amplitude of oscillation so long as the amplitude is small. To see why this law holds, let us work out the dynamics of the simple pendulum.

Let the mass of the bob be m, the length of the chord be l, and at any point of time the angle of the chord with the vertical be θ – which becomes the configuration coordinate. We assume the air friction be represented by a friction coefficient R. We obtain the equation by the Lagrangian method as

$$T = \frac{1}{2}m\left(\dot{\theta}l\right)^2,$$

$$V = -mgl\cos\theta,$$

$$\mathfrak{R} = \frac{1}{2}R\left(\dot{\theta}l\right)^2,$$

$$\mathcal{L} = \frac{1}{2}m\dot{\theta}^2l^2 + mgl\cos\theta.$$

Assuming $q = \theta$, we have the momentum variable as

$$p = \frac{\partial \mathcal{L}}{\partial \dot{q}} = ml^2\dot{q}.$$

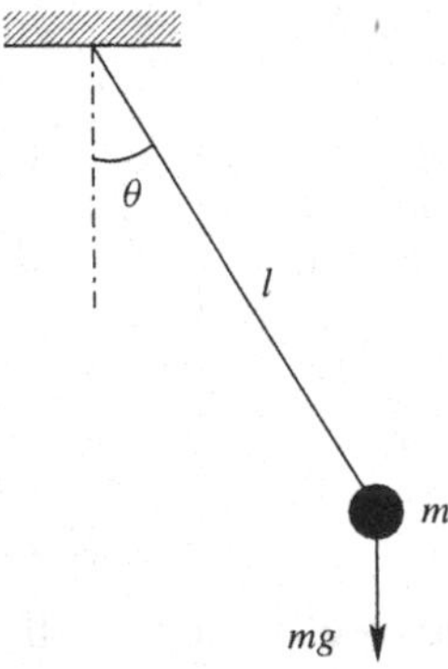

Figure 9.7 The simple pendulum.

This gives the first-order differential equations as

$$\dot{q} = \frac{p}{ml^2}$$

and

$$\begin{aligned}\dot{p} &= -mgl\sin q - Rl^2\dot{q},\\ &= -mgl\sin q - Rl^2\frac{p}{ml^2},\\ &= -mgl\sin q - \frac{Rp}{m}.\end{aligned}$$

By putting $\dot{p} = 0$ and $\dot{q} = 0$, we get $(0, 0)$ as one equilibrium point. Now we proceed to obtain the local linear behaviour around this equilibrium point through the Jacobian matrix:

$$\begin{bmatrix}\dot{x}\\ \dot{y}\end{bmatrix} = \begin{bmatrix}0 & \frac{1}{ml^2}\\ -mgl\cos q & -\frac{R}{m}\end{bmatrix}\begin{bmatrix}x\\ y\end{bmatrix}.$$

At the location of the equilibrium point, $q = 0$. Putting this, we get

$$\begin{bmatrix}\dot{x}\\ \dot{y}\end{bmatrix} = \begin{bmatrix}0 & \frac{1}{ml^2}\\ -mgl & -\frac{R}{m}\end{bmatrix}\begin{bmatrix}x\\ y\end{bmatrix}.$$

The eigenvalues of the square matrix are

$$\lambda = -\frac{R}{2m} \pm \sqrt{\frac{R^2}{4m^2} - \frac{g}{l}}.$$

This implies a few things:

1. If the damping due to air friction is neglected, the frequency of oscillation is $\sqrt{g/l}$. This is independent of the mass of the bob.

2. If the frictional damping is taken into account, the damping factor becomes $-R/2m$, and the frequency of oscillation changes by a very small amount equal to the square of the damping factor.

3. The frequency of oscillation does not depend on the amplitude of oscillation.

This is the basic content of Galileo's observation. We can also see that this observation will be valid so long as the local linear approximation of the nonlinear equations gives a valid representation of the vector field. We know that this is true only in a small neighbourhood of the equilibrium point $(0, 0)$, that is, for small values of p and q.

To appreciate the figures, let us assume some realistic values of the parameters: $m = 0.5$ kg, $l = 1$ m, $R = 0.1$ Ns/m, and let us assume a rounded off value of $g = 10$ m/s^2. With these values, the local linear equation becomes

$$\begin{bmatrix}\dot{x}\\ \dot{y}\end{bmatrix} = \begin{bmatrix}0 & 2\\ -5 & -0.2\end{bmatrix}\begin{bmatrix}x\\ y\end{bmatrix}.$$

The eigenvalues of the square matrix are found to be

$$\lambda = -0.1 \pm j\sqrt{10 - 0.01} \approx -0.1 \pm j\sqrt{10}.$$

We have seen that the solution is given in the form

$$\mathbf{x}(t) = c_1 e^{-0.1t} \begin{pmatrix} -\sin\sqrt{10}t \\ \cos\sqrt{10}t \end{pmatrix} + c_2 e^{-0.1t} \begin{pmatrix} \cos\sqrt{10}t \\ \sin\sqrt{10}t \end{pmatrix}. \tag{9.17}$$

The constants c_1 and c_2 are obtained from the initial conditions. This shows that the oscillations have a frequency of $\sqrt{10}$, which is independent of the initial condition, and the amplitude of oscillation decays as $e^{-0.1t}$. ◄

9.4.4 Eigenvalues real and equal

When the eigenvalues are equal, we have only one real-valued eigenvector $\mathbf{v}$ associated with the eigenvalue λ. So we have only one solution

$$\mathbf{x}_1(t) = e^{\lambda t}\mathbf{v}.$$

In this case, the rule is to look for a second solution of the form

$$\mathbf{x}_2(t) = e^{\lambda t} \begin{bmatrix} A_1 + A_2 t \\ B_1 + B_2 t \end{bmatrix}, \tag{9.18}$$

so that the general solution is

$$\mathbf{x}(t) = c_1 e^{\lambda t} \begin{bmatrix} v_1 \\ v_2 \end{bmatrix} + c_2 e^{\lambda t} \begin{bmatrix} A_1 + A_2 t \\ B_1 + B_2 t \end{bmatrix}. \tag{9.19}$$

► **Example 9.4** Let

$$\begin{aligned} \dot{x} &= 3x - 4y, \\ \dot{y} &= x - y. \end{aligned} \tag{9.20}$$

Here both the eigenvalues are equal to 1. For this eigenvalue the eigenvector equation is $x = 2y$, so that an eigenvector $\mathbf{v} = [2, 1]^T$. Thus, one non-trivial solution is

$$\mathbf{x}_1(t) = e^t \mathbf{v}.$$

We now seek another solution of the form (9.18) with $\lambda = 1$. Substituting this into (9.20), we get

$$\begin{aligned} (A_1 + A_2 t + A_2)e^t &= 3(A_1 + A_2 t)e^t - 4(B_1 + B_2 t)e^t, \\ (B_1 + B_2 t + B_2)e^t &= (A_1 + A_2 t)e^t - (B_1 + B_2 t)e^t. \end{aligned}$$

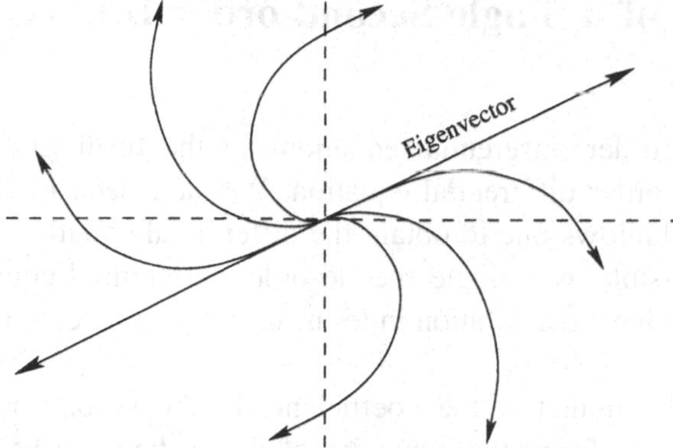

Figure 9.8 The vector field in the state space for the system in Example 9.4, which has only one eigenvalue with eigenvector $[2, 1]^T$.

This reduces to

$$(2A_1 - A_2 - 4B_1) + (2A_2 - 4B_2)t = 0,$$
$$(A_1 - 2B_1 - B_2) + (A_2 - 2B_2)t = 0.$$

Since these equations must hold good independent of t, each term in the above equations must be zero, that is,

$$2A_1 - A_2 - 4B_1 = 0,$$
$$A_1 - 2B_1 - B_2 = 0,$$
$$2A_2 - 4B_2 = 0,$$
$$A_2 - 2B_2 = 0.$$

Solving these, we have $A_1 - 2B_1 = 1$, $A_2 = 2$ and $B_2 = 1$. Since we can take any values of A_1 and B_1 satisfying these equations, we take $A_1 = 1$ and $B_1 = 0$. This gives another linearly independent solution of (9.20) as

$$\mathbf{x}_2(t) = e^t \begin{bmatrix} 1 + 2t \\ t \end{bmatrix}.$$

Hence, the general solution of (9.20) is

$$\mathbf{x}(t) = c_1 e^t \begin{bmatrix} 2 \\ 1 \end{bmatrix} + c_2 e^t \begin{bmatrix} 1 + 2t \\ t \end{bmatrix}.$$

The vector field for this case is shown in Fig. 9.8. ◀

9.5 Solution of a Single Second-order Differential Equation

A set of two first-order differential equations in the form (9.2) is also equivalent to one second-order differential equation. We have seen in Chapter 4 that the Lagrangian method allows one to obtain the differential equations in second order. For systems expressible as a single second-order differential equation, it is sometimes convenient to have the solution in terms of the parameters of the second-order equation.

We shall soon find that if the coefficients in the second-order equation are expressed in a specific form, the dynamics of the system can be predicted simply by looking at the coefficients. Owing to this advantage, it has become customary to express second-order equations in the form

$$\frac{\mathrm{d}^2 q}{\mathrm{d}t^2} + 2\zeta\omega_n \frac{\mathrm{d}q}{\mathrm{d}t} + \omega_n^2 q = 0. \tag{9.21}$$

The two new parameters are ζ (pronounced zeta) and ω_n (pronounced omega n). Expressing it in terms of the differential operator p, we get

$$\left(p^2 + 2\zeta\omega_n p + \omega_n^2\right) q = 0,$$

which gives the characteristic equation as

$$p^2 + 2\zeta\omega_n p + \omega_n^2 = 0. \tag{9.22}$$

One might think this substitution of $\mathrm{d}/\mathrm{d}t$ with p and writing an equation in terms of it, as an ad hoc procedure – a mathematical sleight of hand. But there is an underlying logic that justifies this procedure.

Let us ask: What form must the solution $q(t)$ take so as to satisfy equation (9.21)? Note that in this equation a linear combination of q, $\mathrm{d}q/\mathrm{d}t$, and $\mathrm{d}^2q/\mathrm{d}t^2$ add to zero. Therefore, in order to satisfy (9.21) the three terms must be of the same form and order, differing only in the coefficients. We know that this is true for the exponential function, and so one possible solution is

$$q(t) = ke^{pt},$$

where k and p are constants. There is no restriction on the form of these constants, however. These may be real, imaginary or complex. Substituting this form of the solution in (9.21), we get

$$\left(p^2 ke^{pt} + 2\zeta\omega_n pke^{pt} + \omega_n^2 ke^{pt}\right) = 0,$$

which yields (9.22).

The roots of this characteristic equation are given by

$$p_1, p_2 = -\zeta\omega_n \pm \omega_n\sqrt{\zeta^2 - 1}.$$

How do we obtain the solution in terms of these roots? To see this, let us first express the second-order equation (9.21) in terms of two first order equations – whose solution we already know. Assuming

$$q = x, \quad \text{and} \quad \dot{q} = y,$$

we get

$$\begin{aligned} \dot{x} &= y, \\ \dot{y} &= -2\zeta\omega_n y - \omega_n^2 x \end{aligned}$$

or, in matrix form,

$$\begin{bmatrix} \dot{x} \\ \dot{y} \end{bmatrix} = \begin{bmatrix} 0 & 1 \\ -\omega_n^2 & -2\zeta\omega_n \end{bmatrix} \begin{bmatrix} x \\ y \end{bmatrix}. \tag{9.23}$$

In order to obtain its eigenvalues, we write

$$|\mathbf{A} - \lambda\mathbf{I}| = \begin{vmatrix} -\lambda & 1 \\ -\omega_n^2 & -2\zeta\omega_n - \lambda \end{vmatrix} = 0$$

or

$$\lambda^2 + 2\zeta\omega_n\lambda + \omega_n^2 = 0,$$

which has the same form as (9.22). Therefore, the roots of the characteristic equation (9.22) are identical with the eigenvalues of the **A** matrix in the first-order form (9.23).

We already know that the solution of (9.23) can be of three types, depending on the character of the eigenvalues. We can now relate these to the parameters of the equation (9.21) as follows:

Case 1: If $\zeta > 1$, the roots are real and distinct. The solution is given by

$$q(t) = e^{-\zeta\omega_n t}\left(c_1 e^{\omega_n\sqrt{\zeta^2-1}\,t} + c_2 e^{-\omega_n\sqrt{\zeta^2-1}\,t}\right).$$

Case 2: If $\zeta = 1$, the roots are real and equal. The solution is of the form

$$q(t) = (c_1 + c_2 t)\, e^{-\omega_n t}.$$

Case 3: If $0 < \zeta < 1$, the roots are complex conjugate. These can be expressed as

$$p_1, p_2 = \sigma \pm j\omega,$$

where $\sigma = -\zeta\omega_n$ and $\omega = \omega_n\sqrt{1-\zeta^2}$. In this case, the solution is of the form

$$q(t) = e^{-\zeta\omega_n t}\left(c_1 e^{j\omega_n\sqrt{1-\zeta^2}\,t} + c_2 e^{-j\omega_n\sqrt{1-\zeta^2}\,t}\right),$$
$$= e^{\sigma t}\left(c_1 e^{j\omega t} + c_2 e^{-j\omega t}\right),$$

which can finally be written as

$$q(t) = A\, e^{\sigma t}\sin(\omega t + \theta),$$

where the constants A and θ are to be obtained from the initial condition.

There is a geometrical relationship between the parameters, which is seen when we write

$$\sigma^2 + \omega^2 = \zeta^2\omega_n^2 + \omega_n^2\left(1-\zeta^2\right),$$
$$= \omega_n^2.$$

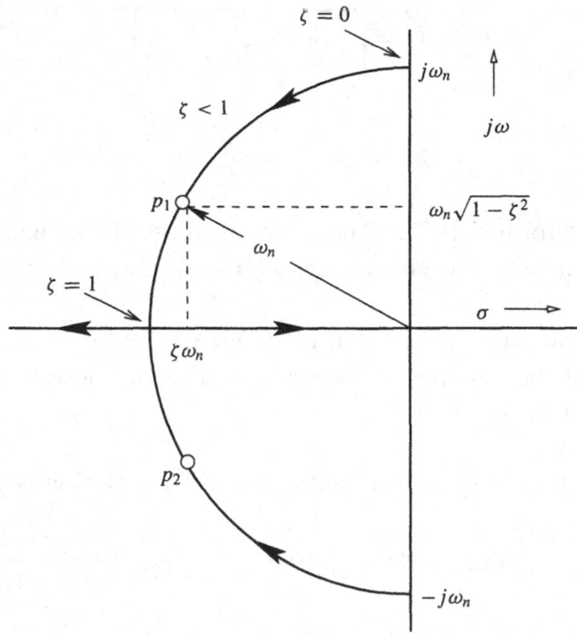

Figure 9.9 The loci of the complex roots, and the relationships among variables as ζ is varied from 0 to ∞.

This implies that in the complex plane the vectors representing the roots of the characteristic equation are of magnitude ω_n. If ζ is varied from 0 to 1 with ω_n remaining constant, the locus of the roots will be a circle of radius ω_n. For different values of ω_n these will be circles of different radii. At $\zeta = 1$ the two roots converge on the real axis. For $\zeta > 1$ the two real roots move away from each other. This is shown in Fig. 9.9.

It is clear therefore, that for $\zeta \geq 1$ the solution is non-oscillatory. If $\sigma = -\zeta\omega_n$ is negative, the solution is an exponentially decreasing function and if σ is positive, the solution is an exponentially increasing function.

For $0 < \zeta < 1$ the solution is oscillatory, and the frequency of oscillation is $\omega = \omega_n\sqrt{1-\zeta^2}$. If $\sigma < 0$, the solution is a damped sinusoid with an exponentially decreasing amplitude, and for $\sigma > 0$, the solution is a sinusoidal function with an exponentially increasing amplitude.

If $\zeta = 0$, the roots are $\pm j\omega_n$, therefore $\sigma = 0$ and $\omega = \omega_n$. In this case, the solution is purely a sinusoid without any damping.

9.6 Systems with Higher Dimensions

So far, we have discussed how to understand the dynamics of two-dimensional linear differential equations in terms of its eigenvalues and eigenvectors. For systems of higher dimensions, these ideas can be extended easily. We illustrate it with an example.

► **Example 9.5** Let the system be

$$\dot{\mathbf{x}} = \begin{pmatrix} 0 & 1 & 0 \\ 0 & 0 & 1 \\ -2 & 1 & 2 \end{pmatrix} \mathbf{x}. \tag{9.24}$$

The characteristic equation is

$$\begin{vmatrix} -\lambda & 1 & 0 \\ 0 & -\lambda & 1 \\ -2 & 1 & 2-\lambda \end{vmatrix} = \lambda^2(2-\lambda) - 2 + \lambda = (\lambda^2 - 1)(2-\lambda).$$

Therefore, the eigenvalues are $\lambda_1 = -1$, $\lambda_2 = 1$ and $\lambda_3 = 2$. For $\lambda_1 = -1$, the eigenvector equation is

$$\begin{pmatrix} 1 & 1 & 0 \\ 0 & 1 & 1 \\ -2 & 1 & 3 \end{pmatrix} \begin{pmatrix} v_1 \\ v_2 \\ v_3 \end{pmatrix} = \begin{pmatrix} v_1 + v_2 \\ v_2 + v_3 \\ -2v_1 + v_2 + v_3 \end{pmatrix} = \begin{pmatrix} 0 \\ 0 \\ 0 \end{pmatrix}.$$

The first two equations give $v_1 = -v_2$ and $v_3 = -v_2$. These two equations make the third equation redundant (the reader may check that). Choosing $v_2 = -1$, we get the eigenvector

$[1, -1, 1]^T$ so that a solution becomes

$$\mathbf{x}_1(t) = \begin{pmatrix} 1 \\ -1 \\ 1 \end{pmatrix} e^{-t}.$$

Likewise, for $\lambda_2 = 1$, we get the eigenvector conditions as $v_1 = v_2$ and $v_3 = v_2$ (the third equation being redundant), and setting $v_2 = 1$, we obtain another solution

$$\mathbf{x}_2(t) = \begin{pmatrix} 1 \\ 1 \\ 1 \end{pmatrix} e^{t}.$$

For $\lambda_3 = 2$, we get the eigenvector conditions as $v_2 = 2v_1$ and $v_3 = 2v_2 = 4v_1$. Setting $v_1 = 1$, we get another solution

$$\mathbf{x}_3(t) = \begin{pmatrix} 1 \\ 2 \\ 4 \end{pmatrix} e^{2t}.$$

Thus, the general solution is

$$\mathbf{x}(t) = c_1 \begin{pmatrix} 1 \\ -1 \\ 1 \end{pmatrix} e^{-t} + c_2 \begin{pmatrix} 1 \\ 1 \\ 1 \end{pmatrix} e^{t} + c_3 \begin{pmatrix} 1 \\ 2 \\ 4 \end{pmatrix} e^{2t}.$$ ◀

Notice that the system is stable along the eigenvector $[1, -1, 1]^T$, and unstable along the other two. We may imagine a plane passing through the eigenvectors $[1, 1, 1]^T$ and $[1, 2, 4]^T$, and the system is unstable in this subspace. The subspaces, which have the property that any initial condition placed on the subspace will always remain on it, are called *eigenspace*. The subspace spanned by the vectors $[1, 1, 1]^T$ and $[1, 2, 4]^T$ is an example of an unstable eigenspace, while the line along the eigenvector $[1, -1, 1]^T$ is a stable eigenspace.

In a given linear system, if some eigenvalues are real and negative while some others are real and positive, the system is stable in the subspace spanned by the eigenvectors associated with the negative eigenvalues, and is unstable away from this.

▶ **Example 9.6** Let

$$\dot{\mathbf{x}} = \begin{pmatrix} 1 & 2 & 0 \\ -1 & -1 & 0 \\ 1 & 0 & -1 \end{pmatrix} \mathbf{x}. \tag{9.25}$$

Here the eigenvalues are $\lambda = +j, -j, -1$. Following the same procedure as in earlier problems, we find that for $\lambda_1 = +j$, an eigenvector is $[1 + j, -1, 1]^T$; for $\lambda_2 = -j$, an eigenvector is $[1 - j, -1, 1]^T$; and for $\lambda_3 = -1$, an eigenvector is $[0, 0, 1]^T$. The three solutions are obtained as

$$\mathbf{x}_1(t) = \begin{pmatrix} \cos t - \sin t \\ -\cos t \\ \cos t \end{pmatrix}, \quad \mathbf{x}_2(t) = \begin{pmatrix} \sin t + \cos t \\ -\sin t \\ \sin t \end{pmatrix}, \quad \mathbf{x}_3(t) = \begin{pmatrix} 0 \\ 0 \\ e^{-t} \end{pmatrix}.$$

The general solution is a linear combination of these three solutions.

Now notice that the system is stable in the z-direction, and therefore there will be no steady-state deflection in this direction. The other two eigenvalues are purely imaginary, implying an undamped oscillatory solution in the eigenplane associated with the complex eigenvalues. This is the plane passing through the two vectors obtained from the real and imaginary parts of the complex eigenvectors. These are $[1, -1, 1]^T$ and $[1, 0, 0]^T$. Therefore, asymptotically the orbit converges on the plane passing through these two vectors, and settles on a cyclic orbit in this plane. The radius is given by the initial condition, and the frequency of oscillation is unity. ◄

Since complex eigenvalues can occur only as complex conjugate pairs, a 3-D linear system can either have all three eigenvalues real or one eigenvalue real and the other two complex conjugate.

Following the logic given in the last two examples, it becomes possible to have an intuitive idea of the dynamics without solving the differential equations explicitly.

For example, if two eigenvalues are complex conjugate with negative real part and the third eigenvalue is real and positive, then any orbit will spirally converge onto the eigenvector associated with the real eigenvalue and will diverge along this direction. The rate of divergence along this eigenvector will be given by the magnitude of the real eigenvalue, while the rate of convergence onto this eigenvector will be given by the real part of the complex conjugate eigenvalues (see Fig. 9.10).

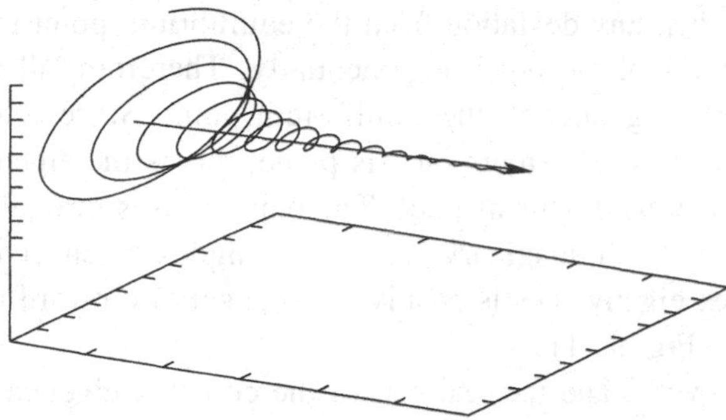

Figure 9.10 The typical orbit for a system with one positive real eigenvalue and two complex conjugate eigenvalues with negative real part.

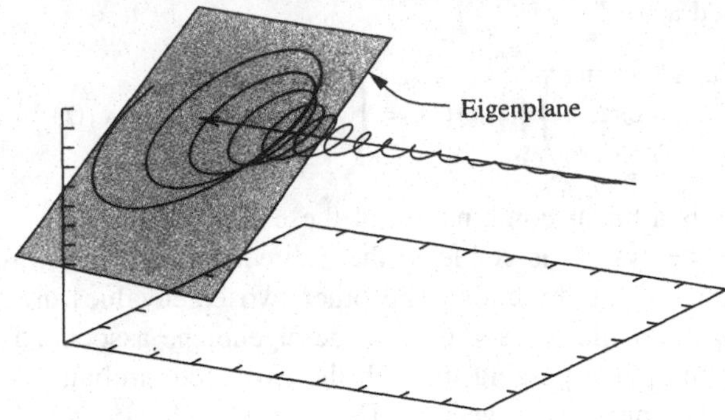

Figure 9.11 The typical orbit for a system with one negative real eigenvalue and two complex conjugate eigenvalues with positive real part.

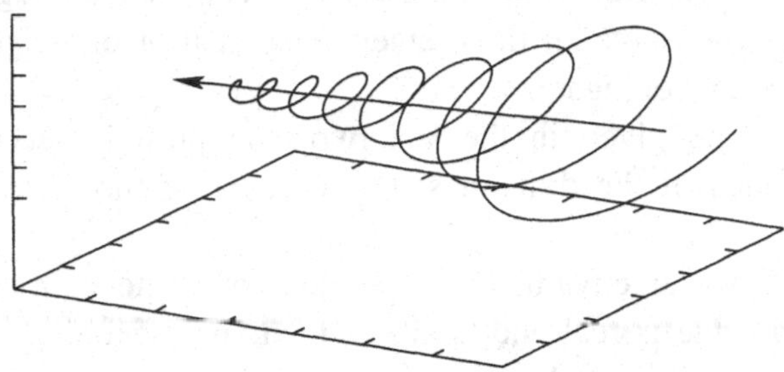

Figure 9.12 An orbit for a system with one negative real eigenvalue and two complex conjugate eigenvalues with negative real part.

If the complex conjugate eigenvalues have positive real part and the real eigenvalue is negative, then any deviation from the equilibrium point in the direction of the real eigenvector will die down exponentially. Therefore, all orbits will move towards a plane passing through the equilibrium point. Since orbits starting from points on this plane forever remain on this plane, this is the eigenspace associated with the complex conjugate eigenvalues. The plane passes through the two vectors obtained from the real and imaginary parts of a complex eigenvector. Since the real part of the complex eigenvalues is positive, orbits spiral outward on this plane and diverge to infinity (Fig. 9.11).

If the real eigenvalue and the real part of the complex eigenvalues are all negative, then all orbits spiral onto the real eigenvector, and the deviation in the direction of this eigenvector decays exponentially (Fig. 9.12).

In a similar manner, one can draw a lot of physically meaningful conclusions about the dynamics of a linear system just by calculating the eigenvalues and eigenvectors. Two such examples are illustrated in Fig. 9.13.

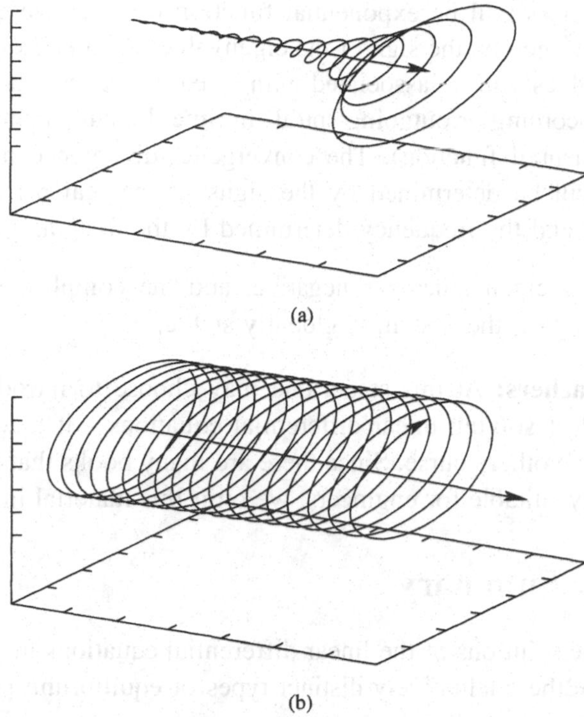

(a)

(b)

Figure 9.13 The typical orbits for a system with (a) one positive real eigenvalue and two complex conjugate eigenvalues with positive real part, and (b) one positive real eigenvalue and two purely imaginary eigenvalues.

For systems of even higher dimension, it might not be possible to depict the trajectories the way we have done so far – because the pictorial representation is possible only for a 2-D system or as the 2-D projection of a 3-D system. Nevertheless, the intuitive understanding we have developed can be extended to such systems also. Following the logic developed so far, we can conclude the following.

1. If there are n number of eigenvalues real and negative, and m number of eigenvalues real and positive, then there is a stable subspace containing the eigenvectors associated with the negative eigenvalues. Any initial condition placed exactly in this subspace will exponentially converge onto the equilibrium point. But even a slight deviation from it will grow exponentially, and therefore for $m \geq 1$ the system is globally unstable.

2. If there are n number of eigenvalues real and m (even) number of eigenvalues complex conjugate, then the dynamics will be combinations of exponential functions and damped sinusoids. The dynamics of the components along the

real eigenvectors will be exponential functions, and convergence/divergence will be determined by the sign of the eigenvalue. Each pair of complex conjugate eigenvalues will be associated with a real subspace where the dynamics will be an incoming or outgoing spiral (in time domain, a sinusoid modulated by an exponential function). The convergence/divergence of orbits in these subspaces will be determined by the signs of the real part of the complex eigenvalues, and the frequency determined by the imaginary part.

3. If all the real eigenvalues are negative, and the complex eigenvalues have negative real part, the system is globally stable.

A note to the teachers: At this stage, one may choose to introduce the Laplace transform method of solving linear differential equations – if that has not already been taught in some other course. Since there are many books that treat the Laplace transform in a way suitable for engineers, we omit this material in the present text.

9.7 Chapter Summary

One can obtain the solutions of the linear differential equations in closed form. The eigenvalues define the qualitatively distinct types of equilibrium points:

- If the eigenvalues are real and negative, the equilibrium point is a non-oscillatory attractor.
- If the eigenvalues are real and positive, the equilibrium point is a non-oscillatory repeller.
- If the eigenvalues are real, with some positive and some negative, the equilibrium point is a saddle.
- If the eigenvalues are complex conjugate with negative real part, the equilibrium point is a spiral attractor.
- If the eigenvalues are complex conjugate with positive real part, the equilibrium point is a spiral repeller.

For systems with higher dimensions, one can identify *subspaces* with the property that if the initial condition is in a subspace, then the whole orbit remains in that subspace. If the eigenvalues associated with a subspace are negative or complex conjugate with negative real part, orbits in that subspace converge on to the equilibrium point. That subspace is said to be a *stable* subspace. Likewise, if the eigenvalues associated with a subspace are positive or complex conjugate with positive real part, the subspace is unstable. The dynamics starting from any arbitrary initial condition is a linear combination of the dynamics in the invariant subspaces.

Further Reading

G. F. Simmons, *Differential Equations with Applications and Historical Notes*, McGraw Hill, New York, 1972.

Problems

1. Derive the solution for the system

$$\begin{bmatrix} \dot{x}_1 \\ \dot{x}_2 \end{bmatrix} = \begin{bmatrix} -1 & a \\ 0.5 & -1 \end{bmatrix} \begin{bmatrix} x_1 \\ x_2 \end{bmatrix}$$

for $a = 0.5$, and initial condition $x_1 = -1$, $x_2 = 1$. Describe the change in the vector field as the parameter a is varied from $+0.5$ through 0 to -0.5.

2. Visualize the change in the vector field of a two-dimensional linear system when its eigenvalues shift in position as shown.

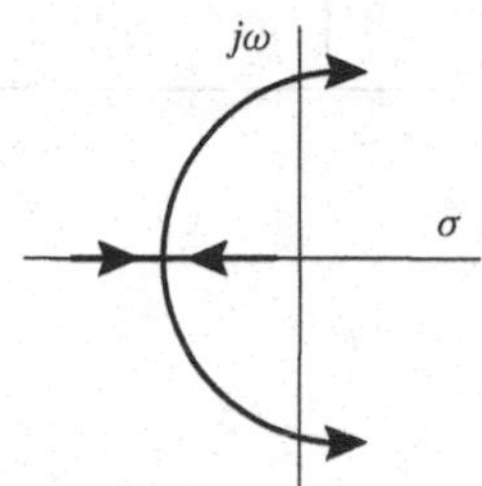

3. Obtain the solutions of the following sets of linear differential equations. Schematically sketch the vector fields

(a) $\dot{x} = -2x + 3y, \quad \dot{y} = 7x - 6y,$

(b) $\dot{x} = x + 2y, \quad \dot{y} = 3x + 2y,$

(c) $\dot{x} = x + y, \quad \dot{y} = 4x - 2y,$

(d) $\dot{x} = 3x - 4y, \quad \dot{y} = x - y,$

(e) $\dot{x} = 54x + y, \quad \dot{y} = -x + y,$

(f) $\dot{x} = x - 2y, \quad \dot{y} = 4x + 5y,$

(g) $\dot{x} = 4x - 3y, \quad \dot{y} = 8x - 6y,$

(h) $\dot{x} = 3x - 2y, \quad \dot{y} = 5x - 4y,$

(i) $\dot{x} = -x - 5y, \quad \dot{y} = x + 3y.$

4. Find the solution of the following second-order differential equations:

   (a) $\ddot{x} + 3\dot{x} + 2x = 0$.

   (b) $\ddot{x} + \dot{x} + 6x = 0$.

   (c) $\ddot{x} + 5\dot{x} + 6x = 0$.

   (d) $\ddot{x} + 5\dot{x} + 4x = 0$.

   (e) $\ddot{x} + 3\dot{x} + 5x = 0$.

5. A third-order linear system has eigenvalues $\lambda_1 = 0.5 + j2$, $\lambda_2 = 0.5 - j2$, and $\lambda_3 = -8.7$. The eigenvector associated with the real eigenvalue is $[-1, 1, 1]^T$. Describe the general properties of the orbits in the state space, starting from any initial condition other than the origin.

6. In the circuit shown, the switch is in position A for a long time, and is moved to position B at $t = 0$. Obtain the equation for the current through the 5 Ω resistance following this instant.

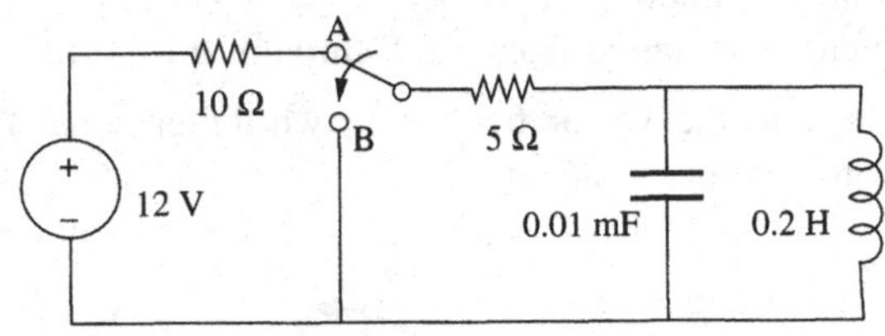

10

Linear Systems with External Input

At the beginning of the last chapter, we said that in general, linear dynamical systems are described by differential equations of the form

$$\dot{\mathbf{x}} = \mathbf{A}\mathbf{x} + \mathbf{B}\mathbf{u},$$

and then proceeded to obtain the dynamics of the unforced system

$$\dot{\mathbf{x}} = \mathbf{A}\mathbf{x}$$

to get an idea about the dynamics in the neighbourhood of the equilibrium point. When an external periodic input is considered, the resulting dynamics will be a combination of the dynamics in the neighbourhood of the equilibrium point, and the position of the equilibrium point.

10.1 Constant External Input

When the external input is a constant – like a dc voltage applied to an electrical circuit or a constant force applied to a mass in a mechanical system – the equilibrium point will not be at the origin. The dynamics can be easily worked out using the knowledge about the position of the equilibrium point and the motion of the state point around the equilibrium.

10.1.1 Constant voltage applied to an RL circuit

Let us illustrate this with reference to the simple RL circuit supplied with a dc voltage as shown in Fig. 10.1. Assume that initially the inductor has no stored

Dynamics for Engineers S. Banerjee

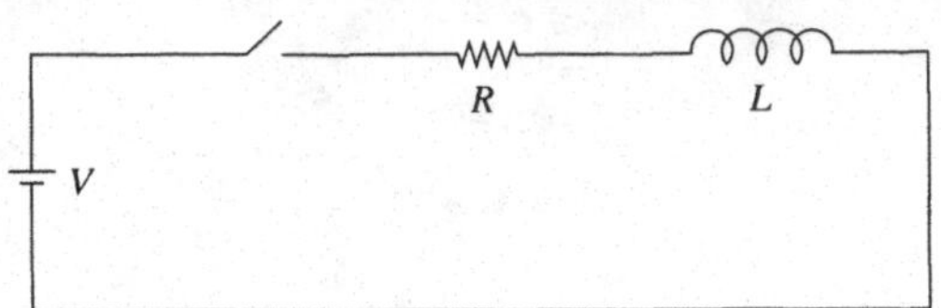

Figure 10.1 The circuit under consideration.

magnetic energy, that is, $i = 0$. By Kirchoff's voltage law the system equation can be written as

$$V = Ri + L\frac{di}{dt}$$

or

$$\frac{di}{dt} = \frac{V}{L} - \frac{R}{L}i. \tag{10.1}$$

We first demonstrate the standard way of solving this equation, as

$$dt = L\,\frac{di}{V - Ri}.$$

Integrating, we get

$$t + k = -\frac{L}{R}\ln(V - Ri)$$

$$\text{or}\quad V - Ri = Ae^{-\frac{R}{L}t},$$

where A is another constant that can be determined from the initial condition. Since at $t = 0$, $i = 0$, we get

$$V - R.0 = Ae^0$$

or $A = V$. Substituting in the solution, we get

$$V - Ri = Ve^{-\frac{R}{L}t}$$

$$\text{or}\quad i = \frac{V}{R}\left(1 - e^{-\frac{R}{L}t}\right).$$

This is the standard way of solving it, found in all textbooks. Now let us take a different look. From (10.1), we see that the equilibrium point obtained by putting $di/dt = 0$, is $i = V/R$. If we now define a new variable

$$x = i - V/R,$$

that is, the deviation of i from its steady-state value, then the equation becomes simply

$$\dot{x} = -\frac{R}{L}x.$$

We have already seen that its solution is

$$x = x_0\, e^{-\frac{R}{L}t}.$$

The initial condition x_0 is the initial deviation of i from the steady-state value, that is, $x_0 = -V/R$. Hence, around the equilibrium point, the dynamics is

$$x = -\frac{V}{R}e^{-\frac{R}{L}t}.$$

This gives the evolution of the current as

$$i = \frac{V}{R} - \frac{V}{R}e^{-\frac{R}{L}t}$$

or

$$i = \frac{V}{R}\left(1 - e^{-\frac{R}{L}t}\right). \qquad (10.2)$$

In this approach, what have we done? We have first reduced the system equation into the form for the unforced system by moving the origin to the equilibrium point. That allows us to intuitively understand the dynamics in terms of the coefficient in the right-hand side (or, for higher dynamical systems, in terms of the eigenvalues) and to obtain the dynamics in the neighbourhood of the equilibrium point. Finally to get the picture in terms of the actual state variable, we substitute the change of coordinate.

From this simple example, the advantage of this approach may not be salient. But when we consider more complicated systems, this gives us a visually intuitive understanding of what can happen in such a system. We shall illustrate that in later sections.

10.1.2 The concept of time constant

It is clear from equation (10.2) that at $t = 0$, $i = 0$ and at $t = \infty$, the final value is V/R. In between, the curve is exponential, as is shown in Fig. 10.2. One would be interested in knowing how fast or how slow the current rises. But this is rather difficult to pinpoint because in all cases it takes infinite time for the current to reach the steady-state value $I_{st} = V/R$.

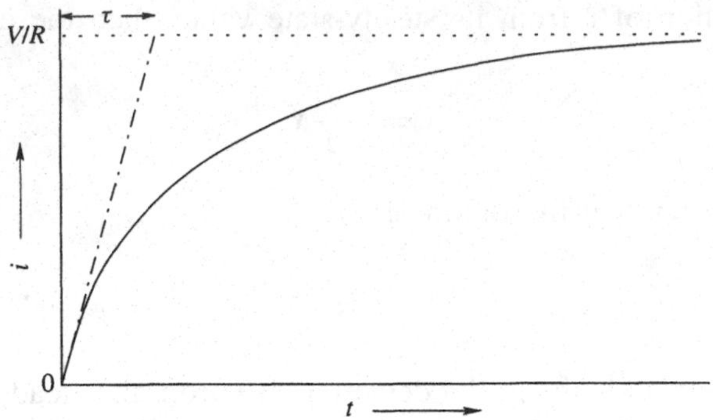

Figure 10.2 The solution given by (10.2).

To overcome this problem, we write the equation (10.2) as

$$i = \frac{V}{R}\left(1 - e^{-t/\tau}\right), \tag{10.3}$$

where $\tau = L/R$, called the *time constant*. Now note that the slope of the current function is

$$\frac{\mathrm{d}i}{\mathrm{d}t} = \frac{V}{\tau R}e^{-t/\tau}.$$

At $t = 0$,

$$\frac{\mathrm{d}i}{\mathrm{d}t} = \frac{V}{\tau R}$$

$$\text{or} \quad \tau\frac{\mathrm{d}i}{\mathrm{d}t} = \frac{V}{R}.$$

This means that if the slope of the current function at $t = 0$ were maintained (as shown by the dash-dot line in Fig. 10.2), then after a lapse of τ seconds it would assume the steady-state value of I_{st}.

It is clear that for small τ the current rises fast, and for large τ the current rises slowly. The definition of the time constant therefore allows us to specify quantitatively how fast or how slow the current rises.

Again, note that at $t = \tau$,

$$i(t) = I_{\mathrm{st}}\left(1 - e^{-1}\right) \approx 0.63 I_{\mathrm{st}}.$$

Thus, the time constant can also be defined as the time after which the current assumes 63% of its final value.

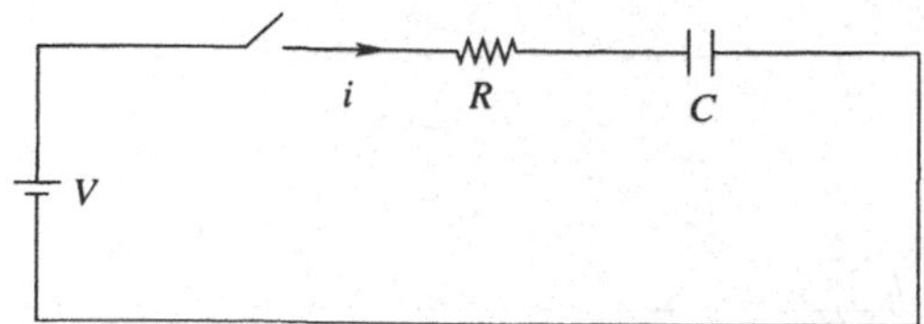

Figure 10.3 The RC circuit under consideration.

We can express the voltages across the resistor and the inductor in terms of the time constant as

$$v_R = iR = V\left(1 - e^{-t/\tau}\right),$$

$$v_L = V - v_R = Ve^{-t/\tau},$$

which means that v_R increases exponentially from zero to V, and v_L decays exponentially from V to zero.

10.1.3 Constant voltage applied to an RC circuit

When an RC circuit as shown in Fig. 10.3 is excited by a constant voltage source, the differential equation obtained from KVL is

$$V = Ri + \frac{1}{C}\int_0^t i\,\mathrm{d}t.$$

Differentiating, we get

$$0 = R\frac{\mathrm{d}i}{\mathrm{d}t} + \frac{1}{C}i$$

$$\text{or} \quad \frac{\mathrm{d}i}{\mathrm{d}t} = -\frac{1}{RC}i.$$

Note that the forcing function does not appear in the current equation. Its solution is

$$i(t) = i_0 e^{-\frac{1}{RC}t}.$$

Now, if we assume that the initial charge stored in the capacitor is zero, then $i_0 = V/R$. Defining the time constant $\tau = RC$, we can write

$$i = \frac{V}{R}e^{-t/\tau},$$

$$v_R = iR = Ve^{-t/\tau},$$

$$v_C = V - v_R = V\left(1 - e^{-t/\tau}\right).$$

The waveforms are shown in Fig. 10.4. The equations also tell that if tangents to the voltage and current waveforms are drawn at $t = 0$, they reach the steady-state

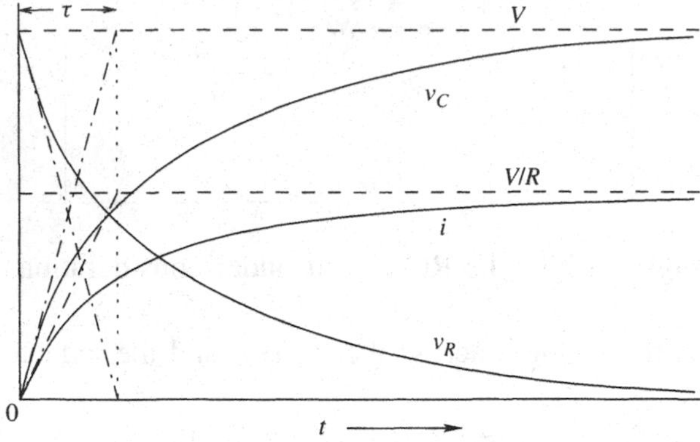

Figure 10.4 The evolution of the variables in the RC circuit.

values of the respective variables after a lapse of τ seconds. It is clear that if the RC time constant is large the system response becomes sluggish, and if it is small, the response is fast.

10.1.4 Constant voltage applied to an RLC circuit

In case of the RLC circuit of Fig. 10.5, the Kirchoff's voltage law gives

$$Ri + L\frac{\mathrm{d}i}{\mathrm{d}t} + \frac{1}{C}\int i\,\mathrm{d}t = V.$$

Differentiating, we get

$$L\frac{\mathrm{d}^2 i}{\mathrm{d}t^2} + R\frac{\mathrm{d}i}{\mathrm{d}t} + \frac{1}{C}i = 0.$$

Thus, in this case also, if the equation is written in terms of i, the equation takes the form of an unforced system. Now, we express this equation in the form

$$\frac{\mathrm{d}^2 i}{\mathrm{d}t^2} + 2\zeta\omega_n\frac{\mathrm{d}i}{\mathrm{d}t} + \omega_n^2 i = 0,$$

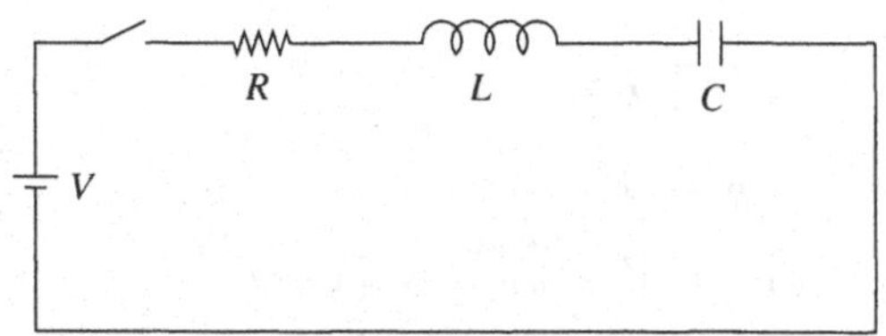

Figure 10.5 The RLC circuit under consideration.

where

$$\zeta = \frac{R}{2}\sqrt{\frac{C}{L}} \quad \text{and} \quad \omega_n = \frac{1}{\sqrt{LC}}.$$

We have already seen in Section 9.5 that when the equation is expressed in this form, the dynamics are completely given by the values of ζ and ω_n.

In particular, we know that if $\zeta > 1$ the eigenvalues of the system are real and hence the response is non-oscillatory. If $\zeta < 1$, the eigenvalues are complex conjugate and the response is a damped sinusoid as given in Fig. 9.5. One can also see that ζ depends on R, and hence if R is progressively increased, the system response may change from being under-damped (oscillatory) to over-damped (non-oscillatory). At which critical value of R does the behaviour change? This is obtained by expressing ζ as

$$\zeta = \frac{R}{R_{\text{cr}}},$$

whence we get

$$R_{\text{cr}} = 2\sqrt{\frac{L}{C}}.$$

10.2 When the Forcing Function is a Square Wave

If the external periodic input is a square wave as shown in Fig. 10.6, then for the period t_1 the applied input is V_1, resulting in a specific location of the equilibrium point. The dynamics in that period will be guided by the matrix **A**, and the initial deviation from that equilibrium point. During this interval, the state point will approach the corresponding equilibrium point; if the eigenvalues of the **A** matrix

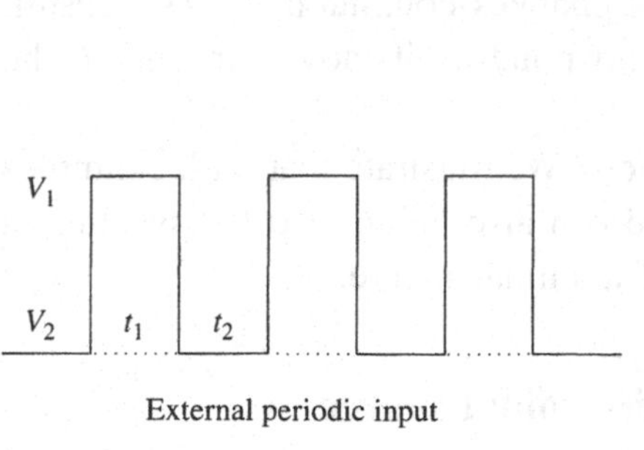

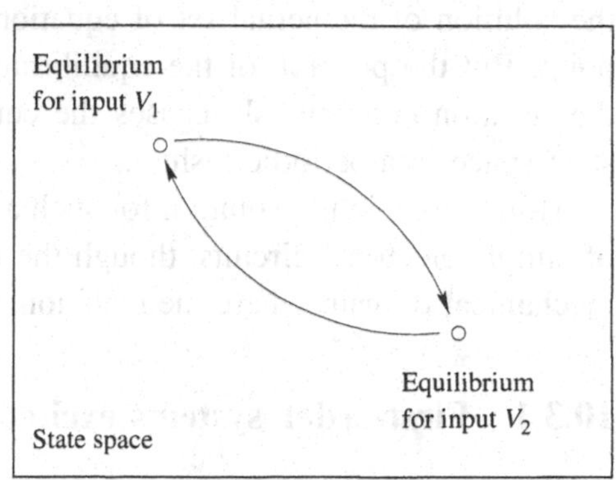

Figure 10.6 The shift in the position of the equilibrium point in the state space as a result of a varying input.

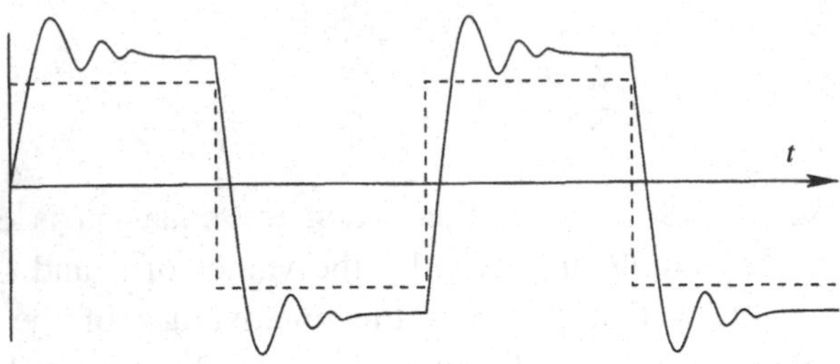

Figure 10.7 The applied voltage (broken line) and the current (firm line) for the RLC circuit with square-wave input.

are complex conjugate, the state will move in a spiral motion, and if they are real the state will follow an exponential function. At time t_1, the applied input changes to V_2, and as a result the equilibrium point moves to a new location. The final condition of the state at the end of the period t_1 becomes the initial condition of the period t_2. If this is now measured in terms of the deviation from the new equilibrium point, the dynamics will be given by the matrix **A** with the new equilibrium point as origin. This process will continue in each period of the forcing function, resulting in a current waveform as shown in Fig. 10.7.

10.3 Sinusoidal Forcing Function

If the forcing function is a continuously varying function like a sinusoidal waveform, the equilibrium point moves continuously as the magnitude of the forcing function changes. If the matrix **A** has eigenvalues with real part negative, that is, if the corresponding homogeneous set of equations gives a stable solution, then the solution of the actual set of equations tends to converge on to the equilibrium point. But the position of the equilibrium point changes continuously. As a result, the solution continuously chases the equilibrium point as it moves around in the state space in a periodic fashion.

How to obtain the solution for such a situation? We illustrate that with examples of simple electrical circuits, though the method can also be applied for systems in mechanical domain where the equations are of a similar nature.

10.3.1 First-order systems excited by sinusoidal source

We first consider a resistance–capacitance circuit supplied by a sinusoidal voltage

$$v(t) = V_m \sin \omega t$$

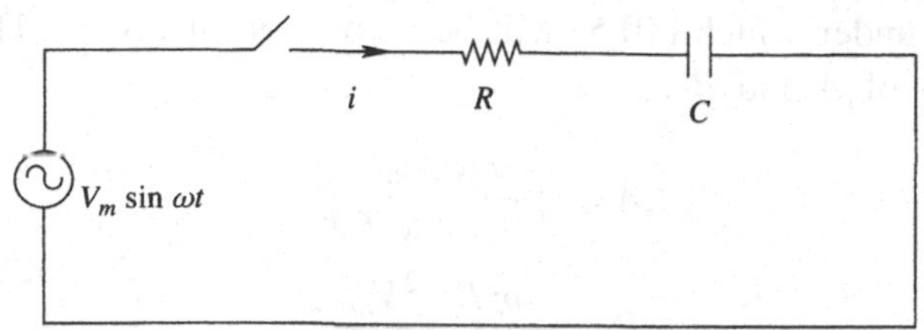

Figure 10.8 The RC network excited by a sinusoidal voltage source.

as shown in Fig. 10.8. Kirchoff's voltage law gives

$$Ri + C\int i\,\mathrm{d}t = V_m \sin\omega t.$$

Differentiating and dividing by R, we get

$$\frac{\mathrm{d}i}{\mathrm{d}t} + \frac{1}{RC}i = \frac{\omega V_m}{R}\cos\omega t. \tag{10.4}$$

How to obtain the solution of this equation? It is not difficult to see that, as per the argument given in Section 10.3, in the steady state the solution (let us call it i_p) will follow the sinusoidal forcing function. A good guess, a so-called "trial solution" can therefore be a general sinusoidal function[1]

$$i_p = A\cos\omega t + B\sin\omega t. \tag{10.5}$$

But is it really a solution of (10.4)? To check, we substitute the functional forms of i_p and $\mathrm{d}i_p/\mathrm{d}t$ into (10.4), to get

$$[-A\omega\sin\omega t + B\omega\cos\omega t] + \frac{A}{RC}\cos\omega t + \frac{B}{RC}\sin\omega t = \frac{\omega V_m}{R}\cos\omega t.$$

Under what condition will this equation be satisfied? This is easily obtained by equating the coefficients of $\cos\omega t$ and $\sin\omega t$ at both sides of the equation. Thus, we get

$$\frac{A}{RC} + B\omega = \frac{\omega V_m}{R}$$

and

$$\frac{B}{RC} - A\omega = 0,$$

[1] A sine term has zero value at $t = 0, \pi, 2\pi, \ldots$, and cosine term has zero value at $t = \frac{\pi}{2}, \frac{3\pi}{2}, \frac{5\pi}{2}, \ldots$, that is, the cosine term is shifted from the sine term by $\pi/2$. An addition of the two yields a sinusoid with any arbitrary magnitude and phase, given by the coefficients A and B.

as the condition under which (10.5) will be a solution of (10.4). This allows us to obtain the values of A and B as

$$A = \frac{\omega C V_m}{1 + \omega^2 R^2 C^2},$$

$$B = \frac{\omega^2 R C^2 V_m}{1 + \omega^2 R^2 C^2}.$$

Thus, we find that

$$i_p = \frac{\omega C V_m}{1 + \omega^2 R^2 C^2} \cos \omega t + \frac{\omega^2 R C^2 V_m}{1 + \omega^2 R^2 C^2} \sin \omega t \tag{10.6}$$

is indeed a solution of (10.4).

As we observed earlier, an addition of two sinusoidal terms is also a sinusoidal function with magnitude and phase given by the coefficients of the sine and cosine terms. Can we find that sinusoidal function? In order to do that, we first express (10.6) in the convenient form

$$i_p = \frac{V_m}{R^2 + (1/\omega^2 C^2)} \left(\frac{1}{\omega C} \cos \omega t + R \sin \omega t \right)$$

by simple algebraic manipulation. Then we define

$$Z \sin \phi = 1/\omega C,$$

$$Z \cos \phi = R$$

from where Z and ϕ are obtained as

$$Z^2 = R^2 + \frac{1}{\omega^2 C^2},$$

$$\tan \phi = \frac{1}{\omega R C}.$$

Note that these variables are geometrically related as shown in Fig. 10.9. R is the *impedance* of the resistor, $1/\omega C$ is the impedance of the capacitor, and the total impedance of the RC network is given by Z.

Substituting, we get

$$\begin{aligned} i_p &= \frac{V_m Z}{R^2 + (1/\omega^2 C^2)} \left(\sin \phi \cos \omega t + \cos \phi \sin \omega t \right), \\ &= \frac{V_m}{Z} \sin \left(\omega t + \phi \right). \end{aligned}$$

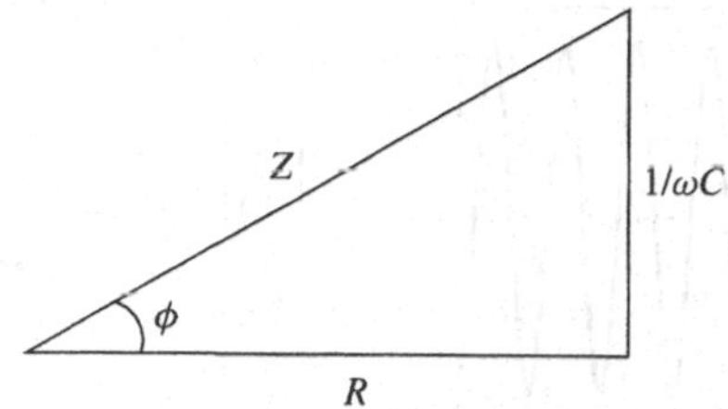

Figure 10.9 The relation between the variables in the RC circuit.

The magnitude is the applied voltage divided by the impedance Z. The current *leads* the applied voltage, and the phase shift is given by the angle ϕ obtained from the impedance triangle in Fig. 10.9.

Note that this is only the steady-state solution. There will also be a transient behaviour that will last for some time after the switch is turned on. What will be the transient behaviour?

It is not difficult to see that the transient will be an exponentially decaying function, and for the RC network its form is

$$i_c = ke^{-t/RC}.$$

The value of k has to be obtained from the initial condition, that is, by putting $t = 0$ in the complete solution and equating it with the initial value of i.

Notice that the initial value of i_p is

$$i_p|_{t=0} = \frac{\omega C V_m}{1 + \omega^2 C^2 R^2}.$$

If the initial charge stored in the capacitor is zero, the initial current in the circuit is the instantaneous value of the applied voltage divided by R. If the switch is turned on when the voltage is zero, the initial current is also zero. Therefore, the value of k should be such that the complete solution $i_p + i_c$ has initial value zero. This gives

$$i_c = -\frac{\omega C V_m}{1 + \omega^2 C^2 R^2} e^{-t/RC}.$$

Therefore, the complete solution of (10.4) is

$$\begin{aligned} i &= i_p + i_c, \\ &= \frac{V_m}{\sqrt{R^2 + (1/\omega^2 C^2)}} \sin\left(\omega t + \tan^{-1}\frac{1}{\omega RC}\right) - \frac{\omega C V_m}{1 + \omega^2 C^2 R^2} e^{-t/RC}. \end{aligned}$$

The part of the solution that refers to the steady-state behaviour is called the *particular integral* and the part that refers to the transient behaviour is called the

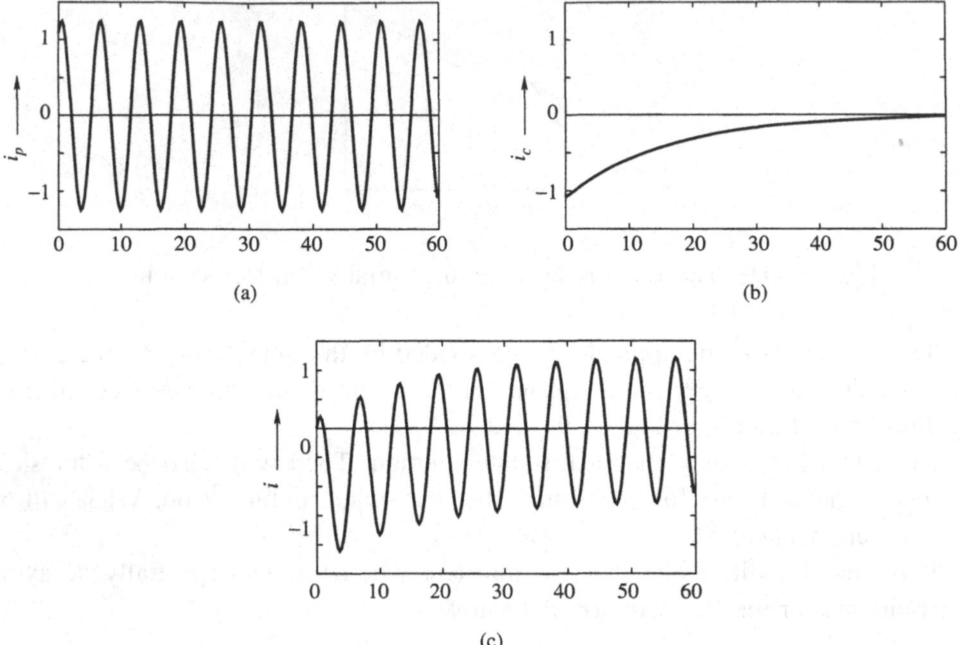

Figure 10.10 The graphs for (a) the particular integral i_p, (b) the complementary function i_c and (c) the complete solution i for $V_m = 10$ V, $R = 4\ \Omega$, $C = 1/4\sqrt{3}$, $Z = 8\ \Omega$, $\phi = \pi/3$, and $\omega = 1$ Hz.

complementary function. Note that at $t = 0$ the value of i exactly matches its initial condition. As time progresses, the complementary function part becomes smaller and the solution settles into the sinusoidal function given by i_p. The waveforms for a specific case are shown in Fig. 10.10.

If a sinusoidal excitation is applied on a series RL circuit, the differential equation becomes

$$L\frac{\mathrm{d}i}{\mathrm{d}t} + Ri = V_m \sin \omega t.$$

Proceeding along similar lines, it can be shown that the particular integral will be

$$i_p = \frac{V_m}{Z} \sin(\omega t - \phi),$$

where

$$Z = \sqrt{R^2 + \omega^2 L^2},$$

$$\phi = \tan^{-1} \frac{\omega L}{R}.$$

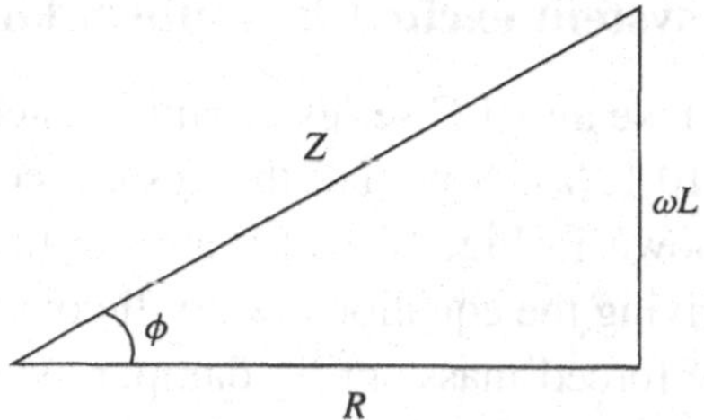

Figure 10.11 The relation between the variables in the RL circuit.

This implies that the steady-state current *lags* the applied voltage by an angle ϕ, obtained from an impedance triangle as shown in Fig. 10.11. The complementary function is

$$i_c = ke^{-(R/L)t},$$

where k has to be obtained from the initial condition.

What will happen if the switch is closed at an arbitrary instant when the value of the sinusoidal forcing function is not zero? This will be reflected by some phase shift in the excitation function, making the differential equation

$$L\frac{\mathrm{d}i}{\mathrm{d}t} + Ri = V_m \sin(\omega t + \theta).$$

Approaching by the same method, we obtain the total solution as

$$i = \frac{V_m}{Z}\sin(\omega t + \theta - \phi) + ke^{-(R/L)t},$$

which implies that the phase of the current waveform in steady state is still shifted from the applied voltage waveform by an angle ϕ. But it reveals another interesting aspect. The current through the circuit was zero before the closure of the switch, and since it contains an inductor, the current cannot change instantaneously after closure of the switch. We can obtain k from the initial condition $i_0 = 0$ as

$$\frac{V_m}{Z}\sin(0 + \theta - \phi) + ke^0 = 0$$

or

$$k = -\frac{V_m}{Z}\sin(\theta - \phi).$$

This means that if $\theta = \phi$, that is, if the angle of the sinusoid at the instant of switching is equal to the angle ϕ, then $k = 0$, and so there will be no transient.

Note that what we have illustrated here with reference to electrical circuits also apply to first-order mechanical systems with sinusoidal excitation.

10.3.2 Second-order system excited by sinusoidal source

To illustrate this case, we take an RLC series circuit excited by a sinusoidal voltage source, as shown in Fig. 10.12(a). Note that the system considered is equivalent to the mechanical system shown in Fig. 10.12(b), and so the two systems will show the same dynamics. In deriving the equations, we will consider the electrical system. Those for the sinusoidally forced mass-spring-damper system can be obtained in a similar manner.

In this case, the system equation is

$$Ri + L\frac{\mathrm{d}i}{\mathrm{d}t} + \frac{1}{C}\int i\,\mathrm{d}t = V_m \sin \omega t.$$

Differentiating, we get

$$L\frac{\mathrm{d}^2 i}{\mathrm{d}t^2} + R\frac{\mathrm{d}i}{\mathrm{d}t} + \frac{1}{C}i = V_m \omega \cos \omega t.$$

Again, let a trial solution be of the form

$$i_p = A\cos \omega t + B \sin \omega t, \tag{10.7}$$

which gives

$$\frac{\mathrm{d}i_p}{\mathrm{d}t} = -A\omega \sin \omega t + B\omega \cos \omega t$$

$$\frac{\mathrm{d}^2 i_p}{\mathrm{d}t^2} = -A\omega^2 \cos \omega t - B\omega^2 \sin \omega t.$$

Substituting these into the differential equation, we get

$$-AL\omega^2 \cos \omega t - BL\omega^2 \sin \omega t - AR\omega \sin \omega t + BR\omega \cos \omega t + \frac{A}{C}\cos \omega t + \frac{B}{C}\sin \omega t = V_m \omega \cos \omega t$$

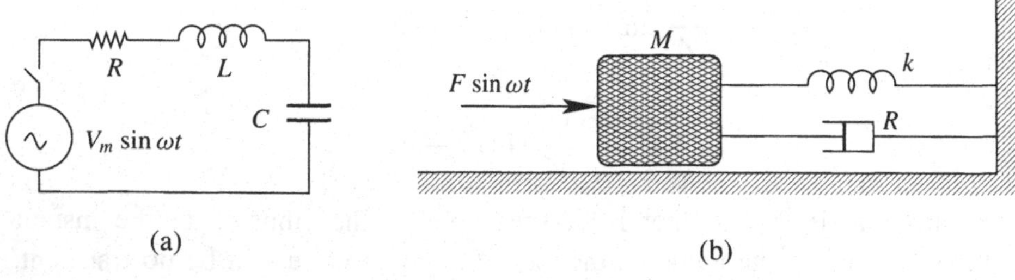

Figure 10.12 The sinusoidally driven RLC circuit and its mechanical equivalent.

or

$$\left(-AL\omega^2 + BR\omega + \frac{A}{C}\right)\cos\omega t + \left(-BL\omega^2 - AR\omega + \frac{B}{C}\right)\sin\omega t = V_m\omega\cos\omega t.$$

This will be true only if

$$-AL\omega^2 + BR\omega + \frac{A}{C} = V_m\omega,$$
$$-BL\omega^2 - AR\omega + \frac{B}{C} = 0.$$

Solving, we get

$$A = \frac{V_m\,(1/\omega C - \omega L)}{(1/\omega C - \omega L)^2 + R^2},$$
$$B = \frac{V_m R}{(1/\omega C - \omega L)^2 + R^2}.$$

Thus, the particular integral becomes

$$i_p = \frac{V_m\,(1/\omega C - \omega L)}{(1/\omega C - \omega L)^2 + R^2}\cos\omega t + \frac{V_m R}{(1/\omega C - \omega L)^2 + R^2}\sin\omega t.$$

To mould it into the convenient form, we define

$$Z\sin\phi = 1/\omega C - \omega L,$$
$$Z\cos\phi = R$$

so that

$$Z^2 = (1/\omega C - \omega L)^2 + R^2,$$
$$\tan\phi = (1/\omega C - \omega L)/R.$$

The variables are geometrically related as shown in Fig. 10.13.

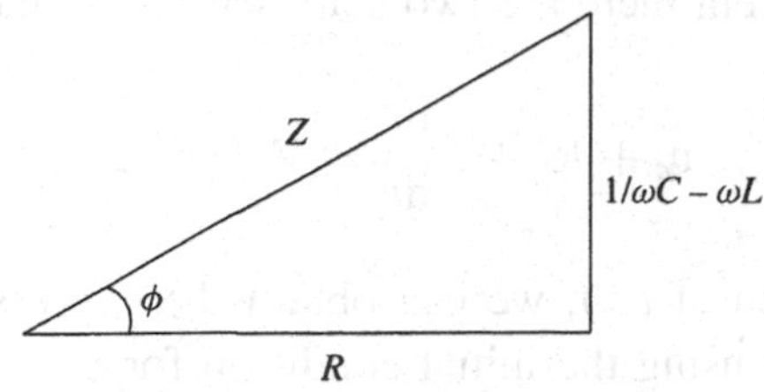

Figure 10.13 The relation between the variables in the RLC circuit.

Substituting, we get

$$i_p = \frac{V_m}{Z}(\sin \omega t \cos \phi + \cos \omega t \sin \phi),$$
$$= \frac{V_m}{Z} \sin(\omega t + \phi).$$

This gives the steady-state waveform of the system. The complete solution is a summation of the steady-state part and the transient part. The form of the transient or the complementary function is obtained by considering the system without external forcing, that is, from the homogeneous equation. We have seen in Section 9.4 that the solution of the homogeneous equation depends on the eigenvalues. If the eigenvalues are real, it is an exponentially decaying function given by

$$\mathbf{x}_c(t) = c_1 e^{\lambda_1 t} \mathbf{v}_1 + c_2 e^{\lambda_2 t} \mathbf{v}_2,$$

where $\mathbf{x}_c$ is the state vector $[v_c, i_c]^T$, $\mathbf{v}_1$ is the eigenvector associated with the eigenvalue λ_1 and $\mathbf{v}_2$ is the eigenvector associated with the eigenvalue λ_2.

If the eigenvalues are complex conjugate, the transient response is a damped sinusoid given by

$$\mathbf{x}_c(t) = c_1 e^{\sigma t} \begin{pmatrix} -\sin \omega t \\ \cos \omega t \end{pmatrix} + c_2 e^{\sigma t} \begin{pmatrix} \cos \omega t \\ \sin \omega t \end{pmatrix}.$$

The constants are to be obtained from the initial condition. For example, if the eigenvalues are complex conjugate (underdamped solution), the total solution of the current waveform $(i_c + i_p)$ will be given by

$$i(t) = e^{\sigma t} (c_1 \cos \omega t + c_2 \sin \omega t) + \frac{V_m}{Z} \sin(\omega t + \phi).$$

If the initial value of i is known, we get one equation involving c_1 and c_2. We require two equations to find both the constants. Which is the other equation?

Note that in this system there are two state variables, and they are related by

$$v_c + Ri + L\frac{\mathrm{d}i}{\mathrm{d}t} = V_m \sin \omega t.$$

Therefore, from the form of $i(t)$, we can obtain the expression for the other state variable $v_c(t)$. Then, by using the initial condition for v_c, we get another equation involving c_1 and c_2. Solving the two equations, the constants appearing in the transient can be obtained.

10.4 Other Forms of Excitation Function

In studying the cases of constant excitation and sinusoidal excitation, we have adopted a specific approach. Let us now state the basic elements of this approach, with which we can work out the solution for some other forms of excitation function.

In this approach, we first worked out the forms of the solution of the corresponding homogeneous equation. Then we argued that when the right-hand side is non-zero, the equilibrium point is not at the origin, and for time-varying right-hand side, the position of the equilibrium point also varies with time. In that case, the state point follows the equilibrium point, and that allowed us to guess the forms of the solution.

The general form of the "trial" steady-state solution will have some coefficients that are not known initially. In order to determine these coefficients, we substitute the trial solution into the original differential equation. By equating terms of the same form at both sides, we obtain a set of algebraic equations. Finally, we obtain the coefficients by solving these equations.

If the form of the excitation function is different from the ones so far encountered in this chapter, the approach will be the same. And the success will depend on our ability to guess the form of the trial solution. If the right-hand side is any arbitrary function of time, it may not be possible to guess it properly, necessitating the application of numerical methods.

Fortunately, in engineering problems we encounter only a few types of excitation function, out of which the constant excitation, piecewise constant excitation (like the square wave) and the sinusoidal excitation are most common. Two other functions that one sometimes comes across are the ramp function and the exponential function. For the ramp function

$$u(t) = at,$$

if the solution follows the equilibrium point, it is natural to guess that the solution will be a general linear function

$$i_p = b_0 t + b_1.$$

Likewise, if the forcing function is exponential of the form

$$u(t) = ae^{\alpha t},$$

a natural guess is that the solution will also be an exponential of the form

$$i_p = be^{\alpha t}.$$

Let us illustrate this with an example.

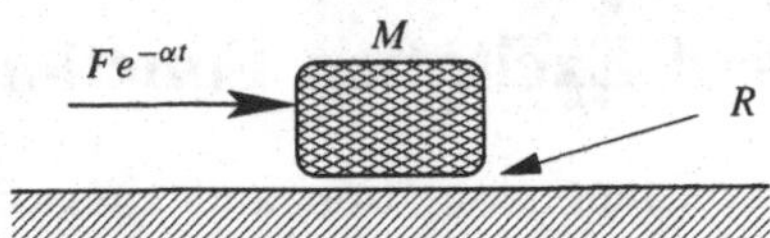

Figure 10.14 The system considered in Example 10.1.

► **Example 10.1** Consider the mechanical system in Fig. 10.14, with exponentially decaying forcing function. The differential equation of this system is

$$M\frac{dv}{dt} + Rv = Fe^{-\alpha t}$$

or

$$\frac{dv}{dt} + \frac{R}{M}v = \frac{F}{M}e^{-\alpha t}.$$

We take a trial solution of the form

$$v_p = be^{-\alpha t}.$$

Substituting into the differential equation, we get

$$-\alpha be^{-\alpha t} + \frac{R}{M}be^{-\alpha t} = \frac{F}{M}e^{-\alpha t},$$

which yields the algebraic equation needed to determine b:

$$b = \frac{F}{R - \alpha M}.$$

Note that this is true for $\alpha \neq R/M$. The characteristic equation obtained from the homogeneous part is

$$p + \frac{R}{M} = 0,$$

therefore the complementary function is

$$v_c = ke^{-Rt/M}.$$

This gives the total solution

$$v(t) = v_p + v_c = \frac{F}{R - \alpha M}e^{-\alpha t} + ke^{-Rt/M}.$$

The value of k has to be obtained from the initial condition. ◄

10.5 Chapter Summary

In a linear differential equation of the form

$$\dot{\mathbf{x}} = \mathbf{Ax} + \mathbf{Bu},$$

the first term determines the dynamics around the equilibrium point while the second term determines the position of the equilibrium point. Therefore, in systems with external forcing, the equilibrium point is placed away from the origin. If the forcing term is constant, the motion of the state point around the equilibrium point is conveniently expressed in terms of the time constant.

For time-varying forcing function, the location of the equilibrium point changes continuously. If the system is stable around the equilibrium point (i.e. if the eigenvalues of matrix **A** have negative real parts), then the solution tends to follow the equilibrium point. This allows one to guess the form of the solution – which in most cases is a generalization of the functional form of the external input. The trial solution and its higher derivatives are then substituted into the original differential equation. By equating terms of the same form at both sides, one obtains a set of algebraic equations – whose solution yields the coefficients. This gives the complete solution of the non-homogeneous linear differential equation.

Further Reading

M. E. Van Valkenburg, *Network Analysis*, Prentice-Hall, Englewood Cliffs, New Jersey, 1974.

Problems

1. For the system

$$\begin{bmatrix}\dot{x}_1\\ \dot{x}_2\end{bmatrix} = \begin{bmatrix}1 & -1.25\\ 5 & -2\end{bmatrix}\begin{bmatrix}x_1\\ x_2\end{bmatrix} + \begin{bmatrix}1\\ 0\end{bmatrix},$$

obtain the state 5 s after starting from the initial state (0, 1). Sketch the vector field, and graph the evolution of each state variable against time.

2. You had obtained the differential equations of this system in Chapter 5. Assuming that the ground vibrates sinusoidally ($y_g = A\sin\omega t$), solve the differential equations and show that this system acts as a vibration isolator.

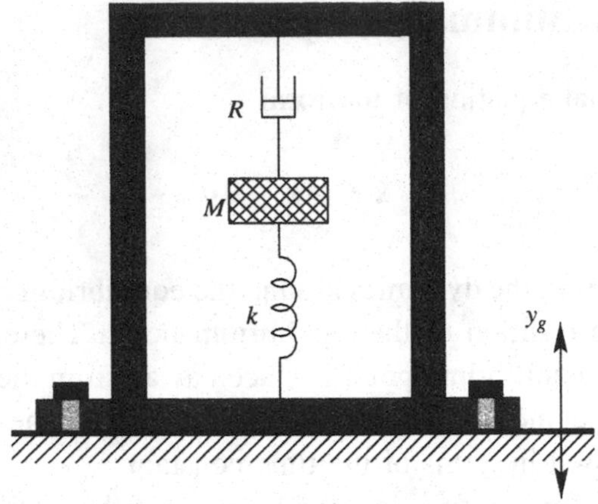

3. The circuit is initially in steady state with the switch closed. Find the evolution of the state variable $i(t)$ after the switch is opened at $t = 0$.

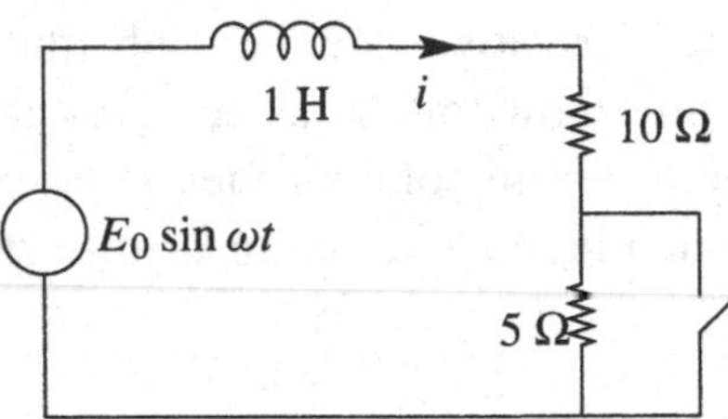

4. In the given circuit, initially the switch is open. At $t = 0$ it is thrown to position 1. After 0.4 s, it is changed to position 2. Find the value of the current at $t = 1$ s. Sketch the current waveform.

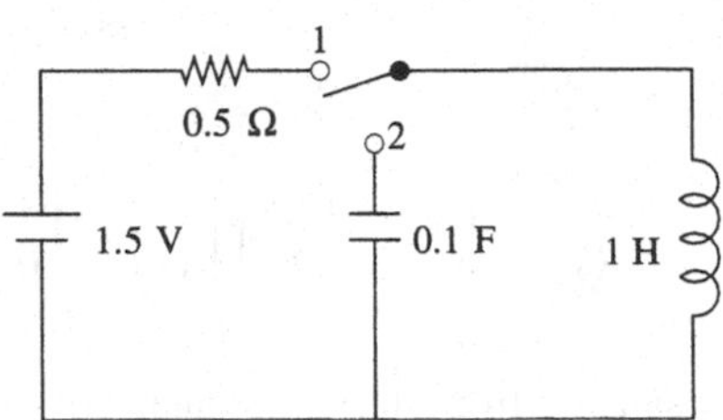

5. The freight car shown in the figure has a mass M cushioned by spring-damper arrangements as shown. The car was moving with a velocity v when it met with an accident

(bumping against a wall as shown). Assuming that the car does not recoil, obtain the equation of motion of the mass $x(t)$.

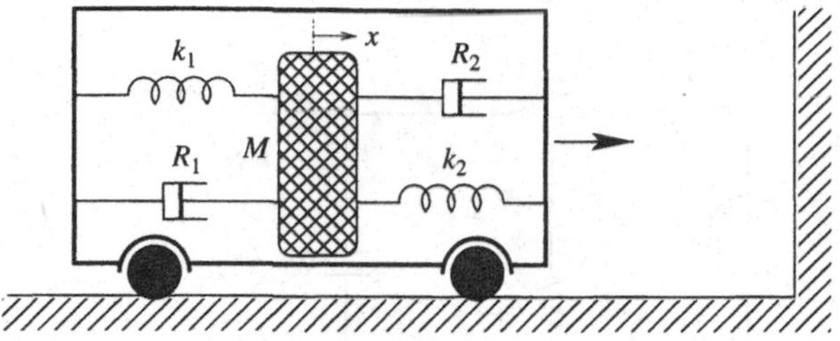

6. In the given circuit, initially the switch is open. At $t = 0$ it is closed. Obtain the equation for the current waveform. What will be the time constant?

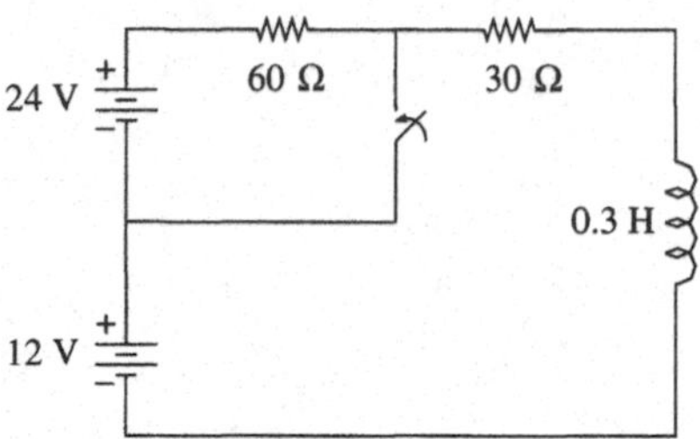

7. In the given circuit, initially both the switches are open. At $t = 0$, S1 is closed while S2 remained open, and at $t = 0.01$ s, S2 is opened and S1 is closed. Obtain the equation for the current waveform and plot till $t = 0.03$ s.

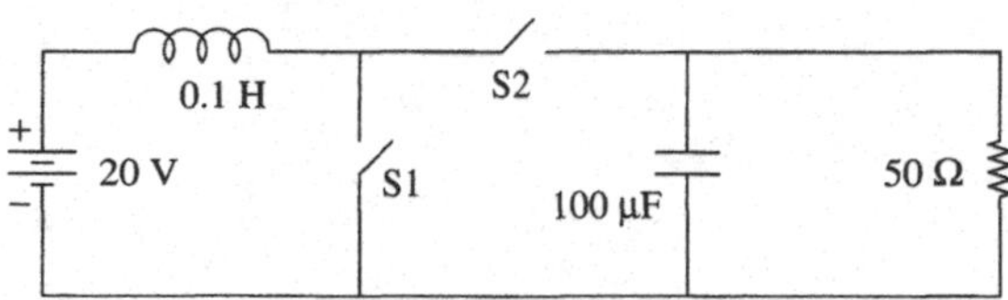

8. In this system, the cart is pulled with a force that varies sinusoidally as $F = 10 \sin 0.1t$. For mass $M = 50$ kg and frictional coefficient $R = 2$ Newtons/m/s, find out by how much time will the phase of the cart's movement lag from that of the applied force. Under what condition will there be no transient?

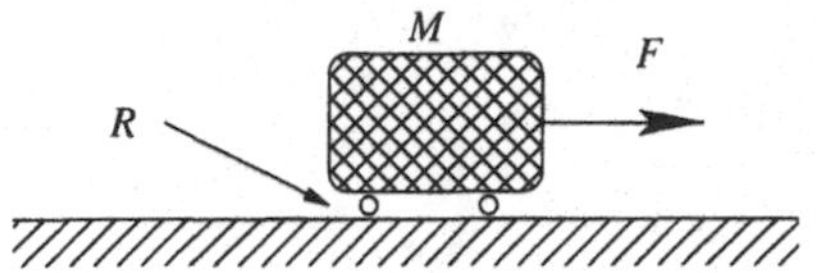

9. The switch was in position 1 for a sufficiently long time, and then it is moved to position 2. For $E = 10\sin 50t$, $R_1 = 10\ \Omega$, $R_2 = 5\ \Omega$, $L_1 = 5$ mH, $L_2 = 1$ mH, $C = 100\ \mu$F, find the capacitor voltage after 0.01 s.

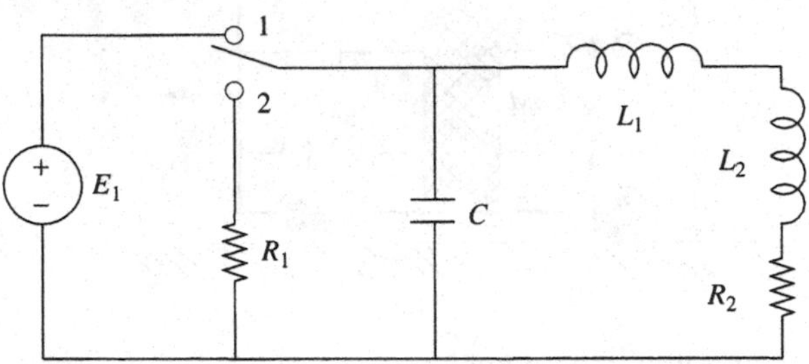

11

Dynamics of Nonlinear Systems

11.1 All Systems of Practical Interest are Nonlinear

In a linear system, there can be only one equilibrium point, and the structure of the vector field over the whole state space is the same – determined by the eigenvalues and eigenvectors of the matrix **A**. Therefore, linear systems are simple to analyse, easy to handle, and many engineering textbooks direct attention only towards such systems.

It is however rare to find a truly linear system in nature or in engineering. We have seen in the earlier chapters while dealing with derivation of system equations that even simple systems yield nonlinear differential equations. Where the derived equations were linear, some approximation was inherent at the modelling stage – like the assumption of linear characteristics of inductors, capacitors, springs and frictional elements – which are approximate simplifications from actual nonlinear characteristics. In general, a linear set of equations is actually a local linear approximation of a nonlinear system in the neighbourhood of an equilibrium point. Let us illustrate this with an example.

► **Example 11.1** The equation for a simple pendulum is

$$l\ddot{x} + g\sin x = 0, \tag{11.1}$$

where x is the angle of the chord with the vertical line. This yields the first-order equations

$$\dot{x} = y, \tag{11.2}$$

$$\dot{y} = -\frac{g}{l}\sin x. \tag{11.3}$$

By locally linearizing this equation, one obtains

$$\begin{bmatrix} \dot{x} \\ \dot{y} \end{bmatrix} = \begin{bmatrix} 0 & 1 \\ -\frac{g}{l}\cos x & 0 \end{bmatrix} \begin{bmatrix} x \\ y \end{bmatrix}, \tag{11.4}$$

which, evaluated in the neighbourhood of the equilibrium point $(0, 0)$, yields

$$\dot{x} = y,$$
$$\dot{y} = -\frac{g}{l}x$$

or, in second order, $\ddot{x} + \frac{g}{l}x = 0$. Most of the students are familiar with this equation from high school level dynamics, but may not have realized that this is actually a local linear representation of a nonlinear system. ◀

The linear system representation is still used widely in engineering because, in general, the nominal operating point of any system is located at an equilibrium point, and if perturbations are small then the linear approximation gives a simple workable model of the dynamical system.

However, if there are large fluctuations or disturbances in a system, the state variables may deviate significantly from the equilibrium point, and the linear approximation no longer remains valid. One therefore has to understand the character of the vector field over a much larger region of the state space, and has to venture out of the range where the linear approximation is valid.

11.2 Vector Fields for Nonlinear Systems

One important feature of nonlinear systems is that the behaviour of the vector field may be different for different parts of the state space, and there can be more than one equilibrium point. In such cases, one can study the *local* properties of the state space around each of the equilibrium points by the same method that we have followed for studying linear systems.

▶ **Example 11.2** Let us consider the equations of the simple pendulum:

$$\dot{x} = y,$$
$$\dot{y} = -\sin x.$$

For the sake of simplicity, we have assumed $l = 9.81$ m, so that $g/l = 1$. This will keep the calculations tidy, and for any other value of l the same procedure can be followed.

By putting the condition $\dot{x} = 0$ and $\dot{y} = 0$, we find that $(0, 0)$ is an equilibrium point. But this is not the only one. $(\pi, 0)$ is also an equilibrium point, so is $(-\pi, 0)$. Owing to the periodic nature of the sinusoidal function, there is in fact an infinite number of equilibrium points located on the x-axis, at $(0, 0)$, $(-\pi, 0)$, $(\pi, 0)$, $(-2\pi, 0)$, $(2\pi, 0)$, and so on.

What are the characters of these equilibrium points? To see this, we locally linearize the system at each of these equilibrium points using the Jacobian matrix in (11.4). At the equilibrium point (0, 0) this yields the **A** matrix

$$\mathbf{A} = \begin{bmatrix} 0 & 1 \\ -1 & 0 \end{bmatrix},$$

whose eigenvalues are purely imaginary, $\pm j$. Therefore, the equilibrium point is a "centre," and the vector field around this equilibrium point will rotate in concentric circles.

Will the rotation be clockwise or anticlockwise? To find out, take the point $(\pi/2, 0)$, and put the values of x and y in the system equations. This gives $\dot{x} = 0$ and $\dot{y} = -1$. This means that the vector at $(\pi/2, 0)$ is pointed downwards, and so the rotation must be clockwise. This allows us to draw the local vector field around the point (0, 0).

Since the elements of the Jacobian matrices of equilibrium points that are 2π apart are identical, the points $(-2\pi, 0)$, $(2\pi, 0)$, and so on, will have the same character of the vector field around them.

At the equilibrium point $(\pi, 0)$ the local linearization yields a different **A** matrix:

$$\mathbf{A} = \begin{bmatrix} 0 & 1 \\ 1 & 0 \end{bmatrix},$$

whose eigenvalues are ± 1. This means that the equilibrium point is a "saddle." For the eigenvalue $\lambda_1 = +1$, the eigendirection is $x = y$. For $\lambda_2 = -1$, the eigendirection is $x = -y$. Therefore, the 45° line represents the unstable direction, and the −45° line represents the stable direction.

Since equilibrium points 2π apart have the same Jacobian matrix, the vector field in the neighbourhood of these points will be identical.

These results are shown in Fig. 11.1.

Let us now understand the physical meaning of this structure of the vector field around the equilibrium points. The points (0, 0), $(-2\pi, 0)$, $(2\pi, 0)$, and so on, correspond to the same configuration, when the bob is hanging vertically below the suspension, and the rotation of the vector field around these equilibrium points represent swinging motion of the pendulum.

The points $(-\pi, 0)$, $(\pi, 0)$, and so on, represent the position when the bob is vertically above the suspension (assuming the chord to be rigid). These points are unstable equilibria or saddles. If the initial condition is exactly on such an equilibrium point, the pendulum

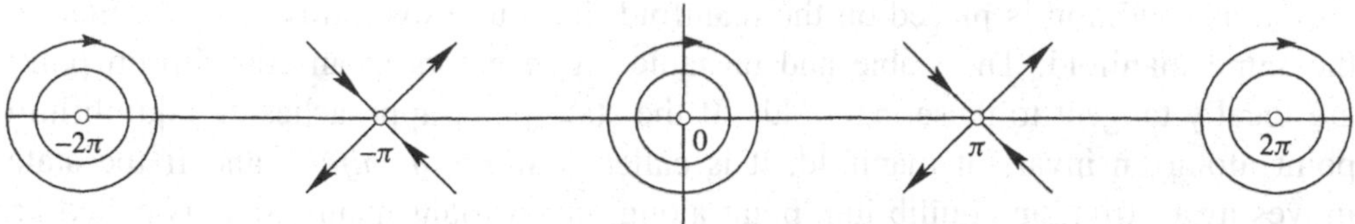

Figure 11.1 The local linear behaviour of the simple pendulum in the neighbourhood of some of the equilibrium points.

should ideally remain in that position forever. But any slight fluctuation from that point will make the state move away from the saddle point along an unstable eigenvector. ◀

This example shows how to work out the character of the vector field in the vicinity of the equilibrium points of a nonlinear system. Does that give any idea about what can happen away from the equilibrium points?

At this point, let us understand some characters of the vector field lines. Notice that we are dealing with differential equation systems of the form

$$\dot{\mathbf{x}} = f(\mathbf{x}).$$

Once such a set of differential equations is defined, for every vector $\mathbf{x}$, $\dot{\mathbf{x}}$ has a unique value. This implies the following properties of vector fields:

1. The vector at any point in the state space is unique.
2. Two vector field lines can never intersect – except at the points where the magnitude of the vector $\dot{\mathbf{x}}$ becomes zero, that is, at the equilibrium points.
3. If the right-hand sides of the differential equations are smooth functions (everywhere differentiable), then there cannot be any sudden and abrupt change in the magnitude and direction of the vector field between two neighbouring points.

In the above sense the vector field lines behave much like magnetic lines of force.

In case of equilibrium points with real eigenvalues, there is one additional property that will be of interest. For each real eigenvalue there will be a specific eigendirection. These lines have the property that if an initial condition lies on an eigenvector, the future evolution of the state must also lie on that eigenvector. And depending on the sign of the eigenvalue, the eigenvector is either stable or unstable.

As one moves away from the equilibrium point, the local linear approximation no longer remains valid. As a result, the lines that started as eigenvectors in the neighbourhood of an equilibrium point will no longer remain straight lines (Fig. 11.2). These curved lines are called *invariant manifolds*, which have the property that if an initial condition is placed on the manifold, its future evolution also remains on the same manifold. The stable and unstable eigenvectors at an equilibrium point are locally tangent to these manifolds. If the state point approaches an equilibrium point along an invariant manifold, it is called a *stable manifold*, and if the state moves away from an equilibrium point along an invariant manifold, it is called an *unstable manifold*.

With the above premise, let us look at the complete vector field of the simple pendulum shown in Fig. 11.3. The vector field in the close vicinity of the

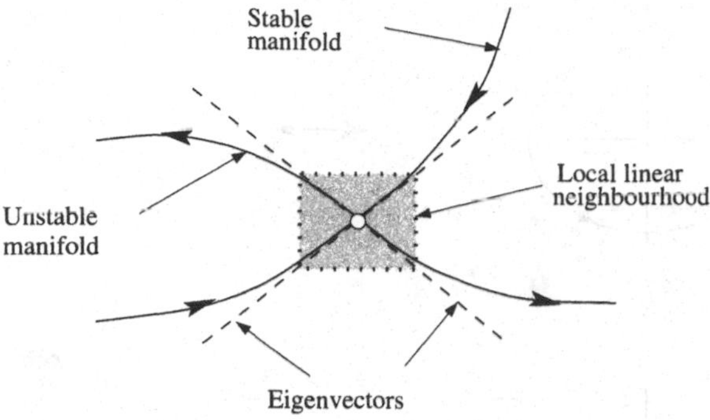

Figure 11.2 The stable and unstable manifolds at a saddle equilibrium point.

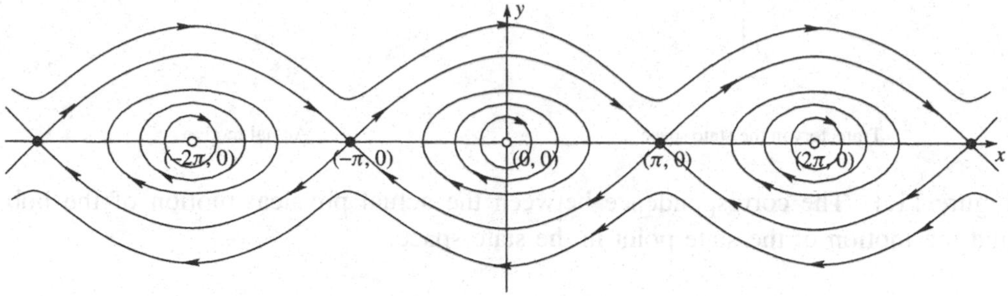

Figure 11.3 The vector field of the simple pendulum.

equilibrium points at $(0, 0)$, $(-2\pi, 0)$, $(+\pi, 0)$, and so on, are concentric circles that become distorted as we move away from the equilibrium points. If we move away along the x-axis, at a distance π we encounter a saddle equilibrium point. What will be the behaviour of the pendulum if the initial condition is placed in the neighbourhood of one of the saddle points?

If the initial deviation is towards the x direction, that is, if the pendulum starts with a small deviation from $x = \pi$ (the vertical position) but with no initial velocity, the resulting motion in the state space will be closed orbits and the bob will move in a swinging motion. If the initial deviation is along the y axis, that is, the bob is given some initial velocity, the vector field indicates that in the state space the motion will be wavy, with the y component always non-zero. This corresponds to continuous rotation of the bob (see Fig. 11.4).

Which of these two kinds of motion – swinging and rotating – will actually take place depends on the initial condition. There is a region of the state space such that if the initial condition is placed in that region, the state evolves in cyclic motion (related to the swinging motion of the bob). There is another region such that if the

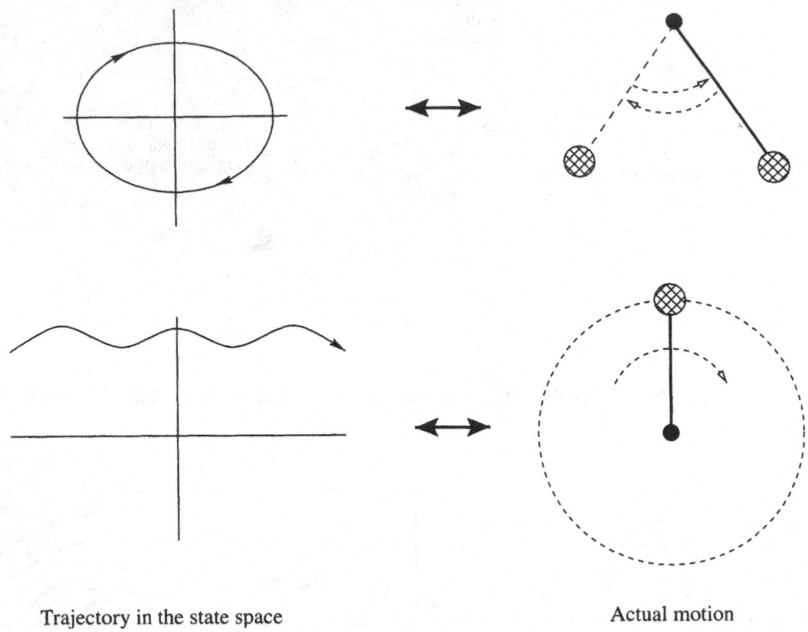

Figure 11.4 The correspondence between the actual physical motion of the bob and the motion of the state point in the state space.

initial condition is placed in that region the state does not show periodic motion in the state space (related to continuous rotation of the bob).

What separates these two regions?

The regions are separated by the invariant manifolds starting from the saddle-type equilibrium points. To understand this, consider the bob exactly in the upright position, that is, the state is at the equilibrium point. Suppose the position is perturbed by an infinitesimally small amount. Since that equilibrium point is unstable, the bob will swing down, and the potential energy will be converted into kinetic energy. Following that, there will be an upswing when the kinetic energy will be reconverted into potential energy. Since in this system there is no dissipation, the bob will rise exactly to the point from where it started, that is, the upright position. Thus, in the state space an initial condition placed at an infinitesimally small distance away from the saddle equilibrium point ends up in the next saddle equilibrium point 2π distance apart.

Geometrically this means that the unstable manifold of one saddle point bends and becomes the stable manifold of the next saddle point, as shown in Fig. 11.3. There will be two such lines joining two consecutive saddle points. These lines separate the regions of the two different types of dynamical behaviour. That is why in some literature these lines are called *separatrix*.

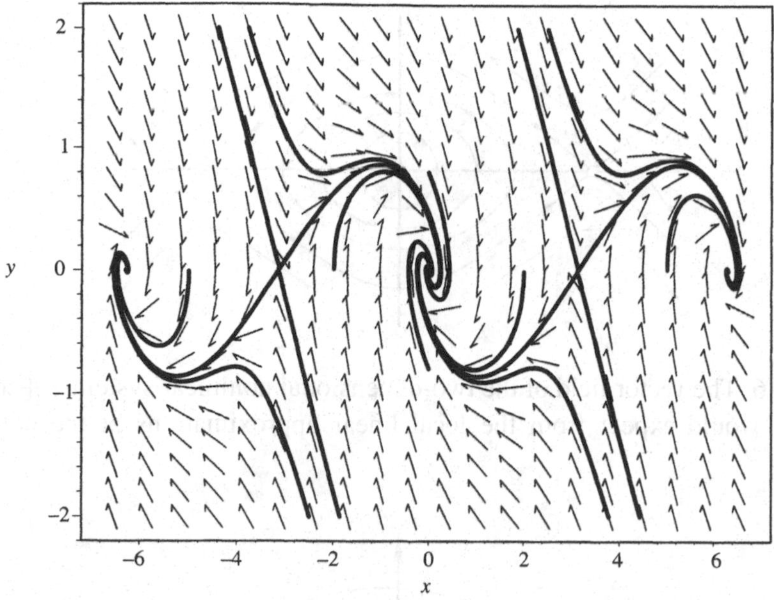

Figure 11.5 The vector field of a damped pendulum.

If we consider damping in the system, the equilibrium point at $(0, 0)$ will no longer be a "centre," with purely imaginary eigenvalues. It will have Jacobian matrix with complex conjugate eigenvalues, resulting in spirally converging orbits. As a result, the separation between the two types of behaviour will be lost (see Fig. 11.5).

Thus, we see that even in this simple system there is much more complexity in the vector field than can be inferred by working piecemeal with local linearization around equilibrium points.

In some cases, the local linear approximation may be misleading, especially where the eigenvalues turn out to be purely imaginary.

► **Example 11.3** Consider the system given by

$$\ddot{x} + x - x^3 = 0. \tag{11.5}$$

There are three equilibrium points, all on the x-axis, located at $x = 0, -1, +1$. The equilibrium point at $(0, 0)$ is a centre (imaginary eigenvalues) and the equilibrium points at $(-1, 0)$ and $(1, 0)$ are saddles. We first draw the vector field around these equilibrium points and logically try to extend the field lines to obtain the behaviour of the vector field over the rest of the state space (Fig. 11.6).

However, there is a catch here. The character of the equilibrium point at $(0, 0)$ – with purely imaginary eigenvalues – is a marginal case between incoming spiralling orbit and outgoing spiralling orbit. And this marginal case is valid only in the immediate neighbourhood of $(0, 0)$ – logically of an infinitesimally small size. There is no valid reason to assume

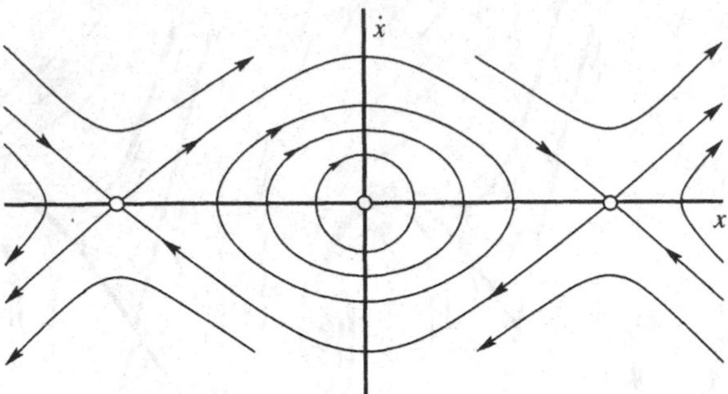

Figure 11.6 The vector field of the two-dimensional nonlinear system $\ddot{x} + x - x^3 = 0$, as one would expect from the local linear approximations at the equilibrium points.

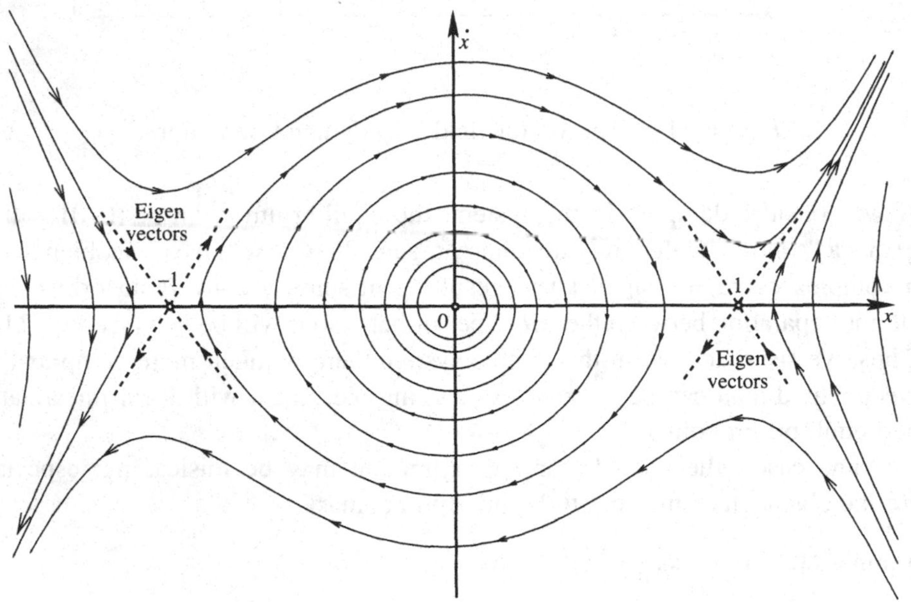

Figure 11.7 Actual vector field of the two-dimensional nonlinear system (11.5), showing that the fixed point at (0, 0) actually has an outgoing spiralling orbit.

that an orbit starting from an initial condition at some distance from the equilibrium point will really close on itself and become periodic, because the state space is actually nonlinear. On following the orbits starting from certain initial conditions, the structure of the state space turns out to be as shown in Fig. 11.7. We find that the equilibrium point at (0, 0) actually has an outward spiralling vector field around it. ◀

In addition to these features, there is also the very important aspect of nonlinear systems: the whole is not just a sum of the parts. There are global properties that cannot be found by looking at the linear approximation around equilibrium points.

11.3 Attractors in Nonlinear Systems

To illustrate a typical feature of nonlinear systems, we take the system given by

$$\ddot{x} - \mu(1 - x^2)\dot{x} + x = 0,$$

known as the *van der Pol equation*. By defining x and $y = \dot{x}$ as the state variables, we get the first-order system of equations:

$$\dot{x} = y,$$

$$\dot{y} = \mu(1 - x^2)y - x.$$

It has only one equilibrium point: at the origin. To obtain the character of the vector field in the neighbourhood of the equilibrium point, we locally linearize this equation to obtain

$$\begin{bmatrix} \dot{x} \\ \dot{y} \end{bmatrix} = \begin{bmatrix} 0 & 1 \\ -2\mu xy - 1 & \mu - \mu x^2 \end{bmatrix} \begin{bmatrix} x \\ y \end{bmatrix}, \tag{11.6}$$

which, evaluated in the neighbourhood of the equilibrium point $(0, 0)$, yields the Jacobian matrix

$$\mathbf{A} = \begin{bmatrix} 0 & 1 \\ -1 & \mu \end{bmatrix}.$$

Its eigenvalues are

$$\lambda_{1,2} = \frac{\mu}{2} \pm \frac{1}{2}\sqrt{\mu^2 - 4}.$$

For $\mu < 2$ the eigenvalues are complex conjugate. If the parameter μ is varied from a negative value to a positive value, the real part of the eigenvalues also changes sign from negative to positive. Therefore, one would expect an incoming spiralling vector field to change into an outgoing spiralling vector field – rendering the system unstable – as μ is varied through zero.

This is where the nonlinearity defeats linear intuition. True, the stable equilibrium point becomes unstable and the field lines spiral outward. But this is true only in the immediate neighbourhood of the equilibrium point. There is no reason to infer that the vector field changes character everywhere in the state space. It is found that the field lines at a distance from the equilibrium point still point inward. As a result, the system does not become globally unstable. Where the two types of field lines meet, there develops a stable *periodic* behaviour. This is called a *limit cycle.*

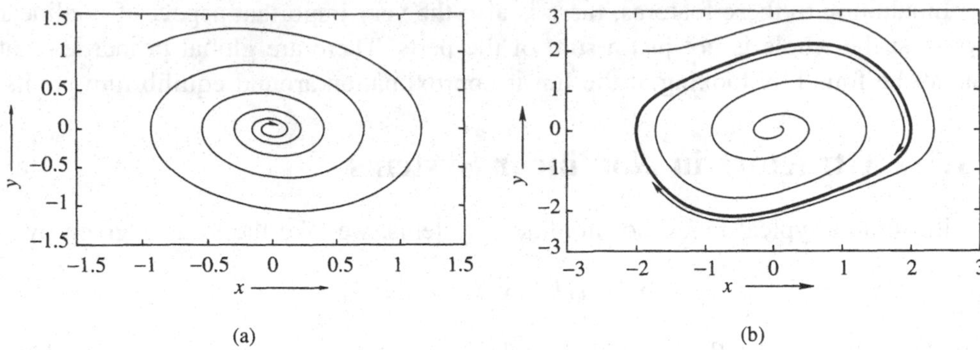

Figure 11.8 The vector fields for the system $\ddot{x} - \mu(1 - x^2)\dot{x} + x = 0$, (a) for $\mu < 0$ and (b) for $\mu > 0$. The thick line shows the limit cycle.

Fig. 11.8 shows the vector field of this system with state variables x and $y = \dot{x}$, as the parameter μ is varied from a negative value to a positive value. Fig. 11.8(a) shows the spirally attracting vector field for $\mu < 0$, and Fig. 11.8(b) shows the birth of a limit cycle for $\mu > 0$. The resulting periodic orbit is a *global* behaviour whose existence can never be predicted from linear system theory.

11.4 Limit Cycle

One point is to be noted here. There is a fundamental difference between the periodic behaviours in a linear system with purely imaginary eigenvalues, and a limit cycle in a nonlinear system. In the first case, different periodic orbits are attained for initial conditions placed at different distances from the equilibrium point. Such orbits are periodic, but are *not* limit cycles. In case of the limit cycle in a nonlinear system, trajectories starting from different initial conditions converge on to the same periodic behaviour. These two situations are schematically shown in Fig. 11.9. The limit cycle appears to attract points of the state space. This is an example of an *attractor*.

Attractors can also exist in linear systems. For example, if there is a linear system with eigenvalues real and negative, all initial conditions would be attracted to the equilibrium point – which will then be the attractor of this system. A similar situation will occur in case of linear systems with complex conjugate eigenvalues with negative real parts. However, note that there can be only one type of attractor in a linear system: a *point attractor*. In a nonlinear system, a periodic orbit can also be an attractor.

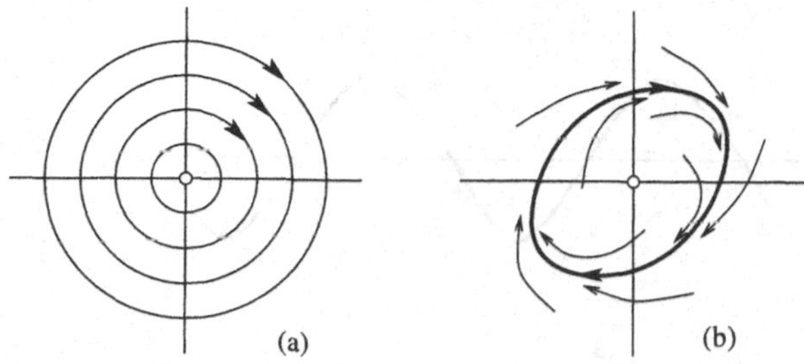

Figure 11.9 (a) Periodic orbits in a linear system with imaginary eigenvalues, and (b) limit cycle in a nonlinear system.

It is not difficult to see that wherever there is any *stable* periodic behaviour in any engineering system, it must be a limit cycle. That is why all oscillators we come across in electronics are nonlinear systems, and the orbit in the state space is a limit cycle. If one were to construct an oscillator as a linear system with imaginary eigenvalues, any disturbance would push it into a different orbit with a different amplitude of oscillation. In contrast, if it is designed as a limit cycle in a nonlinear system, any disturbance would eventually die down and the orbit would settle into the intended periodic behaviour.

For the same reason, all oscillatory behaviour that we come across in nature must also be of the type of limit cycle. A typical example is the human heart – which has a periodic oscillatory behaviour that must be stable. If one is startled by an unexpected sound, the heart starts pounding – the rhythm of the heart is disturbed. In scientific parlance, we would say that the periodic orbit is perturbed. It is vitally important for our survival that the rhythm must come back to its natural state once the disturbance is removed. This can happen because the heart is a nonlinear system, and the orbit in the state space is a limit cycle.

11.5 Different Types of Periodic Orbits in a Nonlinear System

Thus, in a two-dimensional nonlinear system one can come across periodic attractors as in Fig. 11.8(b). When we plot one of the variables against time, it has a periodic waveform as shown in Fig. 11.10(a), corresponding to a state-space trajectory that shows a single loop as in Fig. 11.10(b).

The question now is: What other shapes can a limit cycle take? Can a limit of the type shown in Fig. 11.11 exist?

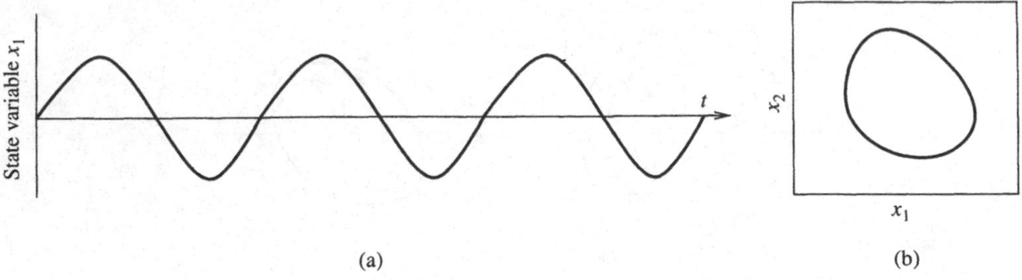

Figure 11.10 (a) The time plot and (b) the state-space trajectory for a period-1 attractor.

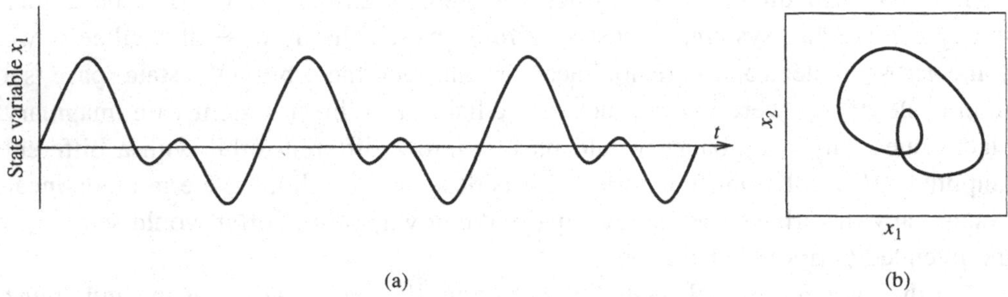

Figure 11.11 A period-2 waveform, (a) in the time domain and (b) in state space.

It is not difficult to see that the state-space trajectory intersects itself at a point, and hence violates one of the important properties of a vector field. There must be a unique velocity vector $\dot{\mathbf{x}}$ associated with every point in the state space. If this trajectory were possible, the vector at the point of intersection would not be unique. Thus, we conclude that an orbit of the type shown in Fig. 11.11 cannot occur.

But wait. Didn't the intersection occur because the state space was assumed to be two dimensional – where a curve wandering freely is constrained to intersect itself? What if the system were three-dimensional? Wouldn't that allow the possibility that at the point where we see an intersection, one line can pass over the other without intersecting?

Thus, we see that a trajectory of the type shown in Fig. 11.11 is possible if the state space has dimension three or more. Such a trajectory, when projected in a 2-D plane as in Fig. 11.11(b), shows two loops. When one of the variables is plotted against time as in Fig. 11.11(a), we see that the same state repeats after every two cycles. This is called a *period-2 orbit*.

In a system of dimension three or more, if a period-2 orbit is possible, why not orbits of higher periodicities? Indeed, there is nothing to prevent the occurrence of orbits of any arbitrary periodicity. In fact, such orbits do occur in specific situations in nonlinear dynamical systems of dimension three or above.

11.6 Chaos

One interesting possibility opens up in systems of order 3 or greater: to get waveforms that do not have any periodicity. In such a case, the system state remains bounded – within a definite volume in the state space, but the same state never repeats. In every loop through the state space the state traverses a new path. This situation is called *chaos* and the resulting attractor is called a *strange attractor*. The system undergoes apparently random oscillations.

As an example, consider the simplified model of atmospheric convection, known as the *Lorenz system*:

$$\begin{aligned}\dot{x} &= -\sigma(x-y),\\ \dot{y} &= -xz+rx-y,\\ \dot{z} &= xy-bz,\end{aligned} \tag{11.7}$$

where r is a variable parameter. Let us assume the other parameters to be set at $b = 8/3$ and $\sigma = 10$.

To analyse the dynamics of this system, we first put $\dot{x} = \dot{y} = \dot{z} = 0$ to obtain three equilibrium points as

$$\begin{aligned}A &= (0,0,0),\\ B &= \left(\sqrt{b(r-1)}, \sqrt{b(r-1)}, r-1\right),\\ C &= \left(-\sqrt{b(r-1)}, -\sqrt{b(r-1)}, r-1\right).\end{aligned}$$

The local behaviour around these equilibrium points will be given by the Jacobian matrix

$$\begin{bmatrix} -\sigma & \sigma & 0 \\ -z+r & -1 & -x \\ y & x & -b \end{bmatrix}.$$

From these we note that for $r < 1$, the equilibrium points B and C do not exist, and the point A is stable. For $r > 1$, the point A at the origin becomes unstable, and the equilibria B and C come into existence. What about their stability?

This can be obtained by substituting their positions into the Jacobian matrix and by obtaining the eigenvalues. We find that for $1 < r < 24.74$, these two equilibria have complex conjugate eigenvalues with real part negative. For $r > 24.74$ the eigenvalues move to the right half of the complex plane (the real part becomes

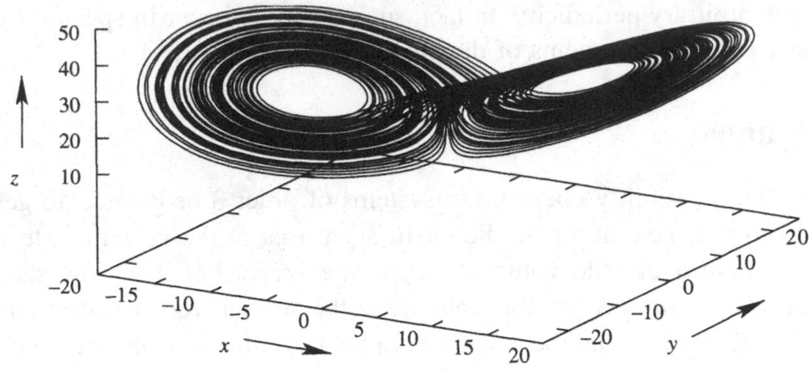

Figure 11.12 The trajectory of the Lorenz system (11.7) for $\sigma = 8/3$, $b = 10$, and $r = 28$.

positive), and so both the equilibrium points become unstable. Orbits near them spiral outward.

This is the maximum that can be inferred from linearization around the equilibrium points. What will be the character of orbits for $r > 24.74$? Will such orbits spiral out to infinity? This can be found out only by solving the equations numerically by the methods outlined in Chapter 7.

The orbit thus obtained is shown in Fig. 11.12. It is noticeable that even though all the equilibrium points are unstable, the orbit does not diverge to infinity. It remains bounded within a definite volume in the state space. But the orbit has no periodicity.

Looking at Fig. 11.12, you can probably see the two eigenplanes associated with the two unstable equilibrium points. Orbits close to the respective equilibrium points spiral outward in these eigenplanes – in one the orbit moves clockwise and in the other it moves counterclockwise. The outward motion along one eigenplane throws the orbit to the other eigenplane and vice versa, thus preventing its escape to infinity.

When one looks at the evolution of one of the variables against time, the waveform looks apparently random and noise-like (Fig. 11.13). But this is not really random noise, because it is a product of a perfectly deterministic process. The differential equations that cause the motion of the state point are known, and simply by following the vector field one gets the motion shown in Fig. 11.12.

Let us now understand another important aspect of a chaotic system. Suppose we compute the evolution starting from two very close initial conditions: one at $x = 0.0$, $y = 7.0$, $z = 7.0$ and the other at $x = 0.0001$, $y = 7.0$, $z = 7.0$. One would normally expect that such tiny difference in the initial condition would not matter.

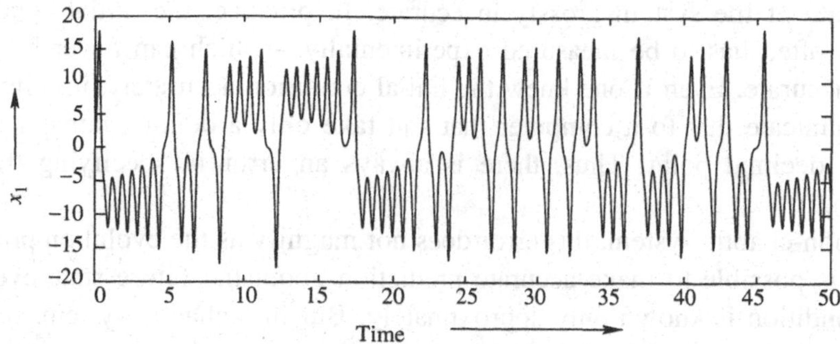

Figure 11.13 The waveform of one of the variables of the Lorenz system plotted against time.

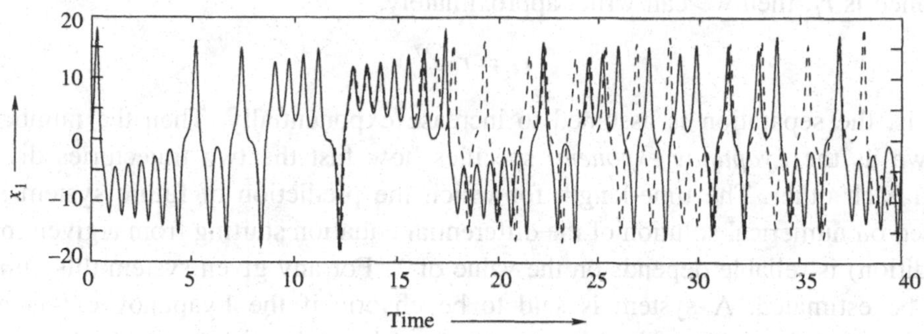

Figure 11.14 The waveforms starting from two very close initial conditions. Firm line: the initial condition is [0, 7, 7], broken line: the initial condition is [0.0001, 7, 7].

But when we actually obtain the trajectories, we find that after some time the two trajectories diverge from each other and then follow different courses of evolution (Fig. 11.14). This is called the *sensitive dependence on initial condition* – a hallmark of chaos.

One might argue that by choosing even more closely placed initial conditions, the two trajectories may be made to remain close to each other. No luck there. It turns out that no matter how small the separation between the two initial conditions, the two orbits would eventually diverge. The reader may try that out by taking initial conditions different only in the seventh decimal place.

This implies something of a grave consequence. In a chaotic system, even a minute error in specifying the initial condition would make prediction about the future state of the system grossly inaccurate. In practice, the initial condition of a system often has to be measured experimentally – which can never be hundred percent accurate. Even if one knew the initial condition accurately, it is impossible to communicate that to a computer that can take only a definite number of digits after the decimal point. Thus, there is always an error in specifying the initial condition.

In a non-chaotic system, this error does not magnify as the evolution progresses. Thus, it is possible to make accurate prediction about the future state even if the initial condition is known only approximately. But in a chaotic system, prediction is impossible beyond a certain time frame.

How long is the time after which prediction is not dependable? Obviously that depends on how fast two nearby trajectories diverge from each other. If, at time $t = 0$ the distance between two points in the state space is r_0, and at time t the distance is r_t, then we can write, approximately,

$$r_t = r_0 e^{\lambda t},$$

that is, the separation is assumed to increase exponentially. Then the number λ, known as the *Lyapunov exponent*, specifies how fast the two trajectories diverge from each other. The time-length for which the prediction of future system state (based on numerical solution of the differential equation starting from a given initial condition) is reliable depends on the value of λ. For any given system this number can be estimated. A system is said to be chaotic if the Lyapunov exponent is positive, and the value of λ gives a measure of how chaotic a system is.

Even though long-term prediction may fail if a system is chaotic, an engineer need not be over-concerned about this failure. Rarely does an engineer need to predict the future state of a system so accurately. An engineer is more concerned with the overall properties of the orbit of a system. Even if one does not know the future state of the system, from the numerical solution of the concerned differential equations one can say with great confidence that the state will not run to infinity, the system will not collapse, and the state will be "somewhere" within a definite volume of the state space. One can also talk in terms of the probability of finding the state in a certain part of the state space. While the states oscillate in an apparently random way, one can still measure the average and RMS values of the state variables. In that sense, even though none of the equilibrium points of the system may be stable, the chaotic attractor itself is stable. Indeed, researchers are finding many applications where such properties can be productively used.

11.7 Quasiperiodicity

In nonlinear dynamical systems, another type of orbit is found to occur – where one periodicity is superimposed on another. We know that the rotation of the Earth

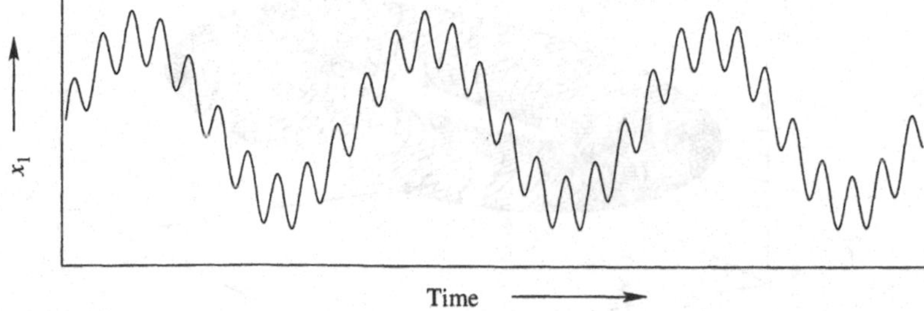

Figure 11.15 A typical waveform of a quasiperiodic orbit in time domain.

around the sun is periodic. We also know that the rotation of the moon around the Earth is also periodic. But what is the character of the orbit of the moon around the sun? It is evident that the orbit will be a combination of two periodicities.

If the two periods are commensurate, that is, if their ratio yields a rational number, then the orbit will be eventually periodic because there is a period that can be obtained by multiplying both the periods by integer numbers. But if the two frequencies are incommensurate, that is, if one can be obtained by multiplying the other by an irrational number, then the same state will never repeat, and the resulting orbit will be aperiodic. This is called a *quasiperiodic* orbit.

► **Example 11.4** Suppose a system has two superimposed frequencies. If we take them individually, in one the same state repeats after every 7 s and in the other the same state repeats every 2 s. When these two periodicities are superimposed, the same state will repeat every 14 s, and hence the orbit will be periodic.

Now consider another system in which one frequency gives a periodicity of 2 s, and the other gives a periodicity of 6.2832... s (which is $2 \times \pi$). Since π is an irrational number, there will be no periodicity in the orbit that has these two incommensurate frequencies. This orbit will be quasiperiodic. ◄

Fig. 11.15 shows a typical waveform of a state variable associated with a quasiperiodic orbit.[1] In the state space, the attractor lies on the surface of a torus, as shown in Fig. 11.16.

Since, the trajectory is aperiodic both in a chaotic system and a quasiperiodic system, what is the difference between the two? The basic difference is that there is sensitive dependence on initial condition in a chaotic system – which is absent in a quasiperiodic system. Indeed, in the latter system the separation between two nearby trajectories remains the same as the system is evolved in time.

[1]Here the two frequencies are clearly identifiable. However, if the two frequencies and their respective amplitudes are close to each other, the waveform may not show apparent signatures of the two periods.

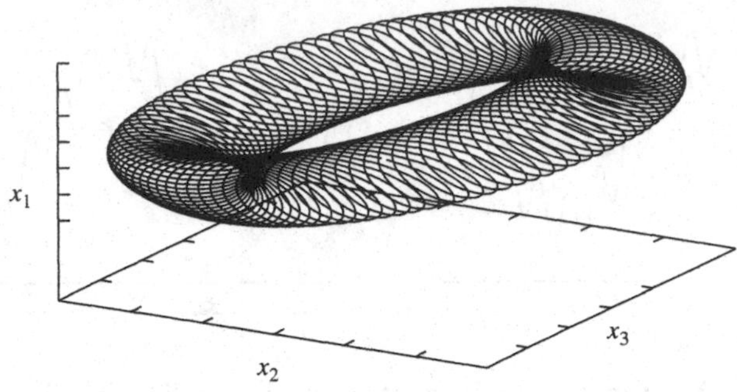

Figure 11.16 The appearance of a quasiperiodic attractor in state space.

11.8 Stability of Limit Cycles

In all systems there are some parameters – like the impedance of an inductor, the mass of a body, a frictional coefficient, a capacitance, a spring constant, and so on. The dynamics depend on these parameters, and if the parameters change, the dynamics will also change.

So long as a periodic limit cycle remains stable, changes in a parameter can only lead to small quantitative changes in the orbit. But if a change of a parameter renders the limit cycle unstable, there is a drastic qualitative change in the behaviour of the system. If another limit cycle of a different periodicity becomes stable, the orbit changes to the new one. If no other limit cycle is stable at that parameter value, the system will collapse.

From an engineering point of view, therefore, it is of vital importance to understand the stability of limit cycles. However, since a limit cycle is a global behaviour of a nonlinear system, its stability cannot be investigated with the help of linear methods. Fortunately, there is a trick – to be discussed in the next chapter – that allows us to understand the stability of a limit cycle in terms of eigenvalues and eigenvectors, much the same way as we understood the stability of equilibrium points through local linearization.

11.9 Chapter Summary

For nonlinear systems, there can be more than one equilibrium point, and the vector field can be very complex. Through local linearization around the equilibrium points, it is possible to get an idea about the character of the orbit if the deviation from an equilibrium point is small. Such linear methods throw little light on the possible system behaviour if there is a relatively large disturbance, pushing the state away from the equilibrium point. Linear methods throw no light on globally stable closed orbits known as *limit cycles* that can exist in a nonlinear system. Limit cycles can be simple single loop, or can have many loops – giving high periodic orbits. In nonlinear systems, there can also be bounded aperiodic orbits with sensitive dependence on initial condition – a phenomenon termed as *chaos*.

Further Reading

K. T. Alligood, T. D. Sauer, and J. A. Yorke, *Chaos: An Introduction to Dynamical Systems*, Springer Verlag, New York, 1996.

R. C. Hilborn, *Chaos and Nonlinear Dynamics*, Oxford University Press, Oxford, UK, 2000.

G. F. Simmons, *Differential Equations with Applications and Historical Notes*, McGraw Hill, New York, 1972.

S. H. Strogatz, *Nonlinear Dynamics and Chaos, with Applications to Physics, Biology, Chemistry and Engineering*, Perseus Publishing, Cambridge, Massachusetts, 1994.

Problems

1. The first-order differential equations of a system are given by

$$\dot{x} = y(x^2 + 1),$$
$$\dot{y} = x^2 - 1.$$

Sketch the vector field over the state space in the range $x = [-2, 2]$, $y = [-2, 2]$. Trace the variations of the x-coordinate with time if the initial conditions are (a) $(0, 0)$ and (b) $(-1, 1)$.

2. Sketch the character of the vector field in the $x - y$ state space.

(a) $\dot{x} = y,\qquad \dot{y} = -x - x^2,$

(b) $\dot{x} = 2xy,\qquad \dot{y} = y^2 - x^2,$

(c) $\dot{x} = x^3 - 2xy^2,\qquad \dot{y} = 2x^2y - y^3,$

(d) $\dot{x} = x - y,\qquad \dot{y} = 1 - xy.$

3. Reduce the following equations to the first-order form by defining an additional variable $y = \dot{x}$, and study the character of orbits in the $x - y$ state space. Where there are parameters in an equation, study the above for different values of the parameters.

(a) $\ddot{x} + x - x^3 = 0.$

(b) $\ddot{x} + x + x^3 = 0.$

(c) $\ddot{x} - ax\dot{x} = 0.$

(d) $\ddot{x} + a\dot{x} - bx + cx^3 = 0.$

(e) $\ddot{x} - a(1 - x^2)\dot{x} + bx + ax^3 = 0.$

(f) $\ddot{x} + e^x = a.$

(g) $\ddot{x} - e^x = a$.

(h) $\ddot{x} = (2\cos x - 1)\sin x$.

4. For the system

$$\dot{x} = -y - x\sqrt{x^2 + y^2},$$
$$\dot{y} = x - y\sqrt{x^2 + y^2},$$

show that the local linearization indicates that the equilibrium point is a centre, while actually it is a spiral point.

5. For the system

$$\ddot{x} + 0.03\dot{x} + x + x^3 = F\cos\omega t,$$

study the change in the character of the orbits as the amplitude and frequency of the external forcing function are varied.

6. The following set of equations are called the *Rössler* equations:

$$\dot{x} = -y - z,$$
$$\dot{y} = x + ay,$$
$$\dot{z} = b + (x - c)z.$$

Write a program to draw the orbit starting from any initial condition, for the parameter choice $a = 0.1$, $b = 0.1$ and $c = 14$.

7. The differential equation of a system is given by

$$\ddot{x} + \left(x^2 - \eta\right)\dot{x} + \omega^2 x = 0.$$

Show that a stable equilibrium point becomes unstable as the parameter η is varied from -1 to $+1$. For $\omega = 1$, at what value of η does the instability occur? What happens to the system after the equilibrium point becomes unstable?

8. The figure shows an RLC circuit with a nonlinear capacitor, whose value depends on the charge across the capacitor, q. Obtain the first-order equations, and sketch the vector field in the state space.

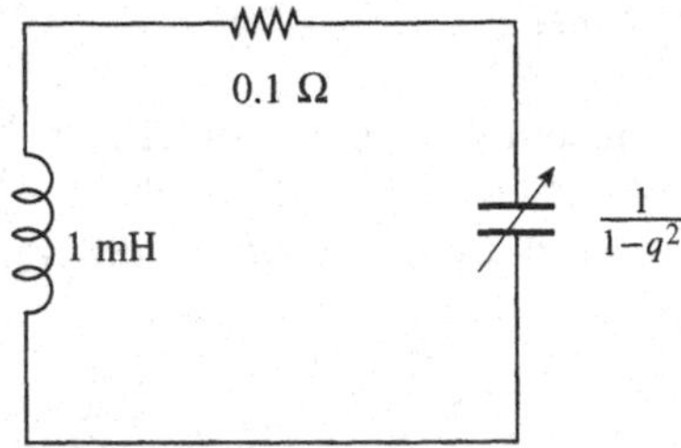

9. Show that orbits in the system

$$\ddot{x} + (x^2 + \dot{x}^2 - 1)\dot{x} + x = 0$$

converge on to a limit cycle.

10. For the mass-spring system with dry friction for which the equation was derived in Section 4.10, obtain the trajectory in the state space for 10 s if the forcing function is $10 \sin t$. Initial condition: $x(0) = 1.0$, $\dot{x}(0) = 0.0$.

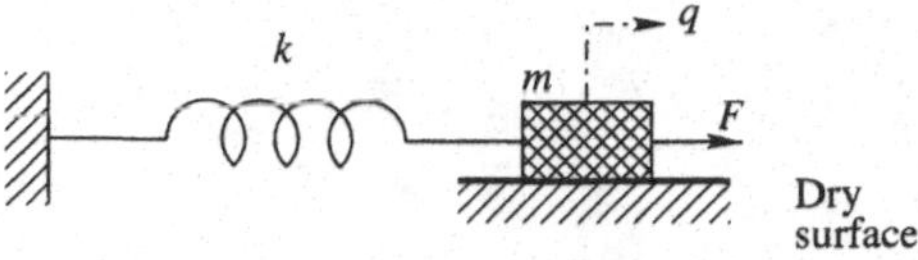

Hint: Evolve the system equations until $\dot{x}$ changes sign. Then change the sign of the friction term in the equations, take the final condition of the last phase as the initial condition of this phase and evolve again. Repeat this process.

11. The equations

$$\dot{x} = a + x^2 y - (1 + b)x,$$
$$\dot{y} = bx - yx^2,$$

represent the dynamical model of a chemical reaction, known as the *Brusselator*. Obtain the position of the equilibrium point and classify its character for

(a) $a = 1$, $b = 2$,

(b) $a = 1$, $b = 1$,

(c) $a = 1$, $b = 5$.

12

Discrete-time Dynamical Systems

We have seen in the last chapter that in nonlinear systems there can be stable periodic orbits or limit cycles. Such orbits, resulting from system nonlinearity, have the desirable property that initial conditions belonging to a large region of the state space are attracted to it, and so perturbations from such orbits tend to die down. That is why this behaviour is frequently used in engineering systems. Indeed, wherever there is a necessity of producing a periodic variation of a state variable (as in an oscillator), one has to deliberately use the nonlinearity of the system to create a limit cycle. And in nature, all oscillatory phenomena are limit cycles.

Since engineers are interested in limit cycles, they also have to worry about the stability of limit cycles. After all, if something is to be applied in engineering, it must be reliable, and therefore stable. Moreover, one needs to understand what is going to happen when a limit cycle loses stability.

We have seen that when the nominal operating condition of a given system is at an equilibrium point, its stability can be probed using the local linear approximation around the equilibrium point. One has to evaluate the Jacobian matrix at the equilibrium point, obtain the eigenvalues, and if the eigenvalues lie in the left half of the complex plane, the system is stable. But since the limit cycle is caused by the behaviour of the vector field over a large region of the state space, local linearization around an equilibrium point says nothing about its occurrence or stability. What can we do in that situation?

12.1 The Poincaré Section

The basic method of addressing the problem was invented by the famous French mathematician, Henri Poincaré. Imagine a surface as shown in Fig. 12.1, called the *Poincaré* section, at a suitable place in the state space such that the orbit intersects it.

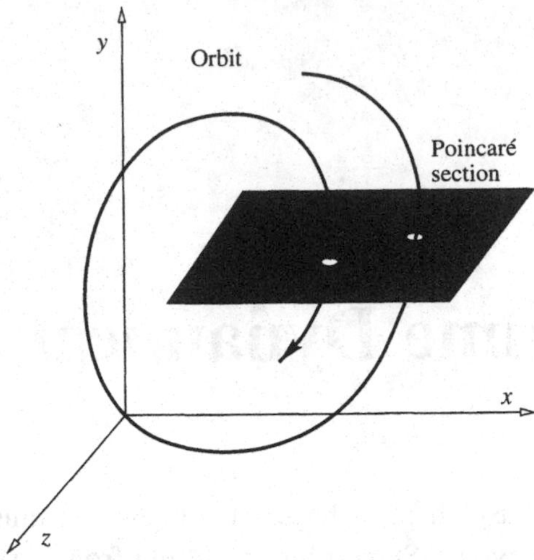

Figure 12.1 The Poincaré section intersecting an orbit. Discrete points are marked on the Poincaré plane as the orbit pierces it from one side.

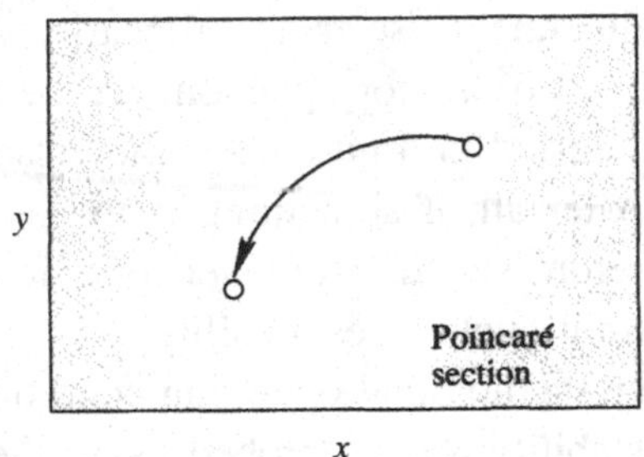

Figure 12.2 The mapping on the Poincaré plane.

Now, if we restrict our attention to what happens *on the Poincaré section*, what will we see? We will see the event of one crossing, and after some time we will see another intersection at some other point. This is as if the state is jumping (or *mapping* in technical language) discretely from one point to another point on that surface (Fig. 12.2). Such a mapping can be defined on the Poincaré section by a function of the form

$$x_{n+1} = f_1(x_n, y_n),$$
$$y_{n+1} = f_2(x_n, y_n),$$

where the position at the $(n + 1)$th intersection is given as a function of the position at the nth intersection. In the form of a compact vector equation, this can be

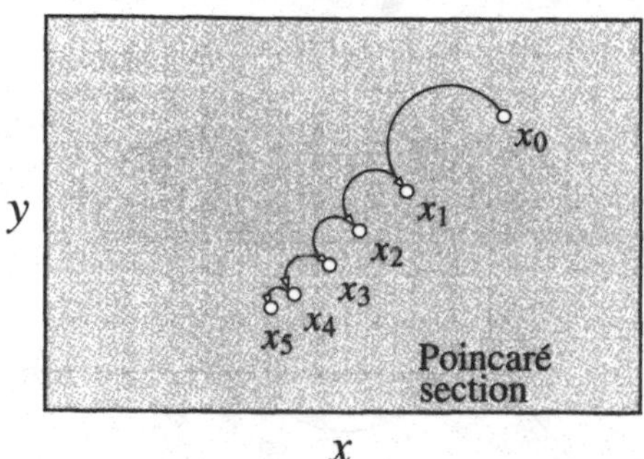

Figure 12.3 A sequence of points on the Poincaré plane, obtained by repeatedly applying equation (12.1).

written as

$$\mathbf{x}_{n+1} = g(\mathbf{x}_n). \tag{12.1}$$

(The subscript on boldface $\mathbf{x}$ denotes the order of occurrence, not a component of the vector.)

We thus obtain a discrete-time dynamical system or *map* from a continuous-time dynamical system

$$\dot{\mathbf{x}} = f(\mathbf{x}).$$

Does it offer any advantage?

As the orbit intersects the Poincaré plane repeatedly, the state will discretely jump from point to point, and this process will continue. On the Poincaré plane, we will thus see a *sequence* of points. This can be obtained by repeatedly applying or *iterating* equation (12.1), as shown in Fig. 12.3.

Now, if the continuous-time orbit is a stable limit cycle, what will we observe on the Poincaré section? We will see just one point if the limit cycle is period-1, as shown in Fig. 12.4(a). If the limit cycle has periodicity 2 as in Fig. 12.4(b), we will see two points as the limit set. Thus, the periodicity of the orbit is exactly reflected in the Poincaré section. If the continuous-time orbit is periodic, there are a finite number of points on the Poincaré section, equal to the periodicity of the continuous-time orbit. Note that for this to work, we have to observe the state only when the orbit intersects the Poincaré plane *from one side*.

In case of quasiperiodic orbits, since the two frequencies are incommensurate, points in the sampled observation will not fall on each other and will be arranged in a closed loop (see Fig. 12.5). For a chaotic system, since the orbit is aperiodic and bounded, there would be an infinite number of points on the Poincaré plane, contained within a finite area and distributed over a region of very intricate structure (we shall see examples of such things later). This is the strange attractor in the

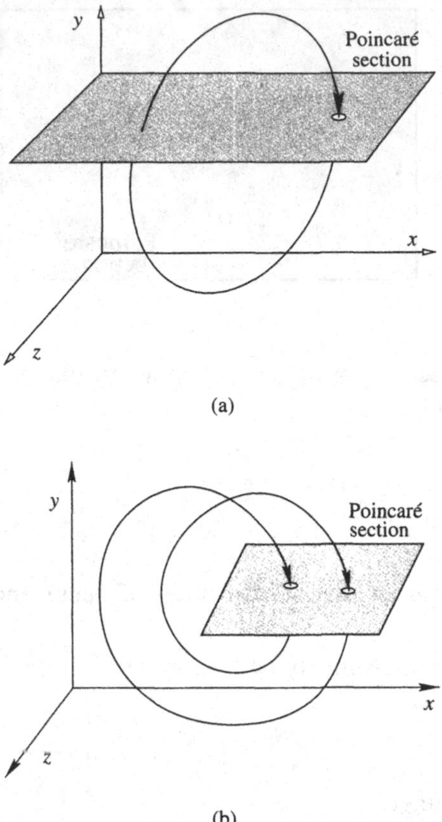

Figure 12.4 The intersections of (a) a period-1 limit cycle and (b) a period-2 limit cycle with a Poincaré plane.

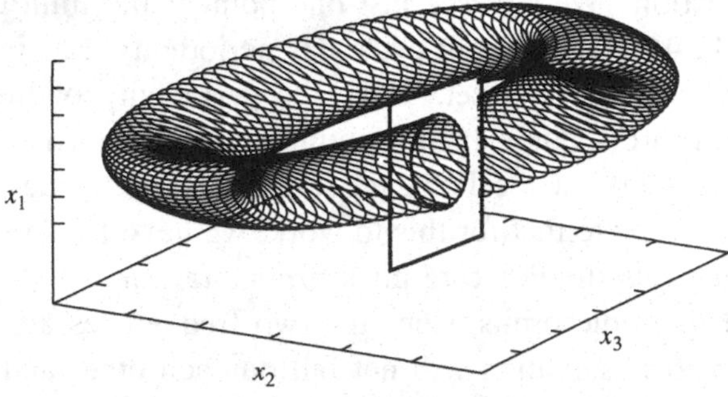

Figure 12.5 A quasiperiodic orbit appears as a closed loop in the Poincaré plane.

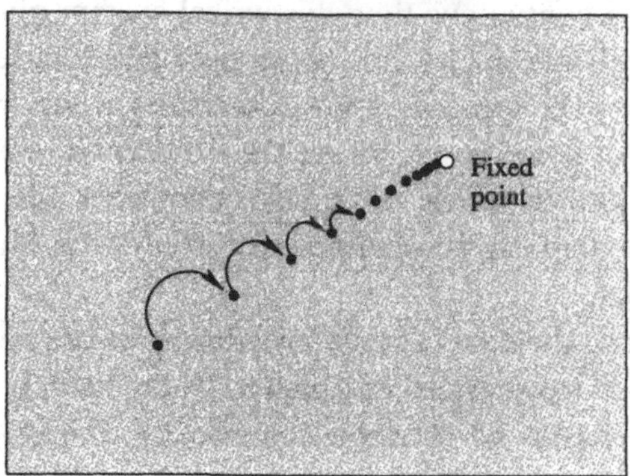

Figure 12.6 The sequence of iterates converging on the fixed point.

discrete domain. Thus, we see that there is a one-to-one correspondence between the character of a continuous-time system and its discrete-time representation.

Now, let us consider the case of a stable limit cycle as shown in Fig. 12.4(a), which shows a single point on the Poincaré section. What will be seen on the Poincaré section if the initial condition is placed away from the limit cycle? Since the limit cycle is an attracting set in the state space, the orbit will progressively move closer to the limit cycle. Therefore, in each piercing of the Poincaré section, the point will land closer to the point corresponding to the limit cycle (Fig. 12.6). Thus, we will obtain a sequence of points limiting onto a point where

$$x_{n+1} = x_n.$$

This point is called the *fixed point* of the map, which corresponds to the intersection of the Poincaré plane with the limit cycle.[1]

If the limit cycle becomes unstable, initial conditions starting away from the fixed point will not converge on it, rather it will move away in subsequent piercings. Thus, we see that the stability of the limit cycle is related to the stability of the fixed point of the Poincaré map. And the behaviour of iterates in the neighbourhood of a fixed point can be probed using local linear approximation of the map. This is the basic idea, which we shall develop in the subsequent sections.

12.2 Obtaining a Discrete-time Model

First, let us understand how the map

$$\mathbf{x}_{n+1} = f(\mathbf{x}_n)$$

[1]Note the nomenclature here. In case of a continuous-time system, the point where $\dot{\mathbf{x}} = 0$ is called an *equilibrium point*. In a discrete-time system, the point where $x_{n+1} = x_n$ is called a *fixed point*.

can be obtained from a continuous-time dynamical system given by a set of differential equations. Since limit cycles can occur only if the differential equations are nonlinear, in most cases the solution of the differential equations cannot be obtained in closed form. Only in some cases, where the differential equations are piecewise linear, the explicit solutions of the differential equations can be obtained, and the functional form of the map can be obtained in closed form. We shall later see some examples of such cases.

But since, in general, the nonlinear differential equations need to be solved using some numerical method (as outlined in Chapter 7), the map $\mathbf{x}_n \mapsto \mathbf{x}_{n+1}$ (the symbol $\mapsto$ means "maps to") has to be obtained numerically. For this, the procedure is as follows:

First, start from any initial condition and solve the differential equations for a long time, and plot the orbit. Then locate a suitable plane such that the orbit intersects it in every cycle. It is generally convenient to choose the plane parallel to one of the axial planes (i.e. either $x = c$ or $y = c$ or $z = c$). Then take an initial condition on that Poincaré plane. This is our $\mathbf{x}_n$ vector. Then, we have to evolve the differential equations until the orbit intersects the plane from the same side. This gives $\mathbf{x}_{n+1}$. Given any initial condition $\mathbf{x}_n$ on the Poincaré section, the program will calculate the value $\mathbf{x}_{n+1}$ at the next piercing. Note that the dimension of the discrete-time state space is one less than that of the continuous-time state space.

The choice of the Poincaré plane in such a system is not unique. But if the system is non-autonomous, that is, if there is an external periodic input, there is a simple and unambiguous way to obtain the discrete-time model. In such systems, we observe the system state in synchronism with the forcing function. That is, if the forcing function has period T, we observe the state at equal intervals – once every T seconds. Thus, the Poincaré map becomes the mapping of the variables at time t onto those at $t + T$ (Fig. 12.7). This is like placing a stroboscope in the state

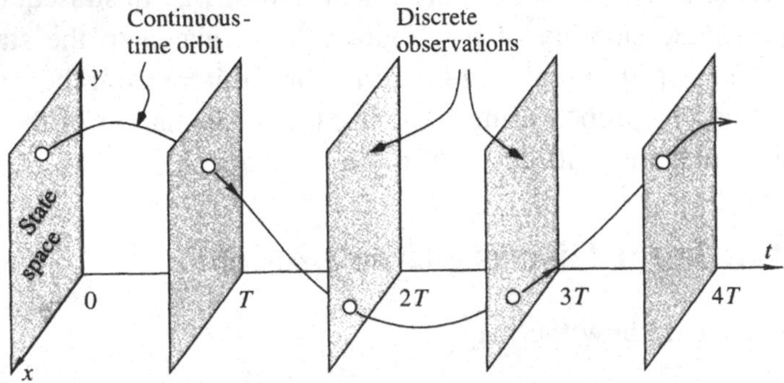

Figure 12.7 Obtaining the discrete model of a non-autonomous system where the external input has period T.

space to illuminate the state at intervals of T. That is why this method of obtaining a discrete-time model is sometimes called *stroboscopic sampling.*

In non-autonomous systems, the periodicity is given by the repetitiveness of the orbit as integral multiples of the period of the external input, and not by the number of loops in the continuous-time state-space trajectory. Thus, if the continuous-time orbit shows one loop in the state space, but there are two cycles of the forcing function within that period, the orbit will be period-2.

► **Example 12.1** Obtain the discrete-time map for the continuous-time dynamical system

$$\dot{r} = br(1-r), \tag{12.2}$$

$$\dot{\theta} = 1, \tag{12.3}$$

where r, θ are polar coordinates.

Solution: It is clear from the system equations that the dynamics occurs in a plane. It is also seen that the motion is rotational, since $\dot{\theta} \neq 0$. Further, $\dot{\theta} = 1$ implies a constant rotational speed of 1 rad/s. $\dot{r}$ has zero magnitude for two values of r: 0 and 1. This implies that $r = 0$ is an equilibrium point, and a cyclic orbit exists at $r = 1$. If b is positive, then for $r < 1$, $\dot{r}$ is positive – which means an outgoing spiral orbit. If $r > 1$, $\dot{r}$ is negative – which means an incoming spiral orbit. This implies that the cyclic orbit at $r = 1$ is stable, that is, a limit cycle.

We place a Poincaré section (in this case, a line) at some arbitrary value of θ, say θ_0. To obtain the discrete map, we observe the state at every 2π s. Now, since the two differential equations are independent, the value of r at the $(n+1)$th instant can be obtained in terms of r_n just by integrating (12.2), as follows:

$$\frac{\mathrm{d}r}{br(1-r)} = \mathrm{d}t,$$

$$\int_{r_n}^{r_{n+1}} \frac{\mathrm{d}r}{br(1-r)} = \int_{t_0}^{t_0+2\pi} \mathrm{d}t,$$

$$\frac{1}{b}\int_{r_n}^{r_{n+1}} \left(\frac{1}{r} + \frac{1}{1-r}\right) \mathrm{d}r = 2\pi,$$

$$[\ln r - \ln(1-r)]_{r_n}^{r_{n+1}} = 2b\pi,$$

$$\ln\left[\left(\frac{r_{n+1}}{1-r_{n+1}}\right)\left(\frac{1-r_n}{r_n}\right)\right] = 2b\pi,$$

$$\left(\frac{r_{n+1}}{1-r_{n+1}}\right)\left(\frac{1-r_n}{r_n}\right) = e^{2b\pi}.$$

From this, by solving for r_{n+1} we get

$$r_{n+1} = \frac{r_n e^{2b\pi}}{1 + r_n(e^{2b\pi} - 1)}. \tag{12.4}$$

This is the functional form of $r_{n+1} = f(r_n)$. ◄

12.3 Dynamics of Discrete-time Systems

As shown earlier, the systems for which the map can be obtained in closed form are rare, and for most systems the map has to be obtained numerically. But in many cases it is possible to find some functional form that qualitatively mimics the behaviour of the system under consideration. By studying such maps one can develop a reasonably good idea about the various dynamical characters of limit cycles, and the ways they can lose stability.

12.4 One-dimensional Maps

In the study of discrete-time dynamical systems, one-dimensional maps have a special position because the graph of the map can be drawn and the dynamics can be graphically visualized. For example, the graph of the map (12.4) obtained in Example 12.1 is as shown in Fig. 12.8.

As a further example, consider the functional form

$$x_{n+1} = \mu x_n(1 - x_n), \tag{12.5}$$

which is known as the *logistic map*. The graph of x_{n+1} versus x_n is shown in Fig. 12.9. The discrete-time modelling of many dynamical systems of practical interest yields continuous functions with a single maximum, and the logistic map serves as a representative of that class of systems.

Notice that the fixed point of the map, obtained by putting $x_{n+1} = x_n$ in (12.5), is simply the point of intersection of the graph of the map with the 45° line, as shown in the figure.

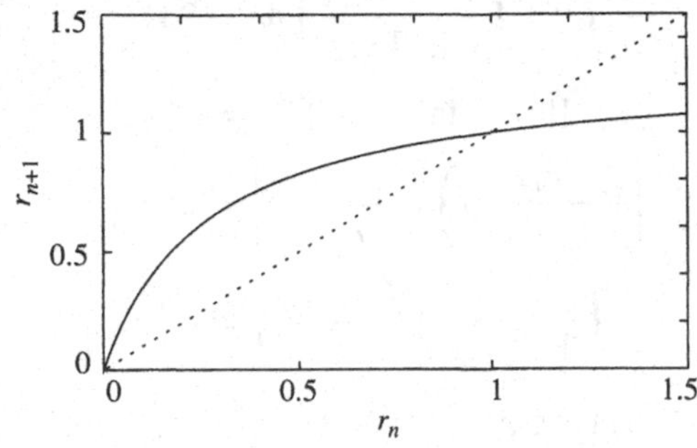

Figure 12.8 The graph of the map (12.4).

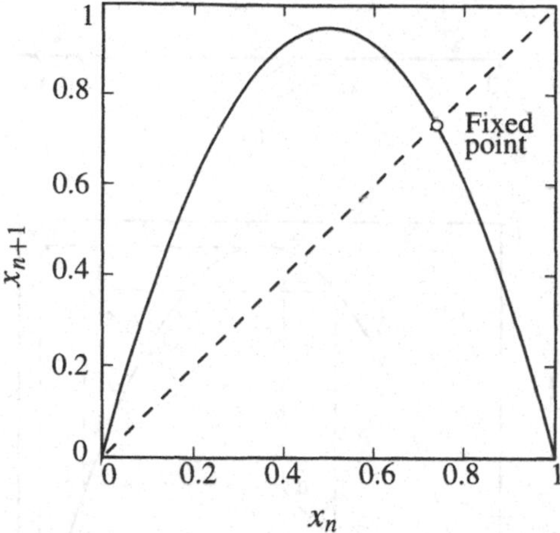

Figure 12.9 The graph of the logistic map (12.5).

The discrete-time orbit is obtained by iterating the map, as

$$x_1 = \mu x_0(1 - x_0),$$
$$x_2 = \mu x_1(1 - x_1),$$
$$x_3 = \mu x_2(1 - x_2),$$

and so on. This sequence of points $x_0 \mapsto x_1 \mapsto x_2 \mapsto x_3 \ldots$, as shown in Fig. 12.10(a), can also be worked out graphically. This is illustrated in Fig. 12.10(b).

First, draw the graph of the map, x_{n+1} versus x_n. Then locate the initial condition x_0 on the x-axis. The value of x_1 is obtained by drawing a vertical line and finding the point of intersection a with the graph of the map. Now, to make the next iteration, this value of x_1 has to be taken along the x-axis. To do that graphically, draw a horizontal line from a to meet the 45° line at point b. The value of x_2 is obtained by drawing a vertical line to meet the graph of the map at point c. To repeat the process to obtain the further iterates, we go horizontally up to point d on the 45° line, and go vertically up to point e on the graph of the map. This gives x_3. This process is repeated.

Can we graphically assess whether a fixed point is stable? If it is stable, an initial condition at some (small) distance from the fixed point should map to a point closer to the fixed point. In Fig. 12.11, we show two situations. In Fig. 12.11(a), the magnitude of the slope of the graph at the fixed point is less than unity. We see that an initial deviation Δx_n from the fixed point reduces at the next iterate

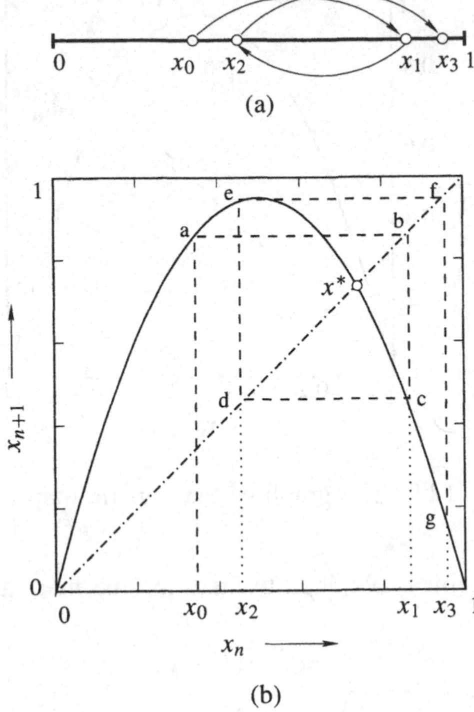

Figure 12.10 Graphical iteration to obtain $x_0 \mapsto x_1 \mapsto x_2 \mapsto x_3 \ldots$.

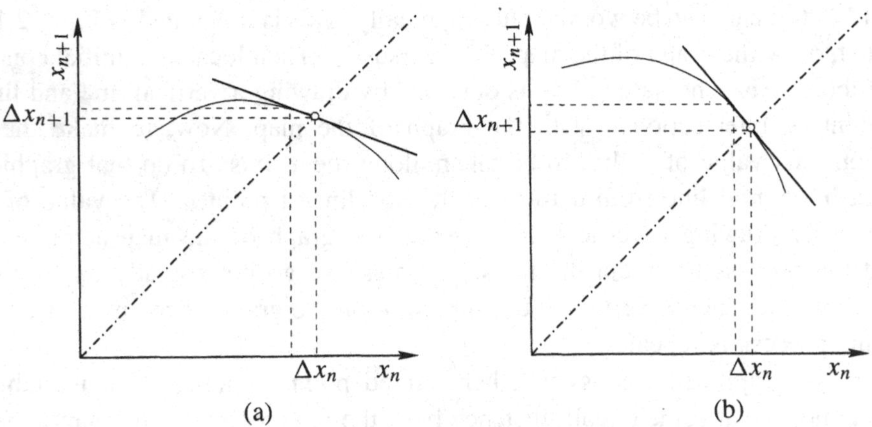

Figure 12.11 Stability of a fixed point is given by the slope of the tangent at the fixed point.

and therefore the point lands closer to the fixed point. On further iteration, the state will converge on to the fixed point. The fixed point is thus stable. In Fig. 12.11(b), the slope of the graph at the fixed point has magnitude greater than unity. In this case, we find that $\Delta x_{n+1} > \Delta x_n$, that is, the point lands away from the fixed point. Hence, the fixed point is unstable.

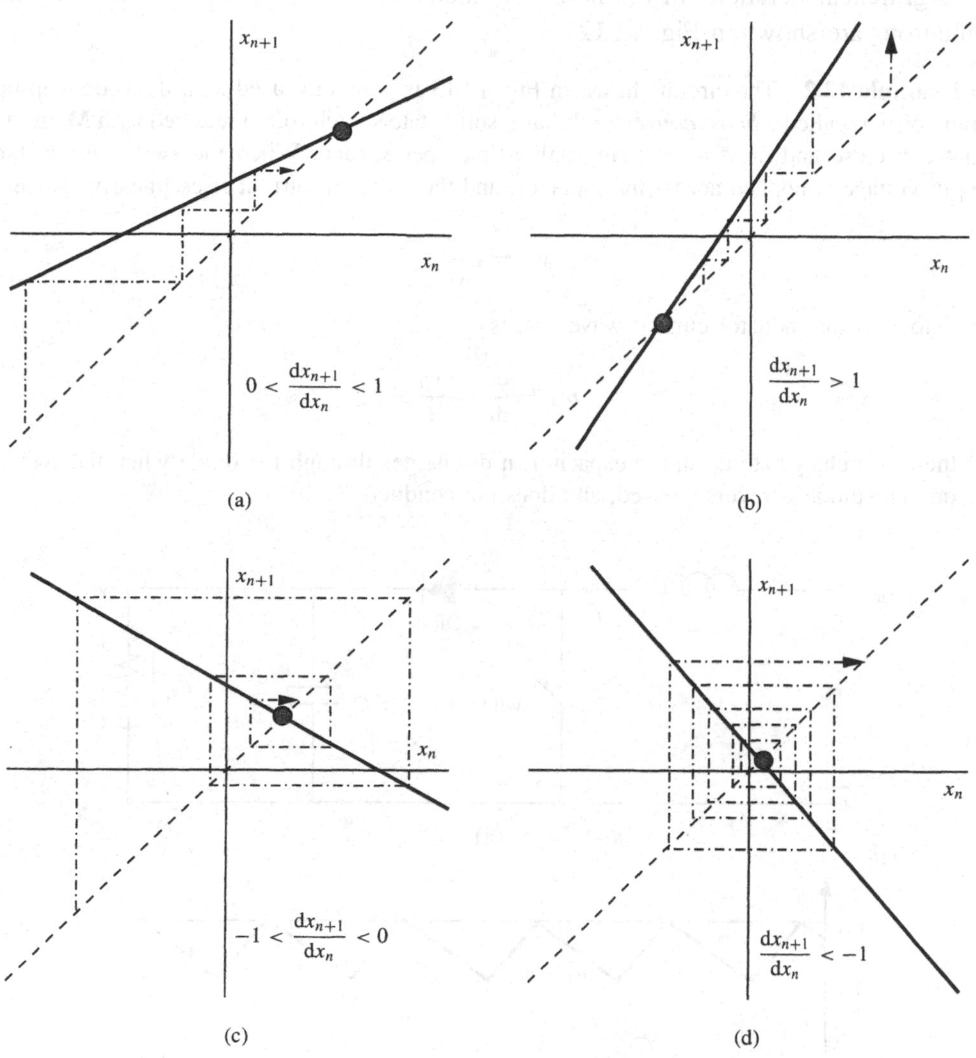

Figure 12.12 The graphical iterations in the neighbourhood of the fixed point (a) when $0 < dx_{n+1}/dx_n < 1$, (b) when $dx_{n+1}/dx_n > 1$ (c) when $-1 < dx_{n+1}/dx_n < 0$, and (d) when $dx_{n+1}/dx_n < -1$. Notice that if the slope is positive, the iterates remain in the same side of the fixed point, and if the slope is negative the state flips from one side to the other in subsequent iterations.

This leads to the conclusion that the stability of the fixed point depends on the slope of the tangent of the graph at the fixed point, and the fixed point is stable if

$$\left|\frac{\mathrm{d}x_{n+1}}{\mathrm{d}x_n}\right|_{\text{at fixed point}} < 1. \tag{12.6}$$

The graphical iterations in the neighbourhood of the fixed point in four possible situations are shown in Fig. 12.12.

► **Example 12.2** The circuit shown in Fig. 12.13 is generally used as a dc-to-dc step-up transformer, called a *boost converter*. It has a solid-state switch (often realized by a MOSFET) that can close and open more than 10,000 times per second. When the switch is on, the input voltage is applied across the inductor, and the inductor current rises linearly. Since

$$V_{\text{in}} = L\frac{\mathrm{d}i}{\mathrm{d}t},$$

the slope of the inductor current waveform is

$$m_1 = \frac{\mathrm{d}i}{\mathrm{d}t} = \frac{V_{\text{in}}}{L}.$$

If there is a charge stored in the capacitor, it discharges through the diode when the switch is on. The diode is reverse biased, and does not conduct.

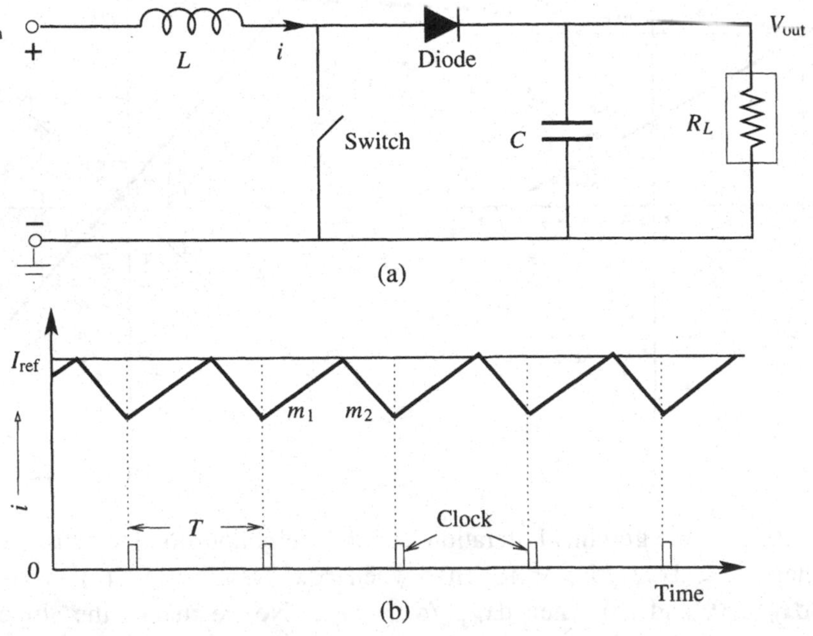

Figure 12.13 The circuit considered in Example 12.2.

When the switch is turned off, the diode becomes forward biased, and the inductor's stored energy passes on to the capacitor and the load resistance R_L. As a result i drops and the voltage across the inductor, given by

$$v_L = L\frac{\mathrm{d}i}{\mathrm{d}t}$$

becomes negative. This voltage is *added* to the input voltage V_{in} to appear across the load. The load therefore sees a voltage greater than the input voltage – hence the name "boost converter."

For this system to work, the switch must be turned on and off periodically, that is, it must establish a limit cycle. One of the commonly used methods of effecting switching is as follows. A periodic "clock" signal is generated separately by means of a commonly available 555 timer chip, and the switch is turned on at every positive edge of the clock, as shown in Fig. 12.13(b). The inductor current is sensed, and compared with a reference current I_{ref}. When the inductor current reaches I_{ref}, the switch is turned off. In case the inductor current does not reach I_{ref} before the arrival of the next clock, the switch remains on. This is known as the *current mode control.*

In practical circuits, a high value of the capacitor is chosen so that the output voltage is approximately a constant dc voltage. For the sake of simplicity, if we assume the output voltage to be constant, then the waveform of the inductor current during the off period will also be linear, with a slope

$$m_2 = \frac{V_{\mathrm{out}} - V_{\mathrm{in}}}{L}.$$

Since the clock is an external periodic input to the system, in order to obtain the map, we have to observe the state (inductor current) at every clock instant. Let the inductor current at the nth instant be i_n. The on period is calculated as follows:

$$\begin{aligned} I_{\mathrm{ref}} &= i_n + m_1 T_{\mathrm{on}} \\ \text{or} \quad T_{\mathrm{on}} &= (I_{\mathrm{ref}} - i_n)/m_1. \end{aligned}$$

The value of the inductor current at the next clock instant can then be calculated as

$$\begin{aligned} i_{n+1} &= I_{\mathrm{ref}} - m_2 T_{\mathrm{off}}, \\ &= I_{\mathrm{ref}} - m_2(T - T_{\mathrm{on}}), \\ &= I_{\mathrm{ref}} - m_2\{T - (I_{\mathrm{ref}} - i_n)/m_1\}, \\ &= \left(I_{\mathrm{ref}} + \frac{m_2}{m_1} I_{\mathrm{ref}} - m_2 T\right) - \frac{m_2}{m_1} i_n. \end{aligned} \tag{12.7}$$

In order to obtain the complete map, we also have to consider the case where the current does not reach I_{ref} before the next clock instant. This happens if

$$i_n \leq I_{\mathrm{ref}} - m_1 T.$$

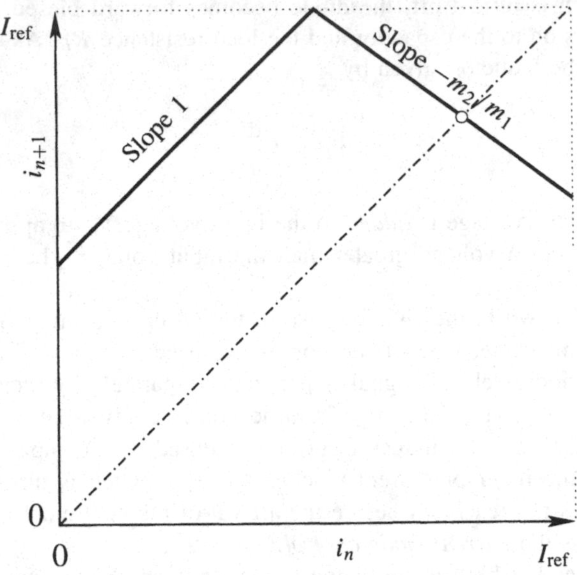

Figure 12.14 The graph of the discrete map corresponding to the system in Fig. 12.13.

In this case, the switch remains on during the whole of the clock period, and hence the current at the next clock instant will be

$$i_{n+1} = i_n + m_1 T. \tag{12.8}$$

Thus, the discrete-time model of the system is given by

$$i_{n+1} = \begin{cases} (12.7) & \text{for} \quad i_n \geq I_{\text{ref}} - m_1 T, \\ (12.8) & \text{for} \quad i_n \leq I_{\text{ref}} - m_1 T. \end{cases}$$

The graph of the map is shown in Fig. 12.14. It has two segments: the one given by (12.8) has slope unity, and the one given by (12.7) has slope $-m_2/m_1$. The part given by (12.7) intersects with the 45° line and therefore the fixed point is located in that segment.

Now we consider the question: Will the periodic orbit be stable? The stability condition (12.6) says that the fixed point will be stable so long as

$$m_2/m_1 < 1.$$

Substituting the values of m_1 and m_2, we get the condition as

$$V_{\text{out}} < 2V_{\text{in}}.$$

Indeed, this result is known to designers of such switching circuits. ◄

12.5 Bifurcations

In any physical system, a change in the magnitude of a parameter leads to some change in the dynamical behaviour. Most of the time these changes are only quantitative in nature, that is, the shape of the trajectory in the state space undergoes some change but the periodicity of the system remains unchanged. There may also be situations where a small variation in parameter results in a major change in steady-state behaviour of a dynamical system. Such events are called *bifurcations*. Naturally, bifurcations are very important dynamical events that may affect the performance of engineering systems.

So long as a fixed point is stable, a change in a parameter can only result in some small quantitative change in the orbit. It is not difficult to see, therefore, that bifurcations are related to the loss of stability of orbits.

There is a convenient way of visualizing the stability status of various orbits as a parameter is varied. Suppose we are varying a parameter of a system in steps, causing the orbit to change. Take one value of the parameter, iterate the map and let the dynamics settle into a stable orbit (i.e. remove the initial transient or "pre-iterates"). Then plot along the x-axis the value of the parameter and along the y-axis about 100 points of the discrete-time orbit. Then take the next value of the parameter and repeat the same procedure. Thus, we obtain a plot with the parameter value in the x-axis and the discretely observed value of a state variable in the y-axis. The resulting plot is called a *bifurcation diagram*. Such a diagram, drawn for the map $x_{n+1} = a - x_n^2$, is shown in Fig. 12.15.

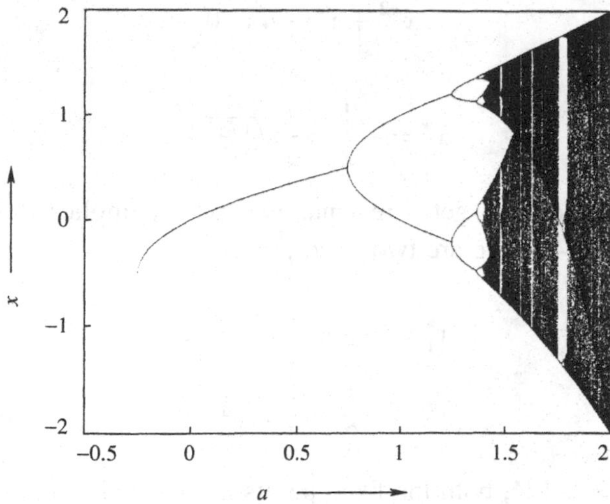

Figure 12.15 The bifurcation diagram for the map $x_{n+1} = a - x_n^2$.

This graph tells a lot of story. For the parameter values where the trajectory is periodic, all the 100 points will fall on the same location, and we shall see just one point. Where the orbit is period-2, 50 points will fall in one position while the other 50 will fall at a different location. We shall thus see two points for that parameter value. When the system becomes chaotic, all the 100 points will fall at different locations (because the system then has no periodicity) and we see a smudge of dots. The graph clearly shows the character of the orbit for every parameter value, and we have a panoramic view of the stability status of various types of orbits as a parameter is varied. The graph also clearly shows where an orbit loses stability, resulting in some qualitative change in the character of the orbit.

Condition (12.6) tells that in a one-dimensional map, a periodic orbit can lose stability in two basic ways – first, if the derivative becomes greater than $+1$, and second, if the derivative becomes less than -1. These result in two different mechanisms of the loss of stability of limit cycles. We shall now consider these two situations.

12.6 Saddle-node Bifurcation

Notice that the slope at a fixed point can become $+1$ only if the graph of the map becomes tangent to the 45° line. To illustrate what happens in that condition, let us consider the map

$$x_{n+1} = a - x_n^2. \tag{12.9}$$

First, we locate the fixed points x^* of this system, where $x_{n+1} = x_n = x^*$. This gives the equation

$$x^{*2} + x^* - a = 0,$$

whose solutions are

$$x^* = -\frac{1}{2} \pm \frac{1}{2}\sqrt{1+4a}.$$

For $a < -1/4$, we do not get a real number, which implies that no fixed point exists. For $a > -1/4$, there are two fixed points

$$x_1^* = -\frac{1}{2} + \frac{1}{2}\sqrt{1+4a},$$
$$x_2^* = -\frac{1}{2} - \frac{1}{2}\sqrt{1+4a}.$$

Notice that at $a = -1/4$, both the fixed points are located at the same place: $x^* = -1/2$. As a is increased from this value, the fixed points move away from each other.

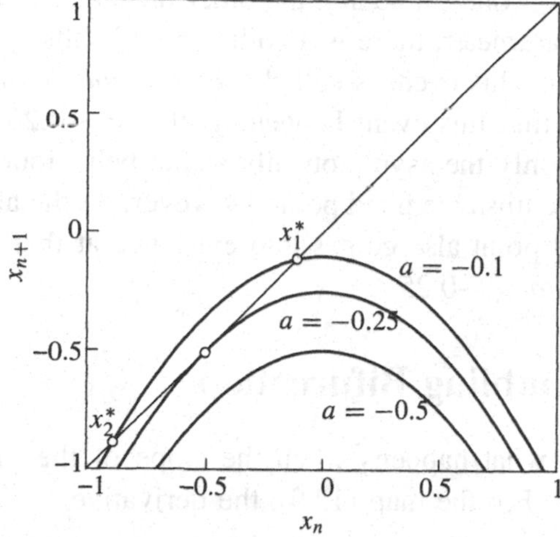

Figure 12.16 The graphs of the map $x_{n+1} = a - x_n^2$ as a is varied.

Now let us look at the derivative of the map (12.9)

$$\frac{dx_{n+1}}{dx_n} = -2x_n.$$

The derivative will be equal to $+1$ for $x_n = -1/2$. The graphical illustration of the sequence of events is given in Fig. 12.16.

It shows that at $a = -0.5$, the graph of the map does not intersect the 45° line, and hence there is no fixed point. As the parameter a is increased, the graph makes contact with the 45° line at $a = -0.25$. At that parameter value, there is only one fixed point, placed at $x_n = -0.5$. As the parameter is further increased to $a = -0.1$, we find two fixed points x_1^* and x_2^*. These two fixed points come into existence at the parameter value where the slope of the graph at the fixed point becomes exactly equal to $+1$.

Notice that the slope at x_1^* is less than unity, that is, this fixed point is stable. On the other hand, the slope at x_2^* is greater than unity, that is, this fixed point is unstable. The two fixed points – one stable and one unstable – come into existence at $a = -0.25$.

If we look at the sequence of events with the parameter a varied in the opposite direction, we see a couple of fixed points existing for $a > -0.25$. They come closer to each other as a is reduced, and the slopes at the fixed points approach unity. At $a = -0.25$ they coincide, and the slope becomes exactly unity. For lower values of a, no fixed point exists.

To summarize: What happens when the derivative of the map at a fixed point becomes $+1$? In one sense of the variation of the parameter, there is a birth of a couple of fixed points – one stable and the other unstable; in the opposite sense of the variation of the parameter, there is a collision and subsequent annihilation of a couple of fixed points. This event is called a *saddle-node bifurcation.*

In Fig. 12.15, we find this event happening at $a = -0.25$. Note that the bifurcation diagram plots only the asymptotically stable behaviour, and hence does not show the branch of the unstable fixed point. However, by the above theory we know that an unstable fixed point also comes into existence at that parameter value, and continues to exist for $a > -0.25$.

12.7 Period-doubling Bifurcation

Now let us consider what happens when the slope of the graph at a fixed point becomes equal to -1. For the map (12.9), the derivative

$$\frac{\mathrm{d}x_{n+1}}{\mathrm{d}x_n} = -2x_n$$

has to be obtained at a fixed point. For $a > -1/4$, there are two fixed points x_1^* and x_2^*. If we substitute the position of x_1^*, we obtain the expression for the derivative at this fixed point as

$$\left.\frac{\mathrm{d}x_{n+1}}{\mathrm{d}x_n}\right|_{x_1^*} = 1 - \sqrt{1 + 4a}.$$

It assumes a value of -1 at $a = 3/4$. For $-1/4 < a < 3/4$, the magnitude of the derivative is less than unity, and hence the fixed point x_1^* is stable. For $a > 3/4$ the slope has magnitude greater than unity and hence the fixed point is unstable. What happens at $a = 3/4$? As the fixed point becomes unstable, does anything else become stable?

To probe this point, we look at the second-iterate map

$$x_{n+2} = f(x_n),$$

which can be easily obtained from the original map

$$x_{n+1} = a - x_n^2$$

as

$$\begin{aligned} x_{n+2} &= a - x_{n+1}^2, \\ &= a - \left(a - x_n^2\right)^2, \\ &= -x_n^4 + 2ax_n^2 - a^2 + a. \end{aligned}$$

In Fig. 12.17, the graph of this "second-iterate map" is shown along with the graph of the map (12.9) for two cases – one for $a < 0.75$ and the other for $a > 0.75$.

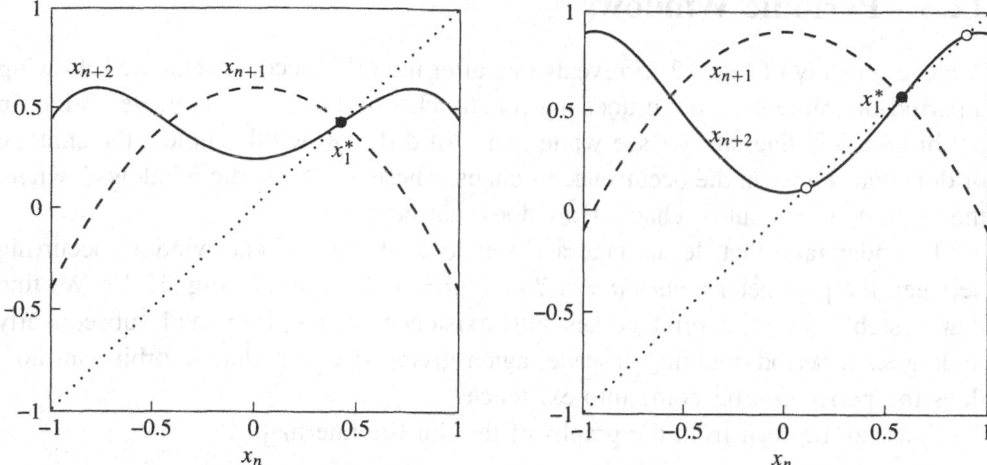

Figure 12.17 The graphs of x_{n+1} (broken line) and those of x_{n+2} (firm line) for the map (12.9) for $a = 0.6$ and for $a = 0.9$. The graphs on the left illustrate the situation for $a < 0.75$, where the period-1 fixed point is stable, and is also the fixed point of the second-iterate map. The graphs on the right show that for $a > 0.75$ the period-1 fixed point becomes unstable, and two stable fixed points appear in the graph of x_{n+2}. This implies that a period-2 orbit becomes stable as the period-1 orbit becomes unstable.

Obviously, the fixed points of the first-iterate map are also the fixed points of the second-iterate map. For $a < 0.75$, the two graphs intersect the 45° line at the same points, implying that they have the same fixed points x_1^* and x_2^*. As the fixed point becomes unstable at $a = 0.75$, two new fixed points appear in the map for x_{n+2}, and the slope at these fixed points is less than unity. As a result, at the critical value of the parameter, we see a transition from a period-1 orbit to a period-2 orbit. This is called a *period-doubling bifurcation.*

As the parameter is further increased, the slopes at the period-2 fixed points change, and at another parameter value these fixed points also become unstable as the slopes become less than -1. Thus, a period-doubling bifurcation occurs again, and if we plot the graph of the fourth-iterate map x_{n+4} we shall find that four new stable fixed points appear. Thus, there is transition from period-2 to period-4. With increasing parameter, there is thus a "period-doubling cascade." Many such period-doubling events are visible in the bifurcation diagram of Fig. 12.15, where an orbit of periodicity n gives way to an orbit of periodicity $2n$. These happen when the slope of the graph at a fixed point of the nth-iterate map attains a value of -1. The cascade finally terminates in an aperiodic behaviour, that is, chaos. This phenomenon is quite common in many practical dynamical systems.

12.8 Periodic Windows

A close scrutiny of Fig. 12.15 reveals that after the orbit becomes chaotic following a period-doubling cascade, it does not remain chaotic for every parameter value. In the bifurcation diagram, we see white bands of different widths amidst the smudge of dots that represent the occurrence of chaos. These are "periodic windows," where the system comes out of chaos. How does that happen?

To understand that, let us take a closer look at the periodic window occurring between the parameter values $a = 1.74$ and $a = 1.80$, given in Fig. 12.18. We find that a stable period-3 orbit comes into existence at $a \approx 1.75$, and subsequently undergoes a period-doubling cascade, again giving rise to a chaotic orbit. But how does the period-3 orbit come into existence?

That can be seen from the graphs of the third-iterate map

$$x_{n+3} = a - \left(-x^4 + 2ax^2 - a^2 + a\right)^2$$

shown in Fig. 12.19, drawn for parameter values before and after the onset of the period-3 window. We find that for $a = 1.73$, there are only two fixed points of the third-iterate map. Indeed, these are the same as the fixed points of the first-iterate map, because if $x_{x+1} = x_n$ is true, $x_{n+3} = x_n$ must also be true. But at $a = 1.76$, we find that the graph of the third-iterate map has crossed the 45° line in three places, resulting in the creation of two period-3 fixed points – one stable and the other unstable. Because the period-3 orbit is stable, all orbits converge on it, and

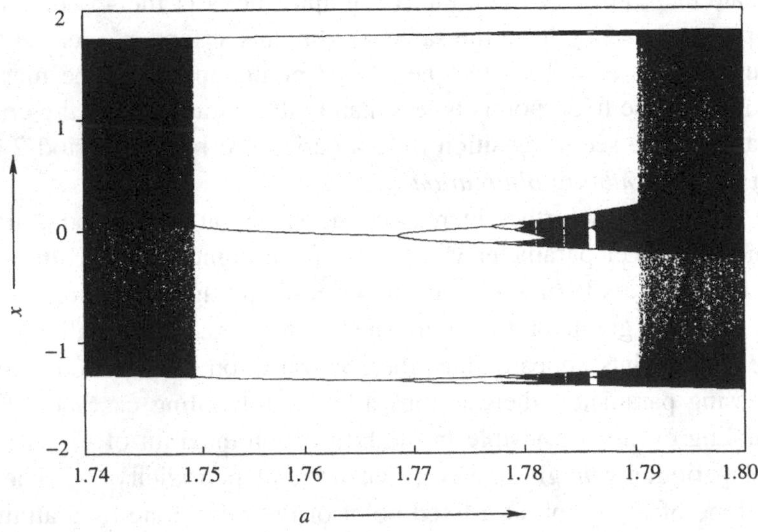

Figure 12.18 An enlarged view of the period-3 window of Fig. 12.15.

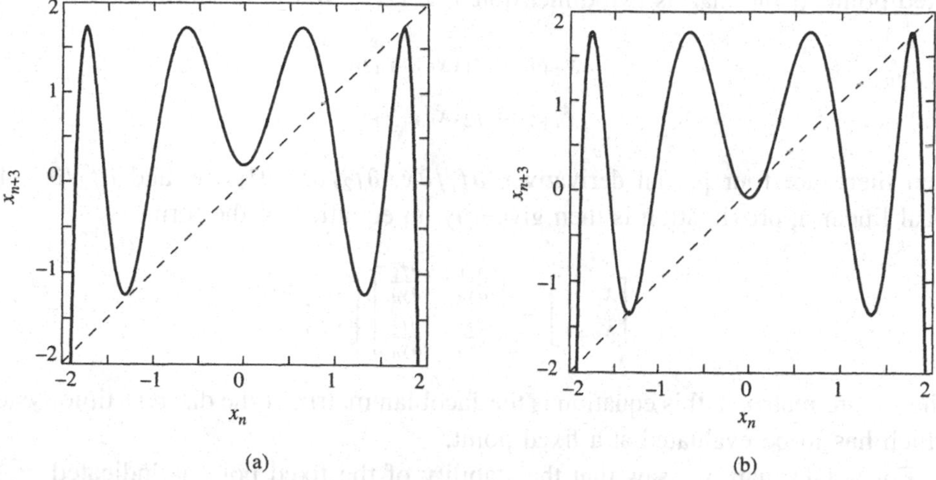

Figure 12.19 The graphs of the third-iterate map, (a) at $a = 1.73$ before the onset of the periodic window, and (b) at $a = 1.76$ after the onset of the periodic window.

hence the chaotic orbit can no longer remain stable. This results in the creation of the periodic window.

What was the value of $\mathrm{d}x_{n+3}/\mathrm{d}x_n$ at the point of bifurcation? At that specific parameter value, the graph must be tangent to the 45° line, and hence the slope must be +1. We see, therefore, that the onset of the periodic window is caused by a saddle-node bifurcation in the third-iterate map.

As the parameter value is increased, at some point the stable period-3 orbit also becomes unstable when its slope becomes equal to −1. This is a period-doubling bifurcation that results in a period-6 orbit. The period-doubling cascade proceeds in the same fashion as seen before, finally giving rise to a chaotic orbit.

Indeed, the same phenomenon – a saddle-node bifurcation followed by a period-doubling cascade – occurs in all the periodic windows seen in a bifurcation diagram. This is also a common phenomenon found in nonlinear systems.

12.9 Two-dimensional Maps

We have seen that in analysing discrete-time nonlinear systems expressed in the form

$$\mathbf{x}_{n+1} = f(\mathbf{x}_n),$$

one first finds the fixed points $\mathbf{x}_{n+1} = \mathbf{x}_n = \mathbf{x}^*$. Then one has to locally linearize the discrete system in the neighbourhood of a fixed point. For the one-dimensional (1-D) map, the local linear approximation was given by the derivative $\mathrm{d}f/\mathrm{d}x$ at the fixed point. If the map is two-dimensional (2-D), given in the form

$$x_{n+1} = f_1(x_n, y_n),$$
$$y_{n+1} = f_2(x_n, y_n),$$

then there are four partial derivatives $\partial f_1/\partial x$, $\partial f_2/\partial x$, $\partial f_1/\partial y$, and $\partial f_2/\partial y$. The local linear approximation is then given by an equation of the form

$$\begin{bmatrix} x_{n+1} \\ y_{n+1} \end{bmatrix} = \begin{bmatrix} \frac{\partial f_1}{\partial x_n} & \frac{\partial f_1}{\partial y_n} \\ \frac{\partial f_2}{\partial x_n} & \frac{\partial f_2}{\partial y_n} \end{bmatrix} \begin{bmatrix} x_n \\ y_n \end{bmatrix}.$$

The square matrix in this equation is the Jacobian matrix of the discrete-time system, which has to be evaluated at a fixed point.

For a 1-D map, we saw that the stability of the fixed point is indicated by the slope of the map at the fixed point. In case of a 2-D map, the eigenvalues of the Jacobian matrix indicate the stability of the fixed point.

But there is a subtle difference between what the eigenvalues indicate in a continuous-time system and what they indicate in a discrete-time system. In a continuous-time system, the Jacobian matrix, when operated on a state vector, gives $\dot{\mathbf{x}}$, that is, the *velocity vector* corresponding to that state. But in a linearized discrete-time system the operation of the Jacobian matrix on a state vector gives the *state vector at the next iterate*.

To illustrate, imagine a two-dimensional discrete-time system linearized around the fixed point $(0, 0)$:

$$\mathbf{x}_{n+1} = \mathbf{J}\mathbf{x}_n,$$

given by

$$\begin{bmatrix} x_{n+1} \\ y_{n+1} \end{bmatrix} = \begin{bmatrix} J_{11} & J_{12} \\ J_{21} & J_{22} \end{bmatrix} \begin{bmatrix} x_n \\ y_n \end{bmatrix}.$$

This implies that for every specific state vector $[x_n, y_n]^T$, pre-multiplication with the matrix $\mathbf{J}$ gives the state vector at the next (discrete) time instant. In other words, the state at any instant is obtained by *operating* the Jacobian matrix on the previous state vector.

Let λ_1 and λ_2 be the eigenvalues of the Jacobian matrix. Imagine an initial condition placed on the eigenvector $\mathbf{v}_1$ associated with a real eigenvalue λ_1. Where will the next iterate fall? First, since $\mathbf{v}_1$ is an eigenvector, the next iterate must remain on the same eigenvector. Second, the magnitude of the vector will be scaled by the factor λ_1 to give the magnitude of the vector in the next iterate. Thus, if a state is on an eigendirection, the state at the next iterate is obtained simply by multiplying the state vector with a scalar quantity λ_1.

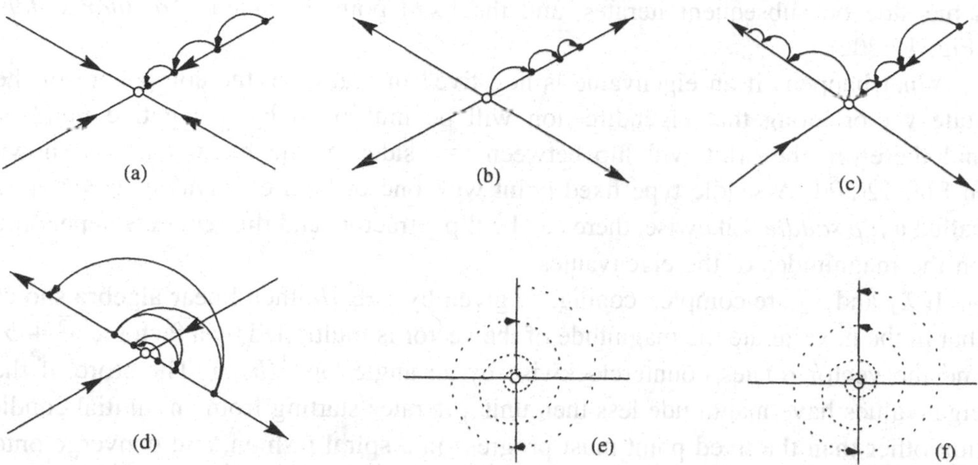

Figure 12.20 Examples of fixed points in a linearized two-dimensional discrete system: (a) An attracting node: eigenvalues real, $0 < \lambda_1, \lambda_2 < 1$. (b) A repelling node: eigenvalues real, $\lambda_1, \lambda_2 > 1$. (c) A regular saddle: eigenvalues real, $0 < \lambda_1 < 1$, $\lambda_2 > 1$. (d) A flip saddle: eigenvalues real, $0 < \lambda_1 < 1, \lambda_2 < -1$. (e) A spiral attractor: eigenvalues complex, $|\lambda_1|, |\lambda_2| < 1$. (f) A spiral repeller: eigenvalues complex, $|\lambda_1|, |\lambda_2| > 1$.

Now, if λ_1 has a magnitude less than unity, the next iterate will fall on another point of the eigenvector closer to the fixed point, and in subsequent iterates the state will converge onto the fixed point along the eigenvector (Fig. 12.20a). The same will happen if the state is on the other eigendirection.

What will happen if the initial state is on any other point? In that case, the state vector can be decomposed into components along the eigendirections, and the state vector at the next observation instant will be given by the vector addition

$$\mathbf{x}_{n+1} = \lambda_1 \times \text{component of } \mathbf{x}_n \text{ along } \mathbf{v}_1 + \lambda_2 \times \text{component of } \mathbf{x}_n \text{ along } \mathbf{v}_2.$$

If both the eigenvalues are real and have magnitude less then unity, any initial condition will asymptotically move towards an eigenvector and will converge onto the fixed point. Thus, the system is stable and the fixed point is called an *attracting node*.

On the other hand, if both the eigenvalues have magnitudes greater than unity, the system is unstable and the fixed point is a *repelling node* (Fig. 12.20b). If these eigenvalues are real, with λ_1 of magnitude less than unity and λ_2 of magnitude greater than unity, the system is stable in one direction (the eigenvector associated with λ_1) and unstable in the other. Such a fixed point is called a *saddle*. If both the eigenvalues are positive, a state on one side of the fixed point remains on the same side on subsequent iterates, and the fixed point is called a *regular saddle* (Fig. 12.20c).

What happens if an eigenvalue is negative? In that case, the component of the state vector along that eigendirection will be multiplied by a negative number, and therefore the orbit will flip between two sides of the fixed point as shown in Fig. 12.20d. A saddle-type fixed point with one or both eigenvalues negative is called a *flip saddle*. Likewise, there can be flip attractors and flip repellers depending on the magnitudes of the eigenvalues.

If λ_1 and λ_2 are complex conjugate, given by $a \pm jb$, then linear algebra shows that in the next iterate the magnitude of the vector is multiplied by a factor $\sqrt{a^2 + b^2}$ and the vector rotates counterclockwise by an angle $\tan^{-1}(b/a)$. Therefore, if the eigenvalues have magnitude less then unity, iterates starting from any initial condition other than the fixed point must progress in a spiral fashion, and converge onto the fixed point. Such a fixed point is said to be *spirally attracting* (Fig. 12.20e). In some literature, a fixed point with a spiralling orbit is called a *focus*. Likewise, if the eigenvalues are complex conjugate and with magnitude greater than unity, the fixed point is a *spirally repelling* focus (Fig. 12.20f).

It is important to note that in a discrete system a fixed point is stable if all the eigenvalues of the Jacobian matrix have magnitude less than unity, while in a continuous-time system an equilibrium point is stable if the eigenvalues have negative real part.

12.10 Bifurcations in 2-D Discrete-time Systems

As we have noted earlier, a bifurcation happens when a fixed point becomes unstable and some other orbit takes its place – resulting in an abrupt change in the character of an orbit. Since we are interested in the stability of fixed points – which in turn signifies the stability of limit cycles in continuous time – let us first ask the question: In how many ways can a fixed point lose stability?

Notice that a fixed point is stable so long as the eigenvalues have magnitude less than unity. If the eigenvalues are plotted in the complex plane, this can be visualized as the eigenvalues being inside a circle of radius unity. Therefore, the above question is equivalent to asking, "In how many different ways can the eigenvalues exit the unit circle?"

It is not difficult to see that there are three possible ways:

1. An eigenvalue may exit the unit circle on the negative real line.

2. An eigenvalue may exit unit circle on the positive real line.

3. A complex conjugate pair of eigenvalues may exit the unit circle.

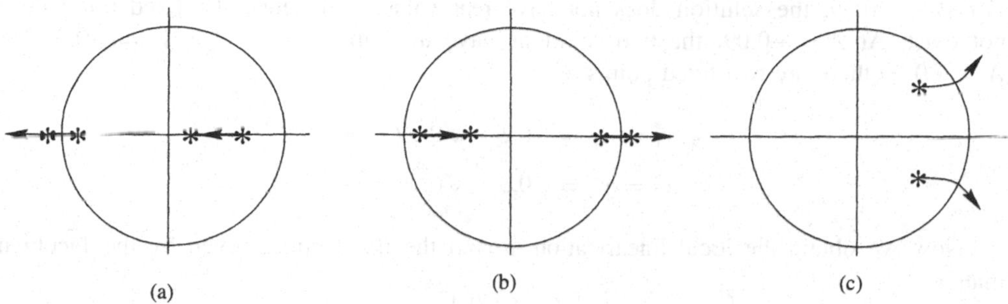

Figure 12.21 (a) A period-doubling bifurcation: eigenvalue crosses the unit circle on the negative real line, (b) a saddle-node bifurcation: an eigenvalue touches the unit circle on the positive real line, and (c) a Neimark–Sacker bifurcation: a complex conjugate pair of eigenvalues cross the unit circle.

The first one is the period-doubling bifurcation and the second one is the saddle-node bifurcation. We have come across these phenomena in 1-D systems. The third one, called the *Neimark–Sacker bifurcation*, cannot happen in 1-D systems since it requires the eigenvalues to be complex conjugate. These three cases are shown in Fig. 12.21 and will now be illustrated with examples.

► **Example 12.3** Consider the following discrete-time dynamical system, known as the *Hénon map*:

$$x_{n+1} = A - x_n^2 + By_n, \tag{12.10}$$

$$y_{n+1} = x_n. \tag{12.11}$$

Let us assume one of the parameter values, $B = 0.4$ and study the dynamics as dependent on the parameter A.

In analysing the dynamics of the system, we first obtain the fixed point (x^*, y^*) by substituting $\mathbf{x}_{n+1} = \mathbf{x}_n$, as

$$x^* = A - x^{*2} + 0.4y^*, \quad \text{and} \quad y^* = x^*.$$

Substituting the second equation into the first, we get the quadratic equation

$$x^{*2} + 0.6x^* - A = 0.$$

Solving, we get

$$x^* = \frac{1}{2}\left(-0.6 \pm \sqrt{0.36 + 4A}\right)$$

and

$$y^* = x^*.$$

For $A < -0.09$, the solution does not have real value, and hence the fixed point does not exist. At $A = -0.09$, the two solutions have the same value, $x^* = y^* = -0.3$. For $A > -0.09$ there are two fixed points

$$x_1^* = y_1^* = -0.3 + \sqrt{0.09 + A},$$
$$x_2^* = y_1^* = -0.3 - \sqrt{0.09 + A}.$$

Now we obtain the local linearization around the fixed point, given by the Jacobian matrix

$$\begin{bmatrix} -2x^* & 0.4 \\ 1 & 0 \end{bmatrix}.$$

The characteristic equation is

$$-\lambda\left(-2x^* - \lambda\right) - 0.4 = 0$$

or

$$\lambda^2 + 2x^*\lambda - 0.4 = 0.$$

Thus, the eigenvalues are

$$\lambda = -x^* \pm \sqrt{x^{*2} + 0.4}. \tag{12.12}$$

Notice that at the point where the fixed point begins to exist, that is, at $x^* = -0.3$, one of the eigenvalues assumes a value of $+1$. As the value of A is reduced from some value above -0.09, the two fixed points $\mathbf{x}_1^*$ and $\mathbf{x}_2^*$ come closer to each other, and at $A = -0.09$ they collide and disappear. At the same time, the eigenvalues at these fixed points approach unity, and at $A = -0.09$ they become exactly unity.

What is the character of the fixed points created at $A = -0.09$? To investigate this, let us take a parameter value for $A > -0.09$, say, $A = 0$. At that parameter value, $\mathbf{x}_1^* = (0, 0)$ and $\mathbf{x}_2^* = (-0.6, -0.6)$. Equation (12.12) shows that the eigenvalues at the fixed point 1 are ± 0.6324, and those at the fixed point 2 are 0.2718 and -1.4718. This implies that $\mathbf{x}_2^*$ is a saddle-fixed point (a flip saddle, because one eigenvalue is negative), and $\mathbf{x}_1^*$ is an attracting node. The event at $A = -0.09$ results in the birth of a saddle-type fixed point and a node-type fixed point. That is why this event is called a *saddle-node bifurcation*.

At what parameter value does one of the eigenvalues become equal to -1? Putting $\lambda = -1$ in (12.12) we find that this happens when $\mathbf{x}_1^* = (0.3, 0.3)$, and this value is attained at $A = 0.27$. Therefore, at $A = 0.27$ the fixed point becomes unstable. What happens then?

If we obtain the second-iterate map $\mathbf{x}_{n+2} = f(\mathbf{x}_n)$, obtain its fixed points[2] and calculate the eigenvalues, we will find that two fixed points appear for $A > 0.27$, and they have eigenvalues with magnitude less than unity. The algebra is a bit cumbersome because it yields fourth-order equations. We leave it to the reader to check.

The result is that as the period-1 orbit becomes unstable, a period-2 orbit becomes stable. This is the period-doubling bifurcation. As the parameter is further increased, there

[2] $\mathbf{x}_{n+1} = \mathbf{x}_n$ also implies $\mathbf{x}_{n+2} = \mathbf{x}_n$, and so the fixed points of the first-iterate map are also the fixed points of the second-iterate map. Therefore, two of the fixed points of the second-iterate map will be the same as the period-1 fixed points obtained earlier.

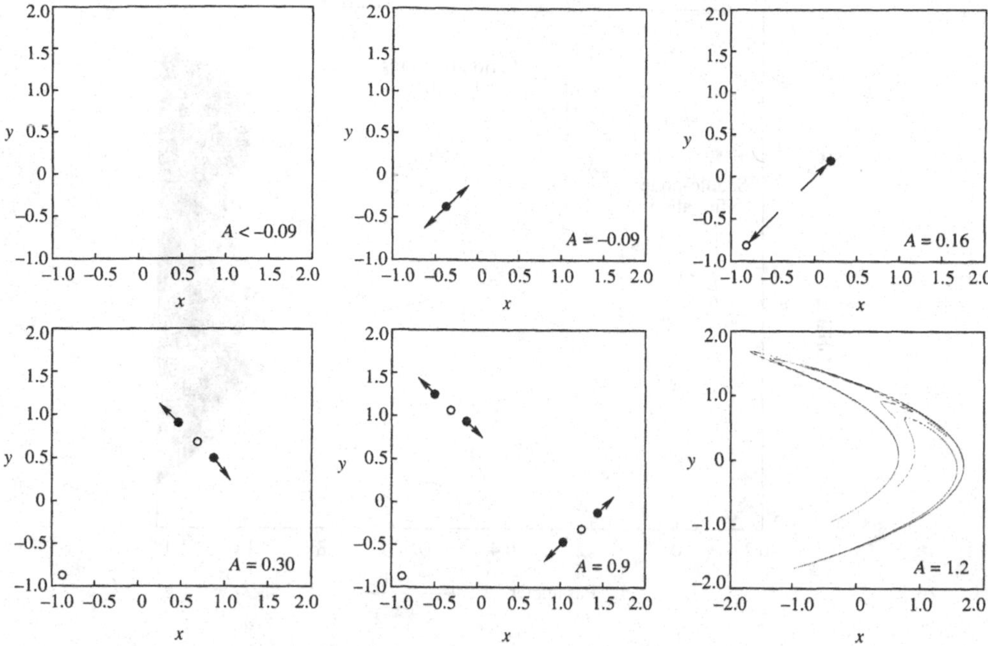

Figure 12.22 Change in the fixed points of the Hénon map as the parameter A is varied. For $A < -0.09$, there is no fixed point. At $A = -0.09$, a pair of stable and unstable fixed points are born. At $A = 0.16$, the saddle and the node have moved away from each other. At $A = 0.3$, $\mathbf{x}_1^*$ has become unstable and a pair of period-2 fixed points have become stable. At $A = 0.9$, the period-2 orbit has become unstable and a stable period-4 orbit emerges. At $A = 1.2$, there is a chaotic attractor.

is a cascade of period-doublings, resulting in chaos. This is shown in Fig. 12.22, and the bifurcation diagram of Fig. 12.23. ◀

Note the remarkable similarity of the bifurcation diagrams in Fig. 12.15 and Fig. 12.23 even though the systems are entirely different. This points to some "universal" aspects of the dynamics of nonlinear systems. Indeed, in many practical engineering systems from electronic circuits to the rocking motion of ships in the sea, such period-doubling cascades have been observed.

Now let us consider the third mechanism of loss of stability, where a complex conjugate pair of eigenvalues cross the unit circle.

▶ **Example 12.4** Consider the map

$$x_{n+1} = y_n,$$
$$y_{n+1} = \mu/2 + (\mu + 1)(y_n - x_n - 2x_n y_n).$$

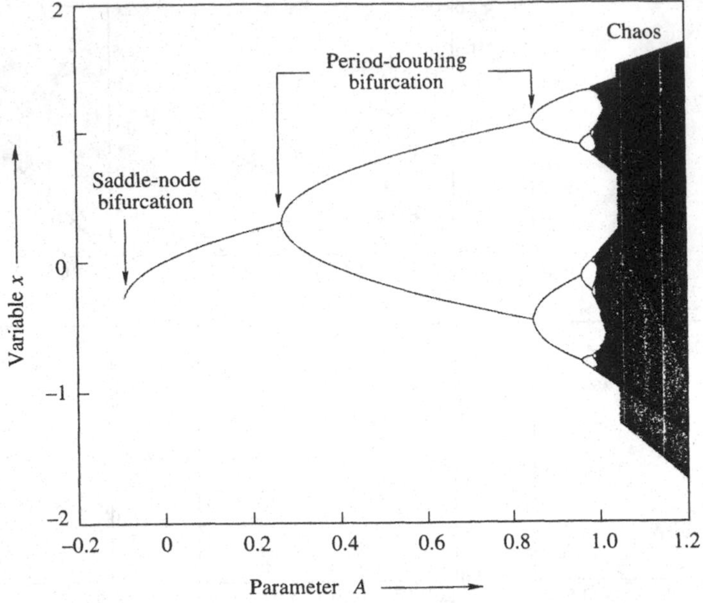

Figure 12.23 The bifurcation diagram of the Hénon map (12.11).

The fixed points of this system are

$$\text{Fixed point 1:}\quad \left(-\frac{1}{2}, -\frac{1}{2}\right),$$

$$\text{Fixed point 2:}\quad \left(\frac{\mu}{2(\mu+1)}, \frac{\mu}{2(\mu+1)}\right).$$

Thus, the fixed point 1 does not move as the parameter μ changes, but the fixed point 2 moves in the state space. Let us now investigate the stability of the fixed point 2. The Jacobian matrix is given by

$$\begin{bmatrix} 0 & 1 \\ (\mu+1)(-1-2y^*) & (\mu+1)(1-2x^*) \end{bmatrix}.$$

Substituting the location of fixed point 2, we get the Jacobian matrix as

$$\mathbf{J} = \begin{bmatrix} 0 & 1 \\ -2\mu-1 & 1 \end{bmatrix},$$

whose eigenvalues are

$$\lambda_{1,2} = \frac{1 \pm j\sqrt{3+8\mu}}{2}.$$

It can be seen that at $\mu = 0$, the eigenvalues are $\frac{1}{2} \pm j\frac{\sqrt{3}}{2}$, whose magnitude is exactly 1. For $\mu < 0$, the magnitude is less than unity and for $\mu > 0$, the magnitude is greater than

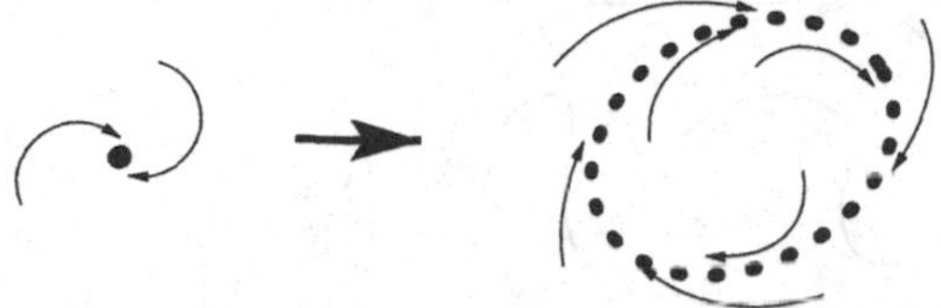

Figure 12.24 The birth of a stable closed loop in the discrete-time dynamical system by a Neimark–Sacker bifurcation.

unity. Therefore, for $\mu < 0$, the fixed point is stable and since the eigenvalues are complex conjugate, the orbits spiral inward. At $\mu = 0$, the fixed point loses stability. For $\mu < 0$, orbits starting in the neighbourhood of the fixed point spiral outward.

However, the outgoing spiral behaviour was inferred by looking at the linearized system. This approximation is valid only in a small neighbourhood of the fixed point. When the excursion from the fixed point is not small, this linear approximation is no longer valid. Because of the nonlinearity in the system, at some distance from the fixed point, iterates still follow inward spiralling orbit. Therefore, the orbit stabilizes on a closed curve – where the outgoing and incoming behaviours meet. Thus, the event occurring at $\mu = 0$ causes the birth of a closed curve on which orbits lie. This is the Neimark–Sacker bifurcation (Fig. 12.24).

Now refer to Fig. 12.5, and recall that a quasiperiodic orbit looks like a closed curve in the Poincaré section. This implies that the Neimark–Sacker bifurcation is the mechanism by which quasiperiodic orbits are created in a dynamical system. ◀

The above examples illustrate the common mechanisms by which oscillatory modes or limit cycles can lose stability in a dynamical system, resulting in an abrupt change in the steady-state behaviour. In these illustrations, we have used specific functional forms of the maps. In practical systems where such phenomena need to be understood, it may not be possible to obtain the map in closed form. In such cases, numerical techniques have to be adopted to obtain the fixed points and the eigenvalues. But because of the universality, the basic phenomenology and the underlying mathematical mechanism causing such events remain the same irrespective of the specific system at hand.

12.11 Global Dynamics of Discrete-time Systems

So far, we were taking the known route of local linearization in the neighbourhood of fixed points in understanding the stability of limit cycles. This method can explain a large number of dynamical phenomena observed in practical systems. There are also some types of dynamical transitions that cannot be understood from the localized approach.

Note that outside the small neighbourhood of a fixed point, the above description of linearized system behaviour no longer remains valid. For example, if the fixed

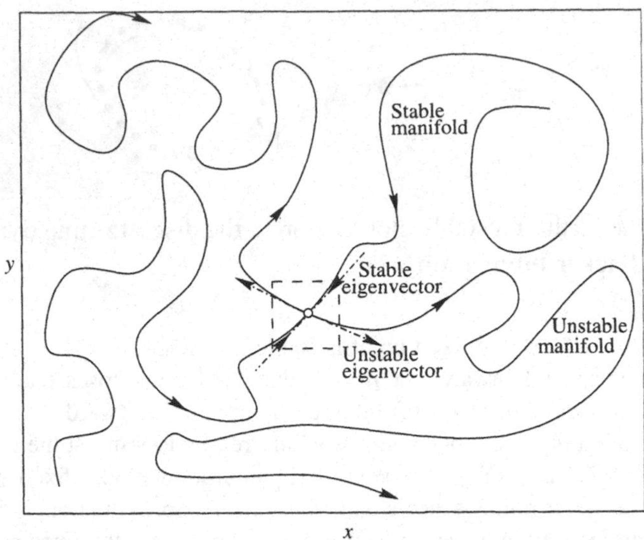

Figure 12.25 Schematic view of the stable and unstable manifold of a saddle-fixed point. The locally linearized neighbourhood of the fixed point is also shown.

point is a saddle, the eigenvectors in the small neighbourhood have the property that if an initial condition is placed on the eigenvector, subsequent iterates remain on the eigenvector. Outside the small neighbourhood the lines with this property no longer remain straight lines. One can therefore identify *curved lines* passing through the fixed point with the property that any initial condition placed on the line forever remains on it under iteration of the map. Such curves are called *invariant manifolds* (Fig. 12.25).

If the iterates of an initial condition approach the fixed point along a manifold, then it is said to be *stable*; if they move away from the fixed point (or move towards it under the application of the inverse map $\mathbf{x}_n = f(\mathbf{x}_{n+1})$), then the manifold is said to be *unstable*. It is clear that the eigenvectors in the linearized model are locally tangent to the stable and unstable manifolds of the fixed point.

It is often found that in a nonlinear system there is more than one attractor existing for the same parameter value. Such a situation is called *multistability* and the attractors are said to be *coexisting.* The collapsing state or "the attractor at infinity" can also be one of the attractors, coexisting along with a stable attractor. In such situations, some initial conditions are attracted to one attractor and some other initial conditions are attracted to the other one. The state space is therefore divided into *basins of attraction* of the respective attractors.

This can be understood in terms of the stable and unstable manifolds. If an initial condition is not placed on one of the invariant manifolds, on iteration of the map the state moves away from the stable manifold and closer to the unstable manifold (Fig. 12.26). The unstable manifold, therefore, "attracts" points in the state space, and the stable manifold "repels" them.

Therefore, if there is a saddle-fixed point in a system, all attracting orbits (or attractors) must lie on the unstable manifold of the saddle. The stable manifold, on

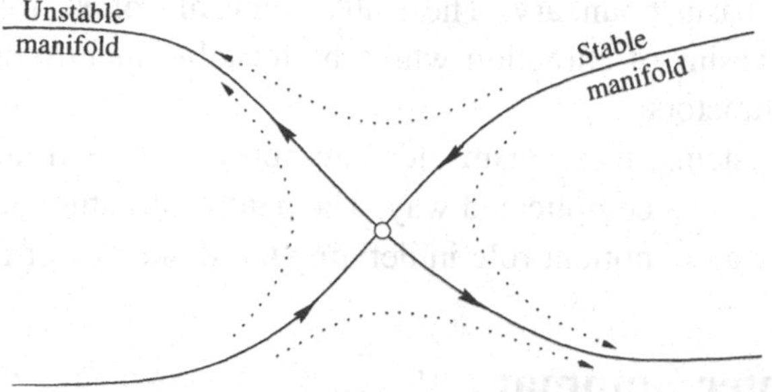

Figure 12.26 The unstable manifold of a regular saddle "attracts" points in the state space while the stable manifold "repels."

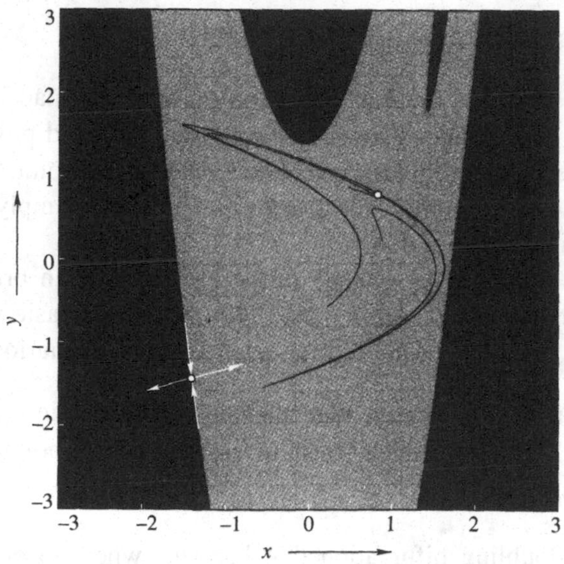

Figure 12.27 The basins of attraction of the Hénon map (12.11) at $A = 1.1$ and $B = 0.4$. One attractor is the stable chaotic orbit and the other is the collapsing state or the "attractor at infinity." The basin of attraction of the chaotic attractor is in light grey while that of the attractor at infinity is in dark grey. A saddle-type fixed point sits in between, whose local eigenvectors are shown. The stable manifold of the saddle forms the basin boundary.

the other hand, repels points in the state space. Therefore, if there are multiple attractors in a system, the stable manifold of a saddle-fixed point acts as the boundary separating the basins of attraction. This is illustrated in Fig. 12.27, where the basins of attraction of the attractors of the Hénon map (12.11) are shown. One attractor is chaotic, and the other is the collapsing state, or the "attractor at infinity." The fixed point $\mathbf{x}_1^* = (0.81, 0.81)$ is placed inside the chaotic attractor, and $\mathbf{x}_2^* = (-1.4, -1.4)$ is located on the basin boundary. The stable manifold of $\mathbf{x}_2^*$ forms the boundary between the two basins of attraction while the unstable manifolds at the two sides lead to the two attractors.

In nonlinear systems, these manifolds can curve and bend and wander around the state space in quite a complicated way. The nature, structure and intersections of the manifolds play an important role in defining the dynamics of nonlinear systems.

12.12 Chapter Summary

The stability of limit cycles in dynamical systems has to be probed by obtaining discrete-time dynamical systems (or maps) of the form

$$\mathbf{x}_{n+1} = g(\mathbf{x}_n)$$

by the method of Poincaré section. The fixed point of the map is uniquely related to the limit cycle in continuous time. The stability of a fixed point is given by the eigenvalues of the Jacobian matrix calculated at the fixed point. If the eigenvalues have magnitude less than unity, the fixed point is stable, implying that the limit cycle in continuous time is stable.

When there is a qualitative change in the character of an orbit as a parameter is varied, the event is called a *bifurcation.* There are three basic ways a fixed point can lose stability, resulting in three different dynamical scenarios. These are

1. The saddle-node bifurcation that happens when an eigenvalue assumes the values of $+1$, resulting in the birth of a pair of fixed points – one stable and the other unstable.

2. The period-doubling bifurcation that happens when an eigenvalue exits the unit circle on the negative real line, resulting in the birth of an orbit of double periodicity.

3. The Neimark–Sacker bifurcation that happens when a pair of complex conjugate eigenvalues exit the unit circle, resulting in the birth of a stable closed loop in the discrete-time system, which represents a quasiperiodic orbit in continuous time.

While the local linear neighbourhood of a fixed point of a discretized system offers a large amount of information, there are many dynamical phenomena that depend on the global character of the map. The behaviour of the stable and unstable manifolds of the fixed points helps us in understanding many such phenomena.

Further Reading

B. Davies, *Exploring Chaos: Theory and Experiment*, Perseus Publishing, Cambridge, Massachusetts, 1999.

E. Ott, *Chaos in Dynamical Systems*, Cambridge University Press, Cambridge, UK, 1997.

Problems

1. Obtain the discrete-time map of the given circuit by observing the state at every positive edge of the input voltage. The waveform of the applied voltage is shown on the right. Draw the graph of the map. What can be inferred from this graph about the behaviour of the circuit?

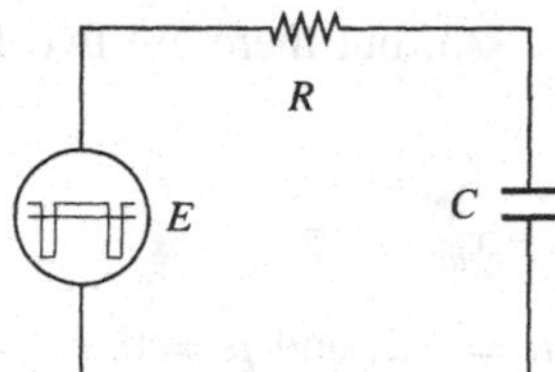

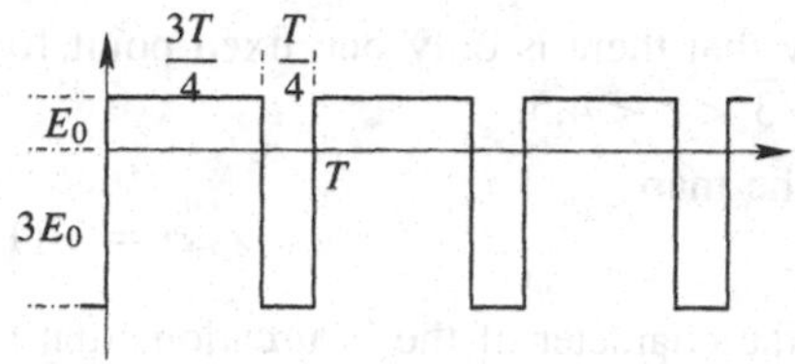

2. Write a program to evolve the Lorenz system

$$\dot{x} = -10(x - y),$$
$$\dot{y} = -xz + 28x - y,$$
$$\dot{z} = xy - 8z/3,$$

numerically. Then obtain the sequence of points by observing the state discretely when the z coordinate reaches a maximum in every cycle. Note that by the last equation, this implies placing a Poincaré surface of section given by $xy - 8z/3 = 0$. Plot the value of z_{n+1} as a function of z_n.

3. For the discrete-time dynamical system given by

$$x_{n+1} = \begin{cases} \mu x & \text{for } 0 < x < \frac{1}{2} \\ \mu(1 - x) & \text{for } \frac{1}{2} < x < 1 \end{cases}$$

the bifurcation diagram of the map is shown for the parameter range [0.8, 2.0].

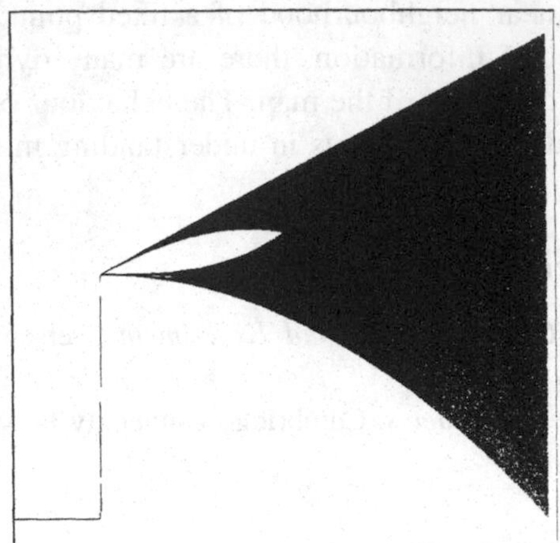

(a) Describe the transition that occurs at $\mu = 1$.

(b) Show that no fixed point is stable for $\mu > 1$.

4. Consider the map

$$x_{n+1} = rx\left(1 - x^2\right)/\sqrt{3}.$$

Show that there is only one fixed point for $0 \leq r < \sqrt{3}$, but there are two fixed points for $\sqrt{3} < r < 4.5$.

5. For the map

$$x_{n+1} = -(1 + \mu)x_n + x_n^3,$$

find the character of the bifurcations that occur at $\mu = -2$, and $\mu = 0$.

6. If the two-dimensional discrete dynamical equations

$$\begin{bmatrix} x_{n+1} \\ y_{n+1} \end{bmatrix} = \begin{bmatrix} 2 & 0.5 \\ 2 & -0.5 \end{bmatrix} \begin{bmatrix} x_n \\ y_n \end{bmatrix}$$

are applied once on the set of points in a circular disk of radius unity centred at (0, 0), find the shape and size of the resulting set. Comment on the shape and orientation of the set after a large number of iterations.

7. Specify what type of bifurcation occurs in the following maps at the parameter values given.

(a) In the map $x_{n+1} = \mu x_n - x_n^3$, at $\mu = 1$.

(b) In the map $x_{n+1} = \mu - x_n^2$, at $\mu = -1/4$.

(c) In the map $x_{n+1} = \mu x_n - \mu x_n^2$, at $\mu = 3$.

8. Let a discrete dynamical system be given by

$$x_{n+1} = \begin{cases} 2 + \sqrt{2}(x - 1) & \text{if} \quad x \leq x \leq c \\ \sqrt{2}(1 - x) & \text{if} \quad c \leq x \leq 1 \end{cases},$$

where $c = 1 - 1/\sqrt{2}$.

(a) Sketch the graph of the map.

(b) Locate the fixed point.

(c) Comment on the character of any typical orbit.

9. For the map

$$x_{n+1} = x_n^2 - 5x_n + y_n,$$
$$y_{n+1} = x_n^2,$$

find the fixed points and specify their type.

Index

9 780470 868447